D0403086

The Making of Mr Gray's Anatomy

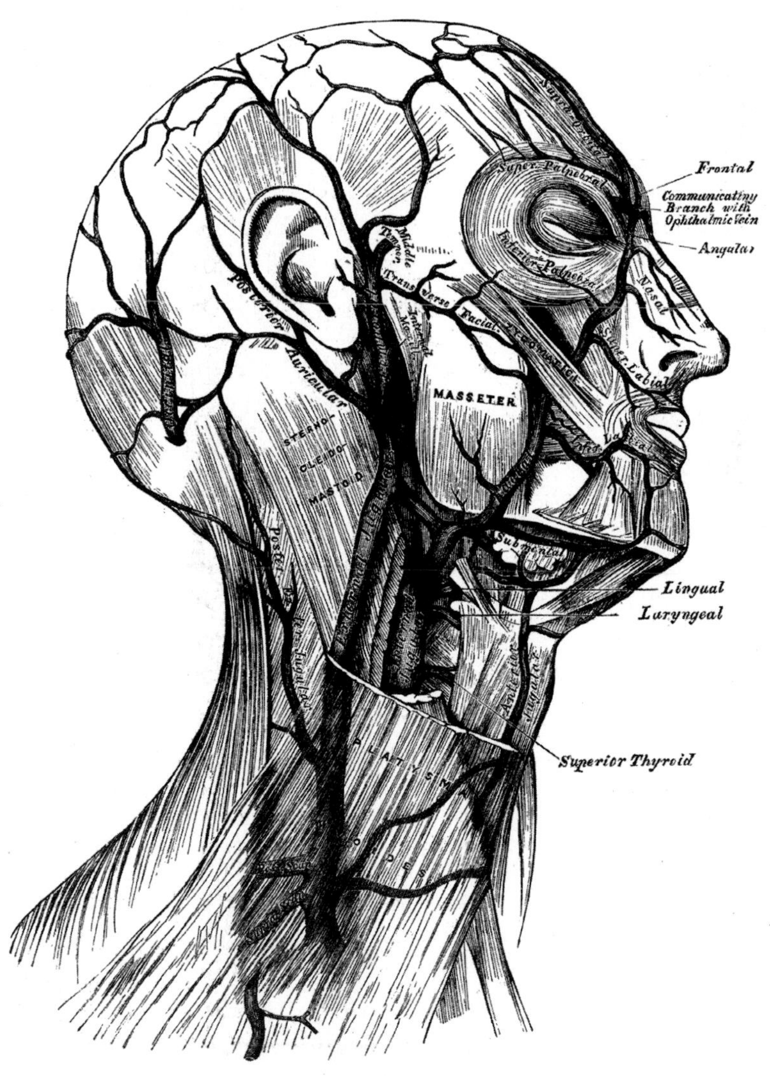

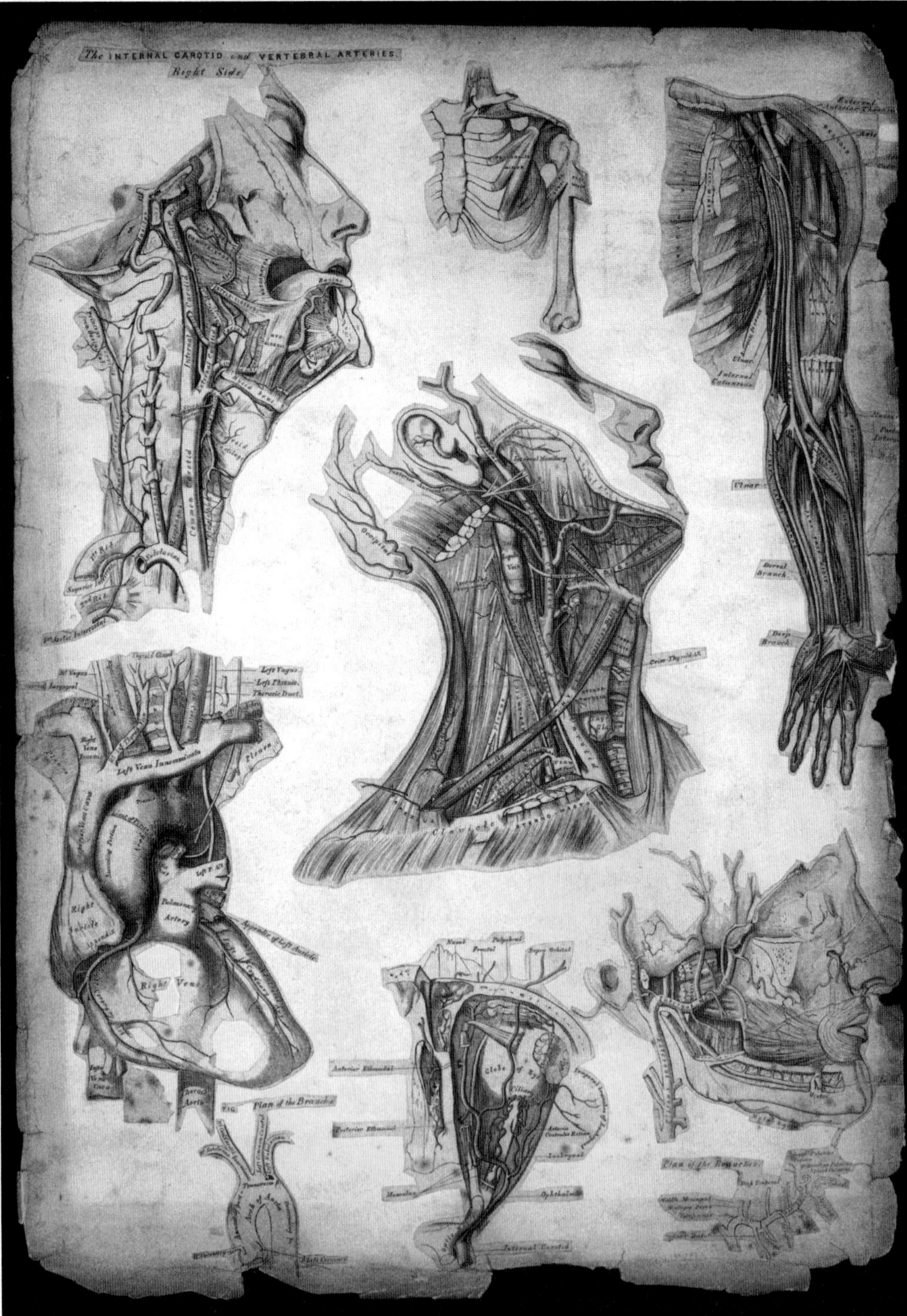
The INTERNAL CAROTID and VERTEBRAL ARTERIES.
Right Side
Left Vagus
Left Phrenic
Thoracic Duct
Left Vena Innominata
Pleura
Pulmonary Artery
Right Auricle
Appendix of Left Auricle
Right Ventricle
Clavicle
Plan of the Branches
Arch of Aorta
Globe of Eye
Anterior Ethmoidal
Posterior Ethmoidal
Muscular
Ophthalmic
Internal Carotid
Frontal
Supra Orbital
Palpebral
Nasal
Plan of the Branches
Dorsal Branch
Deep Branch
Ulnar

The Making of MR GRAY'S ANATOMY

RUTH RICHARDSON

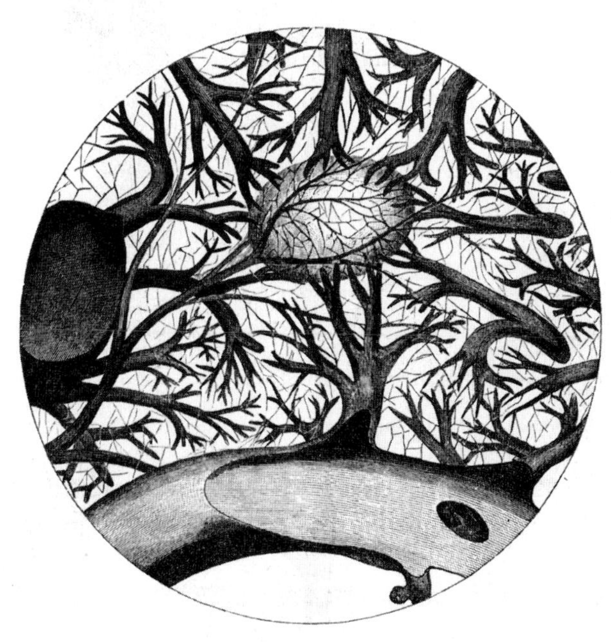

OXFORD
UNIVERSITY PRESS

OXFORD
UNIVERSITY PRESS

Great Clarendon Street, Oxford OX2 6DP

Oxford University Press is a department of the University of Oxford.
It furthers the University's objective of excellence in research, scholarship,
and education by publishing worldwide in

Oxford New York
Auckland Cape Town Dar es Salaam Hong Kong Karachi
Kuala Lumpur Madrid Melbourne Mexico City Nairobi
New Delhi Shanghai Taipei Toronto

With offices in

Argentina Austria Brazil Chile Czech Republic France Greece
Guatemala Hungary Italy Japan Poland Portugal Singapore
South Korea Switzerland Thailand Turkey Ukraine Vietnam

Oxford is a registered trade mark of Oxford University Press
in the UK and in certain other countries

Published in the United States
by Oxford University Press Inc., New York

First published 2008

British Library Cataloguing in Publication Data
Data available

Library of Congress Cataloging in Publication Data
Data available

Printed in Great Britain
on acid-free paper
by CPI Antony Rowe, Chippenham

ISBN 978-0-19-955299-3

1 3 5 7 9 10 8 6 4 2

For Brian

SWEETHEART
COMPANION
FRIEND

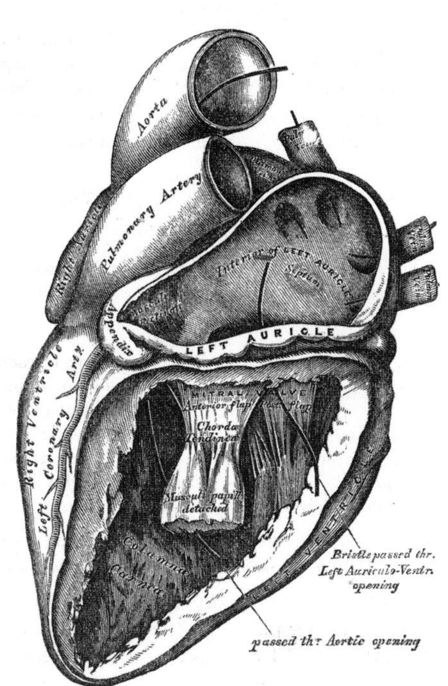

ACKNOWLEDGEMENTS

My first thanks are to Brian Hurwitz, who has encouraged and supported me throughout the effort of making this book; to our Boy; and to my dear Parents, whose encouragement has never failed.

This book would never have been written if Professor Susan Standring of King's College London, had not commissioned me to write a new historical introduction to her own first edition of *Gray's Anatomy*. My warmest thanks and admiration to her.

My Editor at Oxford University Press, Latha Menon, has been a joy and a delight to work with, and I should like also to thank James Thomson, Phil Henderson, Jack Sinden, Bryony Newhouse, Fiona Vlemmiks, and the rest of the team at Great Clarendon Street, who have made the production side of things such a pleasure.

Some of the happiest contacts have been as a result of meeting David Buchanan, of Scarborough City Art Gallery, Professor Gordon Bell and his wife, and Arthur Credland, especially during the run-up to the splendid exhibition on Henry Barlow Carter and Sons, at Hull Maritime Museum and Scarborough Art Gallery. Professor Bell has been a key figure in obtaining the blue plaque for Henry Vandyke Carter in Scarborough. Coulson's, the current owners of Carter's old home, and the Scarborough Civic Society deserve all praise for making it happen. It was also Professor Bell who put me in touch with Sarah Potts, who has so kindly allowed me to reproduce the delightful photograph of HVC with his wife and little son, and to quote from family papers.

I have never met Svetlana Alpers, though I do wish I had. Her book *The Art of Describing* has been an inspiration to me in the writing of this book.

I particularly want to thank the kind and good librarians and archivists who have curated the materials from which it has been possible to unearth the hidden story of *Gray's Anatomy*. So my thanks to:

The staff one and all at the Rare Books Reading Room at the British Library, the Manuscripts Library, the Maps Room, and the old India Office Library, and at Colindale, as well as the staff at the issue desk, and those we readers never get to see, in the stacks, and also the ever good-hearted security staff. Personal help beyond the call of duty has been given by Robin Alston, Edmund King, Des McTernan and Malcolm Marjoram. At the Wellcome Library, I should particularly like to thank Richard Palmer, who catalogued Henry Vandyke Carter's Papers, Julia Sheppard, Richard Aspin, and other members of staff. At St George's Hospital, Tooting, Susan Gove, Nallini Thevakarrunai, Marina Logan-Bruce, and Phil Adds. Dawn Kemp, Andrew Morgan and other staff at the Museum and Library at the Royal College of Surgeons, Edinburgh. At the Royal College of Surgeons, London, Thalia Knight, Tina Craig, Beth Astridge, and Matthew Derrick; and in the Museum, Dr Simon Chaplin and Jane Hughes; Samantha Fairhall at St. Bartholomew's Hospital Archives, Caroline Lam at King's College London Archives, Katie Sambrook and Andrew Baster at King's College London Special Collections; the Golden Archivist at King's College School. Staff at *The Lancet*, Victoria Killick at the London School of Hygiene and Tropical Medicine, Frank James at the Royal Institution, Dee Cook, Archivist at the Society of Apothecaries. Librarians and archivists at the City of London Guildhall Library Archive and Map Room, Dr Williams Library, Family Records Centre, Hull Central Public Library & Archives, London Metropolitan Archives, the Royal College of Physicians and Surgeons of Glasgow, Linnaean Society, London Metropolitan Archives, The National Archives at Kew, The Natural History Museum, Royal Society of Medicine, St Bride Printing Library, Trinity College Cambridge Archives, University of London Senate House Library, University of Reading, The Wellcome Library, Westminster City Archives, the Archivist at Westminster School, staff at Windsor Castle, and at Yale University.

Thanks also to colleagues and friends at the Department of History at the University of Hertfordshire at Hatfield, especially Professor Tim Hitchcock and Dr Sara Lloyd; The Institute of Historical Research, Senate House, and colleagues and friends at the Department of the History and Philosophy of Science at Cambridge, especially Jim and Anne Secord, Tatjana Buklijas, Nick Hopwood, and Nick Whitfield.

I should also like to thank my sisters, and Susan Armstrong, Jane Wildgoose, Phil Adds, Anne Bayliss, Fred Castello, Gordon Cook, Helena Cronin, John Ford, John Foreman, Christian Forsdyke, Denis Gibbs, Chris Hamlin, John Heywood, Madeline Hyde at Elsevier, Monique Kornell, Brian Lake & staff at Jarndyce, Helen MacDonald, Maryon McDonald, Diana Manuel, Diane Middlebrook, Jean Pateman, John Pegington, the Quekett Club, Peter Razzell, Jessica Richardson, John Rickard, Mark Smalley, Tilli Tansey, Tim Whitton and Tony Williams. Mr K E Nicol kindly allowed me access to his private archive and to his stores of knowledge. If I have missed mentioning anyone, forgive me, and have compassion on the little grey cells.

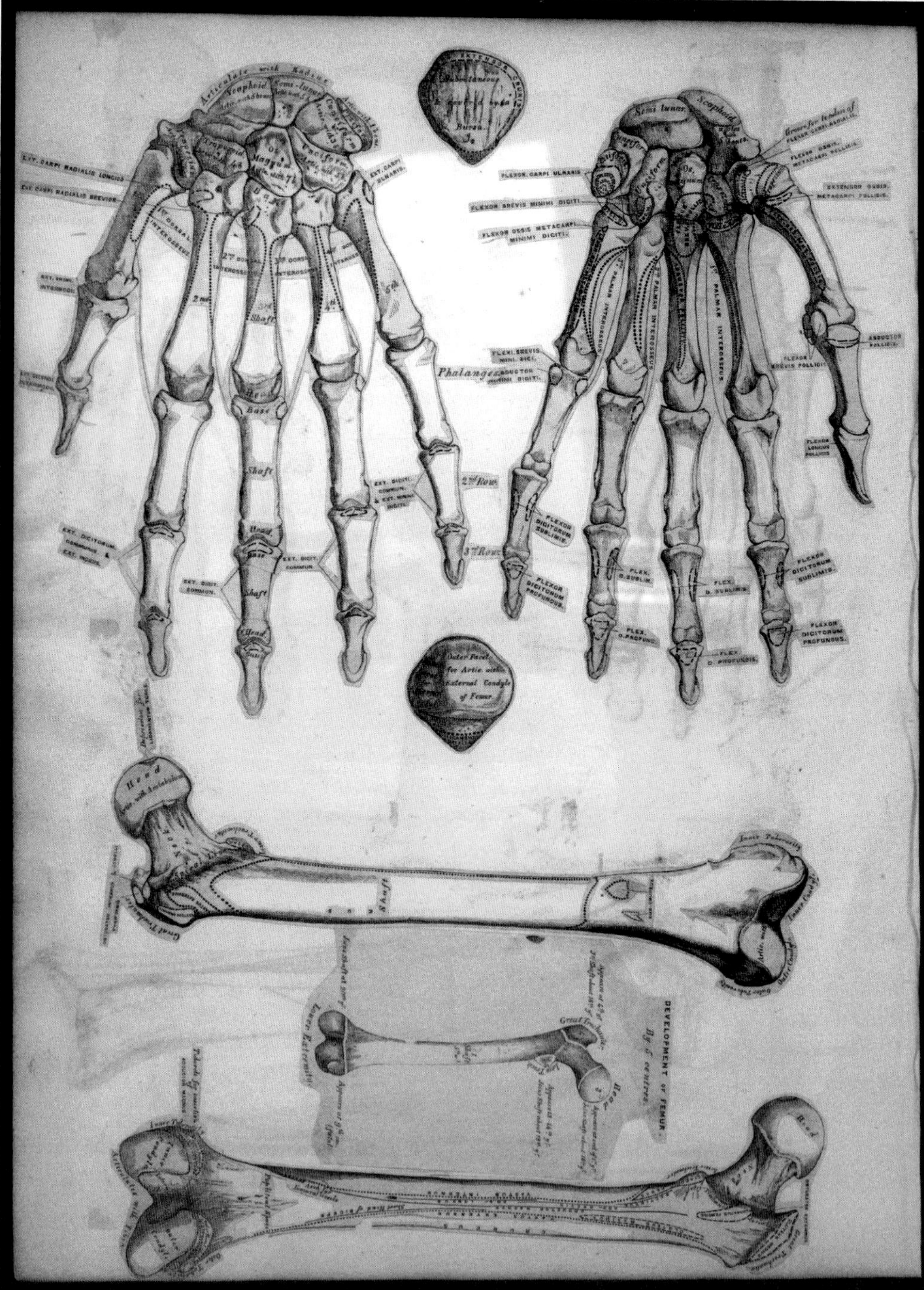

CONTENTS

ILLUSTRATIONS

Picture credits: Royal College of Surgeons of England, London: pp. ii, viii, 142, 172, 175, 190, 192, 203, 211, 219, 251. Guildhall Library, City of London: pp. xii–xiii, 63. Wellcome Library, London: pp. 3, 14, 34, 53, 71 (lower picture). National Portrait Gallery, London: p. 71 (top picture). *Illustrated London News*/Mary Evans Picture Library: p. 131. Royal College of Surgeons, Edinburgh: pp. 198. Sarah Potts Collection: p. 264. St George's Hospital Medical School: p. 248. Unattributed images are from the author's collection.

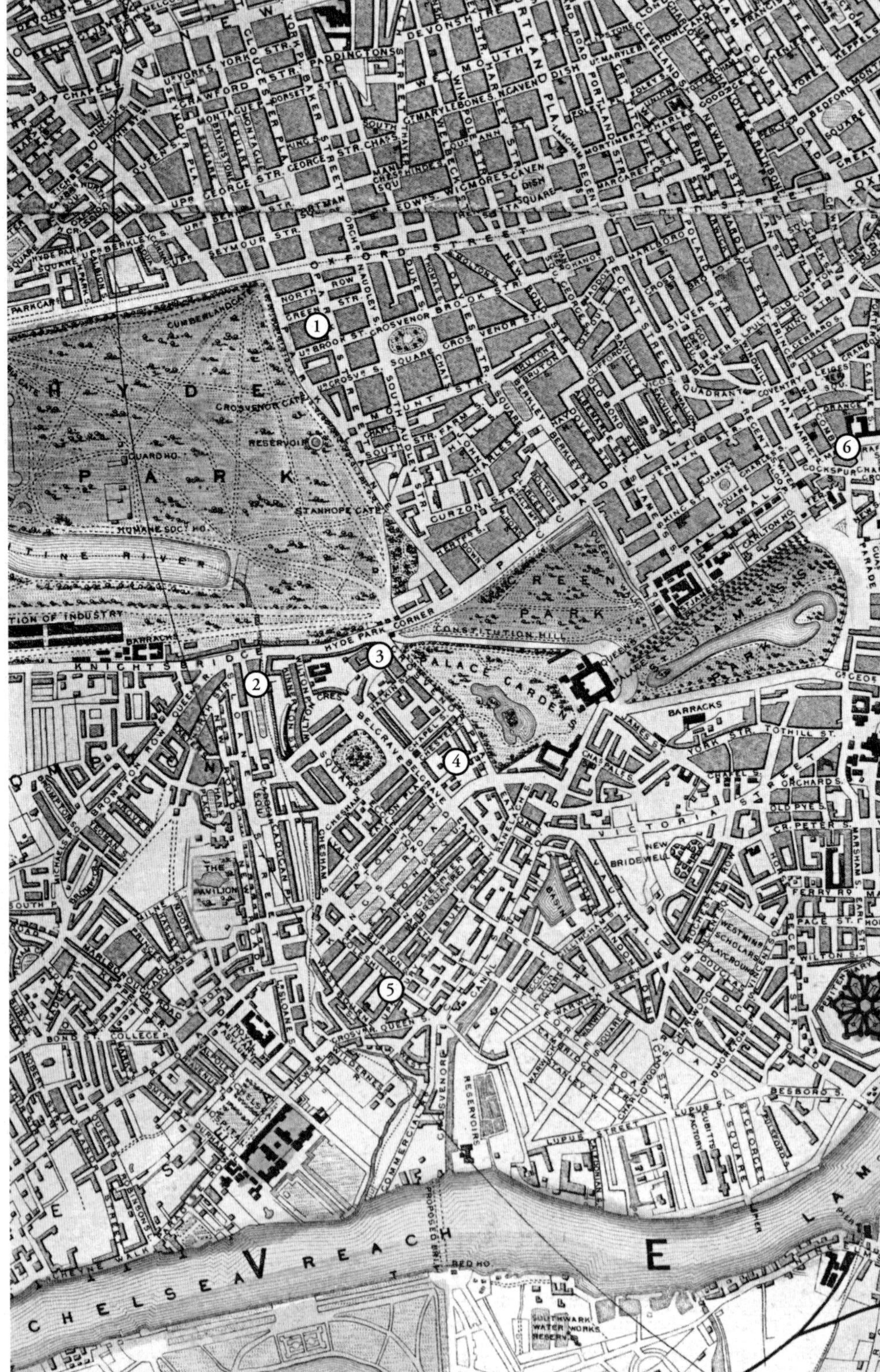

1
2
3
4
5
6
NEW ROAD
PORTLAND PLACE
DEVONSHIRE STR.
PADDINGTON ST.
BAKER STREET
GLOUCESTER PLACE
YORK PL.
CRAWFORD STR.
MONTAGUE SQ.
DORSET STR.
BRYANSTONE SQUARE
MONTAGUE STR.
GEO. STR.
PORTMAN SQ.
SEYMOUR STR.
HYDE PARK SQUARE
UPR. BERKLEY S.
WELBECK STR.
WIMPOLE STR.
HARLEY STR.
CAVENDISH SQUARE
MARGARET ST.
MORTIMER ST.
WEYMOUTH STR.
CLEVELAND ST.
BEDFORD SQUARE
OXFORD STREET
OXFORD STR.
NORTH ROW
CUMBERLAND GATE
GROSVENOR SQUARE
BROOK ST.
GROSVENOR ST.
HANOVER SQUARE
MOUNT STR.
SOUTH AUDLEY STR.
BERKLEY SQUARE
CURZON STR.
REGENT STREET
QUADRANT
JERMYN ST.
KING S. ST. JAMES'S SQUARE
PALL MALL
CARLTON HO.
HAYMARKET
COCKSPUR ST.
HYDE PARK
RESERVOIR
GROSVENOR GATE
GUARD HO.
STANHOPE GATE
HUMANE SOCY. HO.
SERPENTINE RIVER
EXHIBITION OF INDUSTRY
BARRACKS
KNIGHTSBRIDGE
HYDE PARK CORNER
PICCADILLY
GREEN PARK
CONSTITUTION HILL
PALACE GARDENS
ST. JAMES'S PARK
BARRACKS
YORK STR.
TOTHILL ST.
ORCHARD S.
VICTORIA STREET
NEW BRIDEWELL
BASIN
BELGRAVE SQUARE
CHESHAM PL.
EATON SQUARE
SLOANE STREET
CADOGAN PL.
THE PAVILION
BROMPTON ROW
WILTON CRES.
ROYAL MILITARY ASYLUM
HOSPITAL
WESTMINSTER SCHOLARS PLAYGROUND
ECCLESTON SQUARE
LUPUS STREET
BESBORO S.
RESERVOIRS
COMMERCIAL RD.
CANAL
PROPOSED BRIDGE
CHELSEA REACH
SOUTHWARK WATER WORKS RESERV.

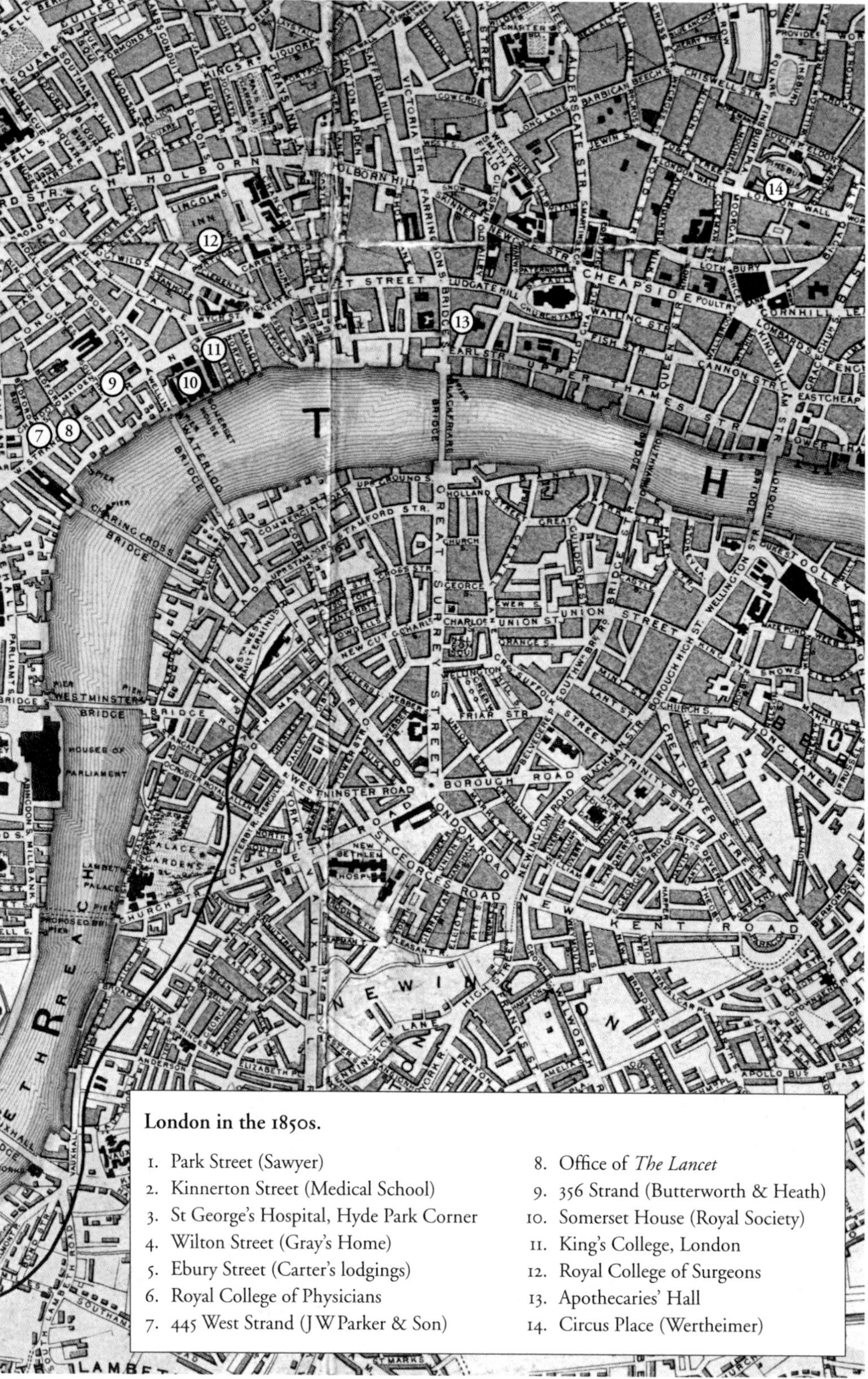

London in the 1850s.

1. Park Street (Sawyer)
2. Kinnerton Street (Medical School)
3. St George's Hospital, Hyde Park Corner
4. Wilton Street (Gray's Home)
5. Ebury Street (Carter's lodgings)
6. Royal College of Physicians
7. 445 West Strand (J W Parker & Son)
8. Office of *The Lancet*
9. 356 Strand (Butterworth & Heath)
10. Somerset House (Royal Society)
11. King's College, London
12. Royal College of Surgeons
13. Apothecaries' Hall
14. Circus Place (Wertheimer)

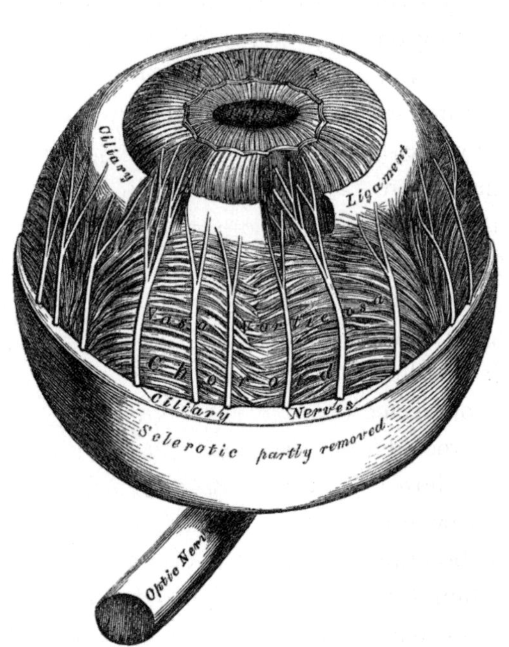
Ciliary
Ligament
Vasa Vorticosa
Choroid
Ciliary
Nerves
Sclerotic partly removed
Optic Nerv

INTRODUCTION

When John Fawthrop, MRCS, bought his copy of *Gray's Anatomy* in 1860, he inscribed it with his name and new status in large-ish letters, followed by the date, in a fine proclamatory script. He had qualified as a Member of the Royal College of Surgeons only the year before, and the purchase was a real statement of success, and intention to treat. Once he was satisfied with the lettering, Mr Fawthrop started drawing curls of foliage around the inscription, little fronds of fern-like stalk and leaf, which went on flowing delicately from his pen until he had enveloped the inscription, and fairly established his occupation of the entire top quarter of the page. The work must have taken him a good while. The leaf of the book he had chosen to inscribe was the dedication page, so there was a fair bit of room for his botanical curlicues.

When one looks now at how the top of the page appeared when he had finished, it feels as if John Fawthrop was beautifying a book he loved, and had every intention to cherish for the rest of his life. There is a sense of pride in the inscription rooted in his own achievement, but it has also a feeling of thanksgiving, somehow expressed privately, towards the book. If it was to *Gray's* that John Fawthrop owed his qualification, if this is why he inscribed his own copy so beautifully, he would not have been alone. This book had been created for him: *Gray's Anatomy* had been designed for medical students, designed to help them go sailing through exams, designed to make anatomy accessible and useful. Fawthrop was among the earliest generation of medical students for whom the book's promise worked its magic.[1]

The great book known to generations as *Gray's Anatomy* is widely thought to have been created single-handedly by its famous author. But in fact it was the work of *two* young men: Henry Gray, anatomist, pathologist, and surgeon, and Henry Vandyke Carter, apothecary-surgeon, microscopist, physician, and artist. Both had trained, and were teachers, at the famous medical school attached to St George's Hospital, at London's Hyde Park Corner. The book—its full title was *Anatomy Descriptive and Surgical*—was a joint project, to which each young man made his own joint and separate contribution. Gray wrote the words, and Carter created the illustrations. The title page credits both young men with the dissections on which the book was based.

Gray's Anatomy is a phenomenon: a textbook which has been in continuous publication ever since 1858. Its publishing history has never been written, so the reasons for its uniqueness are neither properly understood nor appreciated. I focus right down on the 1850s, the decade between the Great Exhibition and the death of Prince Albert, to look at the first edition of *Gray's Anatomy* from gestation to reviews. The nub of the story concerns the small constellation of individuals without whom there would have been no book.

Gray's Anatomy inhabits a vital intersection between the histories of medicine and publishing—and this is a study of *Gray's* as a book: as an intellectual, an artistic, and as a physical artefact. This book is about the making of the book Fawthrop, and many thousands—possibly millions—of medical students worldwide, have used on their way to becoming doctors and surgeons, the book that founded an ongoing institution: *Gray's Anatomy.*

The Dedication Page from the first edition of *Anatomy Descriptive and Surgical* (1858) demonstrating Henry Gray's regard for his mentor Sir Benjamin Brodie. This copy, from the Wellcome Collection, features the beautiful inscription by the book's purchaser, the newly qualified surgeon John Fawthrop MRCS.

TO

SIR BENJAMIN COLLINS BRODIE, BART.,

F.R.S., D.C.L.,

SERJEANT-SURGEON TO THE QUEEN,

CORRESPONDING MEMBER OF THE INSTITUTE OF FRANCE,

THIS WORK IS DEDICATED,

IN ADMIRATION OF HIS GREAT TALENTS,

AND

IN REMEMBRANCE OF MANY ACTS OF KINDNESS

SHOWN TO THE AUTHOR,

FROM AN EARLY PERIOD OF HIS PROFESSIONAL CAREER.

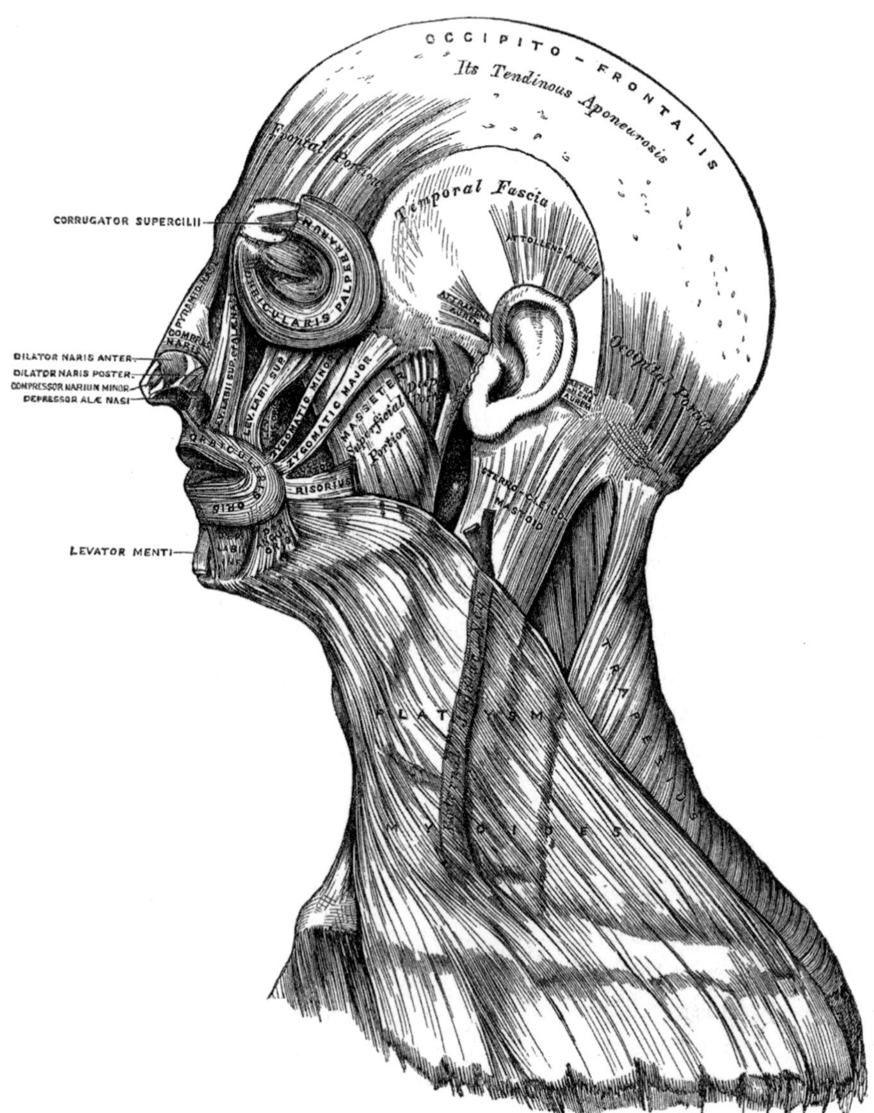

OCCIPITO-FRONTALIS
Its Tendinous Aponeurosis
Frontal Portion
Temporal Fascia
ATTOLLENS AUREM
CORRUGATOR SUPERCILII
ORBICULARIS PALPEBRARUM
DILATOR NARIS ANTER.
DILATOR NARIS POSTER.
COMPRESSOR NARIUM MINOR
DEPRESSOR ALÆ NASI
ZYGOMATIC MINOR
ZYGOMATIC MAJOR
MASSETER
Superficial Portion
RISORIUS
Occipital Portion
LEVATOR MENTI

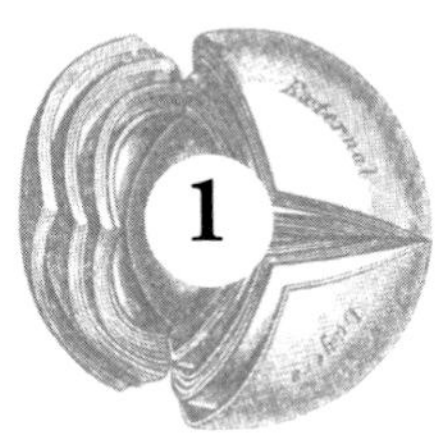

WORDS

Mr Gray of Belgravia

Gray's Anatomy is probably the best-known medical textbook in the world. It is to human anatomy what Mrs Beeton's is to cookery, or Roget's is to thesauri.[1] So highly regarded is Mr Gray's book that it has been placed alongside the Bible and Shakespeare as fundamental to the education of a doctor.[2]

The first edition of *Gray's Anatomy* appeared in London in 1858, and in Philadelphia in 1859, in good time for the start of the medical school year. Unusually, it had two titles: *Anatomy, Descriptive and Surgical* on the title page, but *Gray's Anatomy* in large gilt letters on its spine. It received rapturous reviews in the medical press on its appearance, and was regarded as the standard work within weeks of publication. In the medical world the shorter title served to differentiate it from a number of anatomy textbooks already in existence—such as *Quain's Anatomy* or *Wilson's Anatomy*—so the longer name never stuck. The book became known as *Gray's Anatomy* in dissecting rooms and medical lecture halls, people saw the name on its spine in bookshops and on medical school shelves, landladies glimpsed it on their student-lodgers' mantelpieces, poor students searched for it on second-hand bookstalls.[3]

Gray's Anatomy is so well known today that it's hard to imagine a time when it didn't exist: that is, before its text had been written, before its famous illustrations had been drawn, before the contract with its publisher had even been agreed. But that is what, in this book, we do imagine, so as to be able to see how this most famous of anatomy textbooks came to be created.

The book was the work of two young medical men anxious to make their careers in mid-Victorian London. Their association with St George's Hospital coincided with the London of Charles Dickens—between *A Christmas Carol* (1843) and *Great Expectations (1860–61)*—the London of muddy streets, fog, gaslight, hansom cabs and ragged crossing-sweepers, glittering wealth and desperate poverty cheek by jowl. Henry Gray and Henry Vandyke Carter did their work in the post-mortem room and the dissecting room, on the bodies of the poor taken from the dead-house of their own hospital, and from the mortuaries of the Poor Law workhouses of Victorian London. The story stretches across Victorian London, from the dissecting room in Knightsbridge, via Belgravia and the Strand, to the printing presses in Finsbury, beside the City of London's ancient northern boundary of London Wall.

From start to end—from idea through planning, creation, and production, to advertisement and sale—the book was completed in less than three years. We know this because a dated record exists of the first suggestion that these two young men might work together on the project.

Near the end of November 1855, the book's future illustrator, Henry Vandyke Carter, recorded in his diary a conversation with Henry Gray with this brief note:

> Little to record. Gray made proposal to assist by drawings in bringing out a Manual for students: a good idea but did not come to any plan ... too exacting, for would not be simple artist.[4]

Carter didn't mention where this important conversation took place. Most likely the two young men were somewhere in the medical school in which they both worked, belonging to St George's Hospital at London's Hyde Park Corner. Today, the dissecting room and indeed the hospital in which Gray and Carter met are distant memories. St George's outgrew its site, and in the late twentieth century, relocated several miles south, leaving the old Hospital to be refurbished as a swish hotel, 'The Lanesborough'.

When Henry Gray and Henry Carter were working there in the 1850s, St George's Hospital was a grand neoclassical landmark building, at the entry to the West End of London at the Knightsbridge end of Piccadilly. The Hospital had been at Hyde Park Corner for generations. It had occupied the same site ever since its foundation in the 1730s, when its medical staff had split away from the older Westminster Hospital, and had leased old Lanesborough House, an empty mansion on the road towards Kensington and Brompton, on Grosvenor lands just beside the westernmost margin of Green Park.

The mansion had originally stood in a rather isolated position amid park land and pastures on what was then the western edge of London. In 1733, to the north beyond Hyde Park and the Tyburn Gallows (by what is now Marble Arch) open fields lay all the way through the scattered hamlets to Edgware, and beyond. To the west, the land inclined down towards the watercress beds along the meandering valley of the Serpentine River, and the road we know as Knightsbridge crossed this river at Bloody Bridge, named from some ancient battle. The land to the south and west of Hyde Park Corner we now know as Belgravia was as yet unbuilt: known as the 'five fields', it had been pasturage for grazing sheep and cattle since time immemorial, right down to the swamps of Millbank and the old village of Chelsea. The Hospital's eastern flank overlooked the back wall of the extensive private garden belonging to Buckingham House, not yet a Palace. In the eighteenth century, footpads and highwaymen had been a regular source of fear to passers-by in the area, especially at night.

By the mid-nineteenth century, when Gray and Carter met each another at St George's, 'The Corner' was completely transformed. London had mushroomed in size. The Hospital had been extended, and then entirely rebuilt. It stood solid and self-possessed, with its own raised colonnaded entrance, on the rise at the conjunction of four important roads: Park Lane, Piccadilly, Knightsbridge, and Grosvenor Place. Stucco residences of great opulence had been built over the whole of Belgravia. Fine houses had sprung up to the west of old Westminster towards Kensington, Sloane Square, and Chelsea, and almost the entire way through Pimlico to the River Thames.

The view from the windows of St George's was still the greenest of any hospital in London. The building stood in a pivotal position on a corner, its north front overlooking the vast green space of Hyde Park, and its eastern frontage looking directly over Green Park, the vast royal garden at the rear of Queen Victoria's palace, beyond which could be glimpsed the treetops of St James's Park. Diagonally opposite the Hospital, stood Apsley House, the Duke of Wellington's residence ('Number One, London') which had been porticoed and faced in stone for the Iron Duke. Close by, Rotten Row was a showplace for well-to-do carriage-folk. To the north of Hyde Park, the old gallows at Tyburn had gone, and the crowds on hanging days had moved away east to Newgate, leaving Tyburnia and Westbournia to be built over as comfortable suburbs. The old tree-shaded Tyburn Lane between the Park and May-Fair was now widened to become Park Lane. During the day, Oxford Street, Piccadilly, and Nash's Regent Street were thronged with wealthy shoppers.

Since the demolition of the old turnpike gates right outside the Hospital, the traffic on the hill had become almost incessant: a tide of pedestrians, horses, and the vehicles of all social classes and trades, entering and leaving the West End, in a daily tidal flow.

Despite its burgeoning size, though, London was still traversable on foot. The diarist Arthur Munby, a contemporary of Carter and Gray, strolling up Constitution Hill opposite the Hospital from his office in Whitehall one evening, did not comment on the fine horses and coaches swishing past, or the famously oversized statue of the Iron Duke, but noticed instead a vocal gang of dustwomen, black from their work, trudging home towards the slums of old Westminster from the enormous dustheaps by the Paddington canal.[5]

St George's Hospital is at the very heart of the story of *Gray's Anatomy.* Without that institution, the men who created it would neither have met, nor learned to work together. The idea for their book might never have arisen or become a reality. Both men had learned their anatomy, and later taught it, there. The Hospital's students provided them with the experience of teaching younger men, which was so crucially necessary to the development of such a major textbook. The medical school Library and Museum provided them with essential materials for exploration and research, and the Hospital's Dead-House and its Dissecting Room supplied facilities vital to their labours.

The need for a public hospital on the western fringe of London was very great, and growing as the population of the Victorian metropolis grew. Pressure on space within the building was incessant, and need always greater than provision. The wards could never accommodate the numbers of people applying for relief. Earl Grosvenor's development of Belgravia had raised land values exponentially in the area of the old Five Fields, curtailing any faint hopes the Hospital's governors might have harboured of acquiring adjacent land on which to expand. The stables and much of the garden of Lord Lanesborough's old mansion had been built over when the Hospital was redesigned and rebuilt in phases by William Wilkins, in the 1820s–1830s.

Wilkins's new hospital faced towards Green Park, and embodied hospital reform ideas predating those of Florence Nightingale. It was built on an 'H' plan of twenty-eight wards, with an average of fourteen beds in each, men in the south wings, women in the north. Windows allowed natural light and cross-ventilation, and wards were separated by communications stairs and the innovation of water closets on every floor. There was hot water for movable baths, and an internal heating system. Outpatients was at ground level, and the operating theatre was situated on the top storey, which sounds practical until you discover that it had seating for an audience of 150 and good acoustics (it really was a theatre), and imagine all those students tramping up and down the stairs, and the noise and airborne germs such crowds will have generated. The hospital provided 'every desirable facility for the prosecution of scientific research'.[6,7]

Only much later, and following the impact of Miss Nightingale, and her *Notes on Hospitals*, when audit made hospitals aware that death rates were comparable, open to commentary and competitive improvement, do we see anyone admitting how bad conditions had been there. A report in *The Lancet* on the success of amputations at St George's published in 1860 shows that previously—when Gray and Carter were working there—conditions were pretty shabby, and fatal hospital infections rife:

> For a time, most of the wards of this institution were so unhealthy that a considerable number of capital operations ended badly, as we learnt from personal observation and from the testimony of some of the medical officers. In consequence of improvements carried out last year in most of the wards … the general salubrity has so much increased that pyaemia, which had heretofore been very prevalent, has now almost entirely disappeared.[8]

We have no way of knowing what patients in the wards heard or sensed when an operation was in progress, what it was like to be a patient being operated upon before an audience of 150, or how fearful it may have been to consider oneself vulnerable to pyaemia or the prosecution of scientific research: patients' voices are hard to find in medical and historical sources of this period, preserved to laud institutions and great figures of medicine.

The nearest approach to a patient's-eye view of the wards at St George's I have so far discovered is that of a lay visitor. The poet, Letitia Landon (whose famous pen-name was 'L.E.L.') visited a dying friend there in this era, and afterwards wrote a poem, headed: 'These are familiar things, and yet how few think of this misery!' It continues:

> I left the crowded street and the fresh day,
> And enter'd the dark dwelling, where Death was
> A daily visitant,—where sickness shed
> Its weary langour o'er each fever'd couch.
> There was a sickly light, whose glimmer show'd

Many a shape of misery: there lay
The victims of disease, writhing with pain;
And low faint groans, and breathings short and deep,
Each gasp a heartfelt agony, were all
  That broke the stillness.[9]

Not great poetry, perhaps, but it conveys a sense of the atmosphere on the wards at St George's more genuine and more effectively than do the architect's plans.

In the mid-1830s, while the ambitious rebuilding scheme was in process, premises were rented for medical teaching in a back street to the west of the Hospital. An older private school of medicine already existed close to the Hospital in Grosvenor Place, which had been founded by a St George's man, Samuel Lane. But internal politics at St George's fuelled by a power struggle in which the eminent surgeon Sir Benjamin Brodie played an active role, lead to the Hospital opening its own school in 1835. Its establishment was regarded by *The Lancet* as a financial speculation on Brodie's part, because he had put up the capital to acquire a site at the northernmost end of Kinnerton Street, for a new purpose-built building. The new medical school, with its 'Anatomical Theatre'—reputedly the best dissection room in London at the time—was opened a little later. Henry Gray probably began his studies at St George's in this new building.[10,11,12]

It was at Kinnerton Street that all the dissecting was to be done for *Gray's Anatomy*. The back-street district, and the building Brodie helped the Hospital acquire, still survive only a few minutes walk from old St George's, hidden away behind Knightsbridge. Kinnerton Street runs north–south, between the wealthy streets of Lowndes Square and Wilton Place. The street itself is long and low, mostly of small houses and old stables, with a number of blind-ended closes running off at right angles to the west, all insulated from the main highway by the tall buildings on Knightsbridge itself, and from the rest of Belgravia by the bigger houses of the genteel streets at either side. This secluded warren of humble streets had been designed to house the personnel and livestock servicing the inhabitants of the grander houses close by.

The yards running off Kinnerton Street to the west are dead ends. When they were built, they abutted the valley of one of London's 'lost' rivers, the Westbourne or Serpentine River, the western boundary of the old Five Fields.

The river rose on the northern heights of London, and fed into and drained out of the grand artificial lake known as 'The Serpentine', between Kensington Gardens and Hyde Park, north of Knightsbridge. Apart from the Serpentine, the river itself is now almost invisible above ground, but the conformation of older streets—like the yards off Kinnerton Street—still show evidence of its route. On its way south out of the park at Albert Gate, the river passed in a great conduit underneath the roadway of Knightsbridge (the site of the old Bloody Bridge) then down between the two back streets of Kinnerton Street and William Mews, meandering towards its outfall in the Thames, by the grounds of the Royal Hospital at Chelsea.[13]

The dampness of the old river valley would have made this border region of Belgravia unattractive to affluent inhabitants, so the locality had provided the developer with a conveniently enclosed and largely hidden district to accommodate the necessary labour to service his wealthy tenants. The inferior housing was intended for the stable-hands and servants of the grander mansions, horses and carriages, and the hay on which they fed and slept. Small businesses needing horse-room found space there, like the undertaker at number 24.[14] Such streets of cramped mews housing were a feature of Victorian London: every wealthy street had its shadow. On Charles Booth's extraordinary maps of London's poverty and wealth, they consistently show deep blue or black (poverty, dire poverty) by contrast with the nearby streets coloured gold and red (super-rich, wealthy, substantial middle class).[15]

The Kinnerton Street district is nowadays an area of bijou town houses, painted in pretty pastels. Flower boxes, and potted shrubs and palms cluster everywhere. It is no longer an area of stables, but of wealthy pieds-à-terre, with a sprinkling of more spacious modern mansion flats and town houses. The district has been significantly gentrified, but not beyond all recognition. Despite the hyper-expensive cars parked on the cobbles, its back-to-back yards have an old-fashioned air of the Regency about them, and the older houses still exhibit the mean dimensions of the domestic architecture intended for the working classes and beasts of that era. Even though they stand between the busy hum of Knightsbridge and Harrods, these little streets remain quiet and unfrequented, with the atmosphere of a secret enclave.

The governors of St George's Hospital had been glad of such a position as Kinnerton Street for their new medical school. It was off the beaten track, unfashionable, so cheaper than anywhere else close by, and was both secure and discreet. The northern end of Kinnerton Street was a cul-de-sac, with no passing traffic. The situation was also highly convenient, in that it provided easy access to and from the Hospital itself. When young Mr Gray left the Hospital, he could arrive at the medical school after a brisk walk of less than three minutes.

The new medical school was erected on land lower than the Hospital, backing on to the river valley. Larger in every dimension than the surrounding mews houses in Kinnerton Street, it stood tall above the intervening roofscape. Although the houses on Wilton Place hid its entrance, its upper windows were visible from the top floors at the back of the Hospital. It was an ark of a place, secluded even from Kinnerton Street itself.

The medical school was not a beautiful building, being more akin to the workhouses then being erected for the New Poor Law than to its grand neoclassical parent institution at Hyde Park Corner. It was not designed to be looked at. It was utilitarian in its simplicity, in London stock brick decorated in a very plain Victorian minimalist manner across its facade with a stone string course, wide sustaining arch, and red-brick dressings over windows and doors. A securely gated entrance was formed by the flank walls of two neat houses on either side, whose party wall met at first floor level over the arch. An unadorned stone slab shaped like the wide pent roof of a classical temple broke the horizontal capping of the roofline above these arching rooms. Modern visitors to the now widened Kinnerton Street who look up at it will see that it presents a blank, its old lettering well obscured. But in the days when Gray walked smartly through the archway below, it probably bore a simple legend: 'St. George's Medical School' or more simply still, 'School of Medicine'.

Staff and students alike referred to this place as Kinnerton Street, or just 'KS' for short. Through the archway, modest cottages for the school's caretaker and other servants of St George's Hospital flanked the narrow front yard. At the innermost end of this gated yard, dignified classic columns framed the heavy entrance door of the School itself. The porters' room and other staff offices lay on each side, and the corridor leading to the stairs up to the big Dissection Room at the rear of the building was facing you. In the semi-basement were the chemical and other stores, a kitchen and maintenance facilities, and sluices. The building accommodated a large lecture theatre, and other teaching spaces, a chemical Laboratory, Anatomical Museum, and a technician's room in which bodies and specimens were prepared for teaching and display.[16] The presiding spirit of the place was the genial Professor of Chemistry, Henry Minchin Noad, whose office was close by the front door, and whose lab was his empire. A brilliant chemist and researcher of electrical phenomena, Noad was also an excellent public lecturer, and was soon to become a Fellow of the Royal Society.[17]

In addition to its unostentatious frontage on Kinnerton Street, the medical school seems to have had its own discreet rear entrance. Old maps show there to have been an access way bridging the Serpentine River at the back of the building, by which deliveries and collections could unobtrusively be made. It connected to William Mews, which still joins William Street via an alley. The only other purpose-built anatomy school so far studied—that built in the mid-eighteenth century by John Hunter on Leicester Square—also had a rear entrance, convenient for receiving bodies from the grave-robbers who supplied his school.[18] In the mid-Victorian era, workhouse undertakers had replaced bodysnatchers, but deliveries had still to be made. It was probably thought preferable to keep matters discreet. Today, though, this way across the old river has disappeared, entirely blocked by a high wall and some smart modern town houses facing onto William Mews.[19]

The Kinnerton Street medical school was a large, austere, functional place. Renamed, it still stands, dark and private, enclosed within its own solid walls.[20] Its great teaching rooms have been divided up into more domestic studio spaces, inhabited by artists who value the unobstructed light from its big high-level windows, and the elevated views it still affords over the roofs of the nearby streets.

The crucial conversation in 1855 between Gray and Carter, which initiated their joint negotiations about the possibility of a book, probably took place in one of these well-lit rooms. Mr Henry Gray was twenty-eight, the artist Mr Henry Vandyke Carter twenty-four. The two young men might have been standing in the quiet of the medical school Library, or among the bottles and wax models of the Anatomy Museum. Or, perhaps more likely, Gray had waited until Carter had finished demonstrating anatomical structures under the echoing glass of the high barrel-vault roof of the Dissecting Room, surrounded by teaching materials, his diagrams still chalked on the board.

Gray was a fine looking young man, of medium height. He dressed impeccably well in the dark, slightly dandyish fashion of genteel young men of the day. His young face in surviving photographs has an earnest and determined look which is also thoughtful, giving the impression of gentlemanly intelligence.

This photograph of the young Henry Gray was taken by a medical school friend in the late 1840s or early 1850s, when Gray was in his early twenties. His eyes look intently into the lens, and seem to have dark shadows, perhaps from overwork. This rather intimate photograph shows the youthful Gray as mild, but with a determined quality to his look.

Later ones show a rather brooding seriousness. He probably spoke with a goodish accent. It was known throughout the school that the great ambition of his life was to become the top surgeon at the Hospital. He was diligent and hard-working, focused, clever, ambitious, and had cultivated his bedside manner and teaching technique by observing his elders, and by personal experience. He was more or less a gentleman himself. He lived in Belgravia, only a short stroll south from the Hospital, sharing a house with his widowed mother in Wilton Street, between Eaton Square and the great private garden of Buckingham Palace.

Today, a plaque on the front of 8 Wilton Street informs those who pass by that Henry Gray, Anatomist, lived there, and that he died there in 1861.

Gray died swiftly, tragically, of confluent smallpox at the age of only thirty-four, three years after *Anatomy Descriptive and Surgical* was first published.

He left behind no diaries or reminiscences. A search of the UK National Register of Archives yields a complete blank in this respect. It looks as if the Victorian terror of contagious killer diseases may have led to the stripping of the room in which he died, and that the contents of Gray's study were burnt along with his clothes and bed-hangings. Almost nothing is known of Gray's private life, beyond the story that at the time of his death, he was engaged to be married.[21] Apart from what was briefly recorded by contemporary obituarists, no one who knew Gray during his lifetime seems to have written any kind of personal memoir of him.[22]

The lack of personal biographical information concerning Gray himself means that he exerts a very silent and enigmatic presence in the historical record. We know what he looked like, what he published, a little about his social origins, his career at St George's, and the manner of his death. But very little is known about what kind of person he really was. Any attempt at a biography of him is difficult, as we lack the private inflection of his voice, which in his medical writings he strove to subsume in the formal nondescript 'objectivity' of the authoritative professional medical man.

We have no inside knowledge of Gray's spiritual or ethical certainties or dilemmas, his likes and dislikes, his humour, or the textures of his personality and upbringing, beyond what can be inferred from his handwriting in official documents and scientific work, his appearance in a few photographs, the content of professional writings he left behind. There are glimpses of him in the writings of those who knew him, but where they remain silent we remain ignorant. What can be gleaned from the evidence available will be presented here, but much of what is said has been arrived at by inference. Gray comes over as a clever, competent, and very hard-working surgeon, ambitious for professional success. He was status-hungry, and an institutional stalwart at St George's. His unexpected death left the Hospital bereft.

But after his funeral, his hospital colleagues were most curiously silent. There was no portrait painted or bust erected, and no memorial volume apart from *Gray's Anatomy* itself. It was as if, by writing the book, Mr Gray had been taken to have commemorated himself.

Of Gray's parentage, little is known beyond his parents' names, William and Ann, and the fact that his father had been a functionary at Court, personally serving two kings in the role of what is known as a King's Messenger, a trusted diplomatic and political emissary.[23,24] They had married in 1819, at the church of St Martin-in-the-Fields near Charing Cross, only a short walk from the Lord Chamberlain's office where the Messengers were based, in St James's Palace.[25] Gray's father is said to have relocated the family household from Windsor to Belgravia during the Nash rebuilding of Buckingham Palace for the Prince Regent, so it looks as though the couple had lived in Windsor during the early years of their marriage. Henry Gray was born in 1827, it is said, in the parish of St George Hanover Square.[26]

Where Gray was educated is not known. In the early twentieth century a relative recorded the belief that he had attended Westminster School, the ancient boys' school in the precincts of Westminster Abbey. The present Archivist at the school, however, assures me that this is not the case. Gray may indeed have been schooled within the boundaries of the City of Westminster, but that is not at all the same thing.[27] A twentieth-century American editor of *Gray's*, Charles Mayo Goss, thought Gray had received a good classical education, on the strength of some translations from Latin and Greek authors in one of his essays, but he missed that Gray had later credited that part of the work to a colleague, a graduate in Classics from Cambridge. Goss should have concluded that Gray did not receive a classical education.[28]

Gray had two brothers, one became a lawyer and the other a naval officer, but why Henry Gray himself chose to become a surgeon we do not know. He seems to have had a strong sense of personal heroism, influenced by Britain's naval glories, which would still have been fresh in his parents' memories during his childhood. Gray's personal motto, *'Palmam qui meruit ferat'* (let him who merits bear the palm) was adopted from that of Horatio Lord Nelson, famously killed in the naval victory at Trafalgar.[29]

Negative evidence yields clues. To train as a physician, a young man had first to attend a university: Oxford or Cambridge, or a university further afield, in Scotland, Ireland, or even abroad. But there is no record of Gray having received a university education, and the young age at which he entered medical school renders it unlikely. He looks from the available evidence to have begun his studies at St George's at fifteen years of age.[30] In the 1840s and 1850s, to train as a general medical practitioner, a man needed first to complete an apprenticeship as an apothecary, undertake hospital training in surgery, and qualify in both disciplines. This was the route into medicine taken by many of Gray's student contemporaries at St George's, including Henry Vandyke Carter, as we shall see in the next chapter. But the Archivist at the Society of Apothecaries reports that no record survives of any apprenticeship for Gray, nor of any attempt on his part to apply to qualify as an apothecary, which suggests that he probably didn't train as one.[31]

If this lack of other training was intentional, as it appears to have been, it probably means that from the very outset of his medical education, Henry Gray had neither physic nor general practice in his sights. He seems to have made an early decision to make his mark in surgery alone. Others of his cohort at medical school tended to spread their learning and qualifications to allow some choice of future occupation. Gray chose the risky strategy of putting all his eggs in the one basket, and he was seemingly destined to excel in it. Of course he completed all the pre-clinical courses required of every medical student at St George's, and did well in them, but there is no evidence at all to suggest that he intended to become anything other than a successful surgeon.

The manner in which Henry Gray single-mindedly accomplished his rise through the surgical grades at St George's demonstrates that he had both the diligence and the aptitude to do well: but—perhaps critically—that he also had strategic guidance from an older surgical mentor, one who was able to advise this clever and highly ambitious young man how to make best use of his time. *Anatomy Descriptive and Surgical* was dedicated to the highest star in the hospital firmament at St George's, Sir Benjamin Brodie himself, who was known to be hostile to apothecary training for young men entering the profession.[32,33] Gray's dedication, which was given an entire page, is shown above in the Introduction.

Brodie was one of the great figures of the London medical aristocracy, a nexus of influence within the profession for many years, and an influential figure in the scientific life of mid-Victorian London.[34] He had reached that position from middle-class beginnings, completing his training at St George's under Sir Everard Home, son-in-law of the great anatomist John Hunter, and had risen swiftly to become a royal surgeon. Brodie's life bridged that important period from the Regency to the death of Prince Albert, from grave-robbing to the General Medical Council. He had been awarded the Royal Society's prestigious Copley Medal in 1811 for work on the relationship between the heart, brain, and animal heat, and by 1819 he was Professor of Anatomy, Physiology, and Surgery at the Royal College of Surgeons. Brodie was an operator in both surgical and organizational senses; he succeeded by dint of sharp intelligence, hard work, surgical competence, institutional allegiance, and an acute sense of the workings of power. He was a very big fish at St George's, and at the College of Surgeons. In 1858 (the year *Gray's Anatomy* was published) Brodie would become President of the Royal Society, the first surgeon ever to occupy that office, as well as the first President of the new General Medical Council. One cannot fail to appreciate why Gray held him in the highest regard.

The kindnesses Gray acknowledged from Brodie dated from an early period in Gray's professional career, which is vague enough to mean either that Gray came to Brodie's notice when his meteoric career at St George's was already underway, or, that their mutual acquaintance predated his arrival at St George's. It is the last idea which suggests Brodie may have known Gray's father through mutual connections and attendance at Court. Brodie served George IV when he was Regent, and as King; then William IV, and subsequently, Queen Victoria. A royal messenger and a royal doctor might easily have met on some royal business, and liked one another. It is a slender thread of inference, but it would explain a great deal about Henry Gray's professional strategy, chosen so early on in his career, and the apparent ease with which he made his way.

There may have been specific reasons for Brodie's kindnesses to Gray. It may simply have been that Brodie, who had no children of his own, took a liking to this clever fatherless youngster. Or there could have been some other reason, or connection. For example, if Gray was related to the old Apothecary at St George's (also named Gray), who had been Brodie's partner in his early physiological experiments, Brodie may have promised to keep a kindly professional eye on the young man. Gray is not a rare surname, and even with lengthy genealogical research, familial relationships are not always easy to discern, especially in the years before civil registration when Gray and his forbears were born. Stories suggesting other backdrops to their relationship may have been current in Brodie's day, but so far they have not emerged in the historical record.[35]

Brodie did exercise influence, and intervened to champion individuals. He had done so in 1834 to support Edward Cutler, a surgeon at St George's who owed his job to Brodie. It was under Cutler that Henry Gray trained.[36] In the early nineteenth century, relationships such as these were commonly the key to medical and surgical employment. Although *The Lancet* had long campaigned outspokenly in favour of merit, and against the wielding of personal privilege of this kind, it was still often the way such institutions traditionally worked. But Brodie did not have everything his own way: he had opponents within the institution too, mindful of his power, ready to confound his plans, and wary of those allied to him.

In Gray, Brodie was backing a winner: he was personable, ambitious, hard working, keen to do well, obviously competent, and dependable. Higher fees had been paid for him to become a 'perpetual' student, which meant he was eligible to make his way on the hospital ladder as Assistant House Surgeon and House Surgeon.[37] But this is far from implying that Gray's rising position in the hierarchy at St George's fell into his lap: Gray had to work hard for every step up, had to prove that he merited each promotion. He could not possibly have succeeded without merit in the fiercely competitive environment of hospital medicine at St George's. Other accomplished young men worked assiduously hard, and had to leave St George's to seek careers elsewhere.[38] Gray's dedicated work and fierce personal ambition carried him far, and fast. He had what is nowadays referred to as a turbo-charged medical career: seeming always to know when to do the right thing at the right time, how to do well in the world outside and how to mobilize personal benefit within the institution, how to navigate the system. These varieties of professional knowledge rarely accrue to a young man without the judicious advice and assistance of a mentor, even today.

At the end of the medical school year in the sultry summer of 1846, Gray was awarded prizes in both surgery and clinical surgery. That summer, the Royal College of Surgeons in London advertised a prestigious award. The College Triennial was one of the highlights of the surgical calendar for medical students and graduates alike, an important competition open to all who sought to make a name for themselves within the profession. For the next award, competitors were to submit an essay on the origin of the nerves of the human eye, illustrated by comparative anatomy in other vertebrates. Gray decided to try his hand. Over the next year and a quarter (while Carter was leaving Hull Grammar School, and starting his training as an apothecary in Scarborough) Gray was serving his time on the wards, sleeping at the Hospital in the weeks he was on night duty, studying for his College Membership examination, and working resolutely towards this essay. It was due to be submitted by 25th December 1848.

1848 was a good year for Gray. In January, he was nominated and selected as a member of the Pathological Society of London, a significant scientific medical forum in which serious public conversations were had about scientific medical matters, and professional careers were made. Several senior men from St George's were members. Gray passed his Membership of the Royal College of Surgeons, which was the surgical diploma to practise, in February. Later that same year he was appointed to three posts at St George's: a year's tenure as Post-mortem Examiner, Curator of the Pathology Museum, and Demonstrator of Anatomy.[39] All were important rungs up the ladder.

Henry Vandyke Carter had entered the medical school soon after his seventeenth birthday that May.[40] Although we don't know exactly when the two young men met one another, it is likely to have been during that medical school year, since Gray was probably in the dissection room a good deal working towards his essay, and was subsequently teaching anatomy when young Carter was awarded a certificate of honour in the same subject.

Gray was twenty-one years old in 1848. He had been attached to the hospital and its medical school for five years, since 1843 (when Carter was a twelve-year-old), and had completed his junior and senior years there, so he was well settled at St George's. With his sharp intelligence, and all the confidence of an insider, he may have seemed rather charismatic to Carter, who would have found Gray's seriousness, and his interest in microscopy, attractive. It would not have taken long for Gray to notice the shy new student whose eye was well-trained for the microscope.[41]

Gray's Triennial essay was a prodigious piece of work. He had immersed himself in comparative anatomy, very much the scientific vogue at the time. 'Anatomists of the present age,' he argued in the essay, 'are not content like their predecessors, with confining their researches to Man alone'. After an aside which dismissed the erroneous views generated by the older approach, Gray spoke of the 'great light which Comparative Anatomy furnishes'. He went on to explain that his method would be to work from 'Microscopic Examinations which bring to light what the scalpel never can reveal, and by the facts which Embryology furnishes'.

The thoroughness with which Gray threw himself into the subject is evident: he examined the anatomical origins of the optical nerves in an impressively long series of creatures, from invertebrates like cuttlefish and octopus, to a wide range of vertebrates representing all parts of the animal kingdom, fish, reptiles, birds, and mammals. Among these were all sorts of edible creatures (cod, halibut, skate, goose, duck, pigeon, stag, horse), most of which could have been acquired, with ready money, fresh from the vast storehouses of London markets and fairs. There were also some quite unusual animals, such as boa constrictor, turtle, owl, and porpoise, which probably could not be obtained in such a way, and which he probably found displayed at the Museum at the Royal College of Surgeons.

After detailing his findings from this menagerie, Gray focused right down on the microscopic detail of the brain and optical nerves in the infant chick, the human embryo, and the mole (interesting choices, given the inexperience of sight in each), drawing together his findings in separate sections devoted to the optic nerve itself, the 3rd, 4th, and 6th nerves, and the nerves serving the appendages of the eye—the 5th and the Facial nerve. The long essay ended without much of a conclusion, as though Gray had run out of time and energy, or could hardly summarize (or even perceive?) the significance of what he had learned. With an apology for the essay's length, and a mention of the amount of work he had put in (especially in personally conducted dissections) Gray ended with a grateful genuflexion towards the 'numerous elaborate preparations … in the Museum of the College … amassed by the labours of Hunter, Owen and Swan'. The essay was submitted ahead of the appointed day. Gray fully deserved a rest during the short Christmas break.[42]

Gray's truncated conclusion did not mar the essay in the examiners' eyes: the following July (1849), he was acclaimed the winner of the fifty guineas Triennial prize, a splendid achievement.[43] There is nothing to suggest Gray had curtailed the conclusion for strategic reasons, but the omission subsequently benefited him enormously. After the prize essay had been submitted, Gray wrote up the real meat of his findings at greater leisure as a scientific paper, probably polishing it up during the summer break, after the Royal College of Surgeons' celebration at which the College President presented Gray with the Triennial Prize in July 1849, and before the new medical school year began again in October.[44]

At the end of January 1850, this paper was read at a meeting of the Royal Society, the foremost scientific forum in England, where convention required scientific contributions by outsiders to be presented by an insider. The man responsible for communicating Gray's work to this august body was William Bowman, FRS. He was Professor of Anatomy and Physiology at King's College London, co-author of the best textbook of physiology on the market, and soon to be the most prestigious eye-doctor in London. If not through Brodie's good offices, Gray and Bowman were probably aware of each other through the Licensed Teachers of Anatomy (the professional anatomists' grouping which controlled the distribution of human bodies in the London region), socially, or possibly via membership of one of the medical societies active in London at the time. The two men may have met formally at Gray's Triennial prize presentation.[45]

Henry Gray was still only twenty-three years old at this stage. To witness his own work imparted at the nation's most prestigious scientific forum by such an eminent figure was a tremendous accolade for a young doctor.[46] Bowman had his own reasons, apart from the gratification of bringing on new talent, for putting Gray's paper on the map in such a public way. Gray's findings had unexpectedly precipitated him into the forefront of an argument between big beasts of the scientific world. Gray's work spoke to an argument currently raging between specialist anatomists—or what we would now call physiologists—at an international level, concerning the detailed histology of the anatomical structures of sight. Crucially, Gray had confirmed the findings of Bowman and Karl Ernst von Baer, championing them rather pugnaciously against Emil Huschke and Friedrich Henle. Gray's paper embodied the latest science: applying microscopic investigations in embryology, to confirm investigations made on the same parts in the mature animal. The imprimatur of the Royal College of Surgeons' prize award meant elements of national pride were involved in delivering it at the Royal Society. It later appeared in the Royal Society's printed *Philosophical Transactions*, a gratifyingly high profile publication for Gray's curriculum vitae, and for Bowman's vindication.[47]

This was an auspicious beginning to another significant year for Gray. He twice delivered research papers at the Pathological Society of London, and was appointed House Surgeon at St George's. Gray was also proposed as a Fellow of the Royal Society, with an impressive list of supporting Fellows, including some key figures from Victorian medicine and science, such as Henry Acland, William Bowman, Charles Brooke, Marshall Hall, C. Handfield Jones, Henry Bence Jones, Robert Lee, Richard Partridge, Francis Sibson, and Joseph Toynbee.[48] Sir Benjamin Brodie probably supported from behind. Gray's application was a success, and he became an FRS in 1852, at the young age of 25.[49]

Besides his regular responsibilities at St George's, Gray was now involved in a new research project. Another prize had been advertised, this time by trustees appointed under the will of the famous surgeon Sir Astley Cooper of Guy's Hospital. At his death in 1841, Cooper had endowed a prize of £300 to be given every three years after his death, and a series of difficult topics.[50] After the winner of the 1850 award had received his prize, the trustees had advertised the next topic. For the 1853 Astley Cooper Prize, the subject was to be 'The Structure and Use of the Human Spleen'. Gray would have been aware that the competition was likely to be fierce. The Cooper Prize was highly prestigious, and the reward munificent.

So while Hyde Park was being prepared for the Great Exhibition of 1851, and Joseph Paxton's great Crystal Palace was being erected, filled with its myriad exhibits and thronged with thousands of visitors, not far off, in Kinnerton Street and in Wilton Street, Belgravia, Henry Gray was working away on this essay. Along the way he wrote up a new paper for the Royal Society, on the development of the ductless glands of the chick, which was probably an offshoot from the work on the spleen.[51] In 1852 Gray applied for, and was awarded, a grant of £100 for research on the spleen from the Royal Society, which no doubt helped him cover the costs of the work the essay involved. The finished essay, like his submission on the retina, is impressive. It is more like a doctoral thesis than an essay, nearly 350 pages, handwritten, on blue laid paper. Gray himself referred to it on the title page as a 'dissertation'.[52]

The final work is simply structured in four parts: the embryology (called 'development' at the time) of the spleen, the structure of the organ, its comparative anatomy, and its physiology. Margin notes refer to over fifty illustrations to accompany the essay, and a similar number of anatomical preparations (dissected and preserved tissues, mostly bottled), which under the rules of the competition were to go to the Museum at Guy's Hospital Medical School, where Astley Cooper had taught. Gray's manuscript has survived, but sadly, none of the original illustrations have come down to us.

The essay opens with an impressive historical introduction, reviewing medical ideas concerning the spleen from what looks like an exhaustive list of anatomists, stretching from classical to modern times: from Hippocrates right up to the most important living European scientists. Gray's introduction concluded with the recent reconsideration by modern anatomists of the ideas of the anatomist William Hewson, concerning the close relationship of the spleen and the lymphatic system.[53] The picture presented was of international professional puzzlement concerning the spleen's function.

This detailed introduction was intended to forestall the need to interrupt his own subsequent exposition, Gray explained, by having to cite or dispose of earlier research findings (or errors) in the main body of his own essay. He made clear that he was moving on, on his own. This he did by looking at the minute anatomy and embryological development of the spleen, using chicks and rabbits. Then follows an exhaustive study of the weight of the spleen, using measurements taken after death from animals and from humans from embryo to aged adult, detailing structure both in a gross naked-eye sense and microscopically, and then the physiology, measuring and discussing blood intake and outflow, the varying size of the organ, and experimental work on animals, to ascertain the action of the spleen during life.

As in Gray's retina essay, a wide variety of creatures emerged in the comparative anatomy. Gray had worked on the spleens of dead patients in the dead-house of St George's Hospital, on preserved human spleen specimens in a number of museum collections, and on the spleens of other vertebrates—mammals of many kinds, including horses and rabbits immediately after slaughter, reptiles—a gecko, an alligator, and snakes among them, a variety of fish, and birds: cormorant, puffin, oyster bird, spoonbill, Virginian owl, ostrich, and moorhen. He created an extensive tabulation showing the relative weights of the spleen among vertebrates. His interest focused down on 'the minute structure, and chemical composition, of the most essential elements of this Gland' and most particularly, the 'entering & emerging blood … and of the Lymph'. Most importantly, Gray provided a new account of the embryological origin of the spleen, not previously described. He demonstrated by measurement that the spleen had a strong association with the digestion of food, and bodily nutrition, and argued that it served as a blood reservoir. He concluded that the function of the spleen is to 'regulate the quantity and quality of the blood'.[54] This time, Gray did a better job of pacing himself, and the work feels complete, and fully written up.

Under the same motto as before, Gray won the 1853 Astley Cooper Prize with this essay, and was awarded the £300, and the laurels that came with it. Because all the prize submissions were anonymous, neither he nor we are privy to the identity of his competitors. The competition was open to all comers, and there had been rumours circulating at St George's that one of them was to be the well-known Swiss histologist Albert Kölliker, so the win must have felt like an absolute triumph for Gray. In the following year, 1854, he had the gratification of seeing his work published as a book, *The Structure and Use of the Spleen*.[55,56]

In the twentieth century, the anatomist Charles Mayo Goss—Editor of the American edition of *Gray's Anatomy* at the time of its first centenary in 1958—wrote an admirable essay on Henry Gray, in which he analysed the importance of his anatomical findings concerning the spleen. Goss has shown that Gray was in the forefront of discovery, but that his contributions towards scientific understanding of the spleen were hardly known a century later because subsequent anatomical researchers had not noticed his importance. In particular Goss says that Johannes Müller, the German anatomist, used Gray's work, but did not credit Gray's priority in a number of key respects, including the embryological origins of the spleen, its lymphatics, nerves, and biochemistry. Goss clearly regarded the lack of citation as a culpable omission. He says that Müller's failure to cite Gray's priority was compounded when subsequent anatomists working in the area of the spleen cited Müller, rather than Gray. He had been written out of history.[57]

Goss points out that Gray must have had a fine microscope, because the things he was seeing and reporting upon were up to the standard of the best contemporary histologists of the day, such as Bowman and Kölliker. There is significant contemporary corroboration for Goss's inference, because at the time, Henry Vandyke Carter, who knew a thing or two about microscopy, commented upon Gray's microscope with admiration in his diary, though (sadly for us) without saying what make it was, or what power.[58]

The decade of the 1850s was a very exciting time for microscopy. The discipline had come on in leaps and bounds since the publication (in 1829) of Joseph Jackson Lister's discovery of achromatic lenses, which led to dramatic improvements in the compound microscope, and greatly improved the instrument's capabilities.

The field was so vibrant that places like the Apothecaries' Company opened their halls to host an extraordinary series of what were known as '*Microscopical Conversaziones*'. Keen microscopists and instrument makers (frequently one and the same) arrived with their instruments, to share their enthusiasm and mutually compare and prove the power of their microscopes on identical specimens, hastening with their competitiveness the development of instrumental design. Innovations, new discoveries, pre-prepared microscopic samples, and specially prepared exhibits revealing new fields of enquiry, were displayed and demonstrated at these events. The anatomy of the tiniest creature was said to be rendered as patent as that of the elephant. An interested audience was invited to these *conversaziones* from learned societies in the fields of science, medicine, and photography, including many eminent figures such as Charles Darwin, Michael Faraday, Thomas Hodgkin, Thomas Huxley, and James Paget, but they were also open to a wider public, including women. So busy were these 'banquets of science', and public interest was so great, that at Apothecaries Hall, a detective and six policemen were engaged to control the crowds.[59]

Henry Gray may well have been among the throng attending events such as these. So too, might Henry Vandyke Carter, who had an experienced microscopist's eye long before his arrival at St George's. Being much less financially fortunate than Gray though, Carter could not yet afford his own instrument.[60] But he would doubtless have been interested to see the varying capabilities of the microscopes on show, and probably dreamed of buying one.

Carter was attending ward-rounds and post-mortems with Gray at this period, so they were seeing a lot of each other at the Hospital. It was during a conversation between the two of them on the 14th June 1850, that Gray apparently discovered that Carter was an accomplished artist. Within a week, and for the next two months or so, the two of them worked closely on the creation of a large number of images in a variety of media. Carter mentions creating drawings, paintings, watercolours, and wood-engravings of the anatomical and microscopical structure of the spleen. After the summer break—during which Carter was keenly dissecting fish and invertebrates back home in Scarborough—he continued working with Gray again for another two months, from September to early December 1850. Gray was working on a new paper for presentation (again by Bowman) at the Royal Society.[61]

For some reason Gray omitted to credit Carter with the exquisite illustrations in his prize-winning essay on the spleen—and indeed, in the book that came out of it. Nevertheless, we shall see there can be no question that the best of them were from Carter's hand, as they reappeared credited as original to Carter in *Gray's Anatomy.*

Over the next few years Gray and Carter saw less of each other. Carter was busy making his way through medical school and studying towards his own qualifications, and Gray was busy at the Hospital, his spare time devoted to the creation of the Spleen essay, and then his first book. But there are tantalizing moments when we glimpse Gray in the post-mortem room, at the meetings of learned societies, and attending meetings of the governors at St George's—he had been appointed a Hospital Governor in 1852—so we know he was leading an eventful life. Just occasionally, Carter mentions creating new images to Gray's commission, but as yet it is not certain for what they were intended.[62]

There is a strong likelihood that Gray was also busy working in private practice, possibly with Sir Benjamin Brodie, and certainly with other surgeons from St George's. During the debacle over the appointment of Cutler to the surgeoncy at St George's in 1834, Brodie had defended himself from a public attack in the *Morning Chronicle* newspaper by a spirited address to the Governors, printed in *The Times*, in which Brodie argued that he supported Cutler for the same reason he had asked him to work as his own assistant in his private practice—because he thought him the best candidate.[63] As we shall see, as Cutler became more established himself, Gray assisted in his private practice. Private surgical assistantships were part of the process of pupilage at St George's.

Like microscopy, surgery was undergoing an extraordinary surge of excitement as its age-old barbarity was passing away before the swift development of anaesthesia. Operations on the relaxed unconscious patient were done with less trauma than on the conscious one, to the relief of patients and surgeons alike.

Gray would have witnessed much of this momentous historic change at first hand in surgical operations at St George's. The use of sulphuric ether for anaesthesia was reported in *The Lancet* in January 1847, when Gray was a senior medical student, working on his retina essay, and studying for the Membership examination at the Royal College of Surgeons. The enormous public excitement about its arrival from America might have been one of the reasons Carter was so keen to complete his apprenticeship in London. The Edinburgh doctor James Young Simpson announced his new discovery of chloroform in *The Lancet* in November that same year.[64]

One of the best-known of the early anaesthetists in England was John Snow, more famous now for his work on the transmission of cholera. Snow worked quite extensively as an anaesthetist at St George's. He also worked in private practice for Gray's elders at St George's, Brodie, Caesar Hawkins, Cutler, and Prescott Hewett, and for Gray's Royal Society mentor William Bowman.[65]

It was at a private operation for bladder-stone undertaken by Cutler that Snow recorded having met Henry Gray, at the home of the patient, a Mr Felton, in Eccleston Square, a wealthy area just south of Belgravia.[66] This operation was conducted at the end of 1853, another momentous year for anaesthesia, because that April, Queen Victoria herself had received chloroform during childbirth, and was 'much gratified' by it.[67] The anaesthetist in the case had been John Snow. Victoria's courage in using the new treatment strongly influenced its more widespread use, confounding the large number of male commentators (including *The Lancet*) who believed pain-control both unnecessary and inappropriate in childbirth.[68] What Gray thought of Snow, a quiet and clever man of provincial working-class origins (Snow's father was a labourer in York) who had served the Queen, is anybody's guess. Snow's comments in his casebook are sparing, so we are equally in the dark about what he thought of Gray, but his entry concerning the operation on Mr Felton grants us a welcome glimpse of Gray, working his way to becoming surgeon to London's high society.

A remarkable thing about the medicine of this era is the extent to which the medical men who feature in the world at which we are looking seem able to navigate extremes of wealth and poverty. Snow is one of these figures, passing from the cholera-impregnated districts of Soho to the wealthy comfort of Belgravia and Buckingham Palace. He occasionally notes having worked in the workhouse infirmary belonging to the parish of St James, in Golden Square, Soho, a district of some of the poorest housing and worst poverty in London at the time, and a locus of high mortality during the 1832 and later cholera epidemics.

Gray is another such transitional figure. Besides serving in Belgravia as a surgical assistant, in 1854 Gray was appointed Surgeon to the St George and St James's Dispensary, a charitable out-patient facility close to the Golden Square workhouse. The district around Golden Square was one of the poorest areas of the West End of London at the time. The Dispensary was an establishment of some size, and was tolerably well staffed to cover rotas: two other surgeons were employed, physicians, a cupper, dentists, and accoucheurs (male midwives/obstetricians), but there were no beds for inpatients.[69] It was allied to St George's Hospital, and doubtless referred interesting patients there.[70] Several of the senior staff members at St George's had served a stint at this dispensary. Gray certainly gained a great deal of clinical experience as a result of his work there.[71]

Gray's first book, *The Structure and Use of the Spleen*, was reviewed in *The Lancet* and other medical journals in the June of 1854, so at the time of the operation in Eccleston Square in December 1853, Gray had probably already delivered the manuscript of his new book to the publishers, J W Parker and Son, of West Strand.[72]

The Parkers are key figures in the story of *Gray's Anatomy*, for it was they who were to publish the first edition of the great book itself in 1858. Parker, father and son, will be discussed at greater length in Chapter 3, but here it is important to mention that Gray's arrival on their list of authors is both significant, and unexplained. Its significance resides in the fact that the relationship between author and publisher in 1854 was probably the most important single event influencing the commissioning of Gray's more famous book in the following year. It is unexplained, because how Gray and the Parkers actually met is not known.

A number of possible eventualities present themselves. The younger Parker, who had a keen interest in scientific matters, may well have witnessed Gray's debut at the Royal Society, and may have kept an eye on him thereafter. Benjamin Brodie is known to have been a contributor to *Fraser's Magazine*, which was published by the Parkers, and could have introduced him, or it may have happened that Gray's Royal Society mentor William Bowman—already a longstanding Parker author—recommended Gray to Parker, or Parker to Gray. The introduction of Gray to his new publishers might possibly have been made through the mediation of George Pollock, a teacher of Gray's at St George's who probably knew young Mr Parker through their mutual interest in photography; or perhaps George Kingsley, who had trained for a while at St George's alongside Henry Gray, introduced them. George's brothers were the writers Henry and Charles Kingsley, both of whom were personal friends of Parker junior. George Kingsley certainly helped Gray's professional advancement at a later stage, and could well have been important here, too.[73]

Certainly the publishers had reasons of their own for an interest in the prize essay. A decade earlier, in 1843, they had published Bransby Cooper's enormous biography of the originator of the prize, his uncle Sir Astley Cooper: two fat octavo volumes, with a beautifully engraved frontispiece of Sir Astley after a portrait by Sir Thomas Lawrence.[74] The Parkers knew about the huge prize, because the details stipulating Cooper's wishes appeared at the end of the second volume. They would have known Astley Cooper had particularly specified the spleen as a subject for the competition.[75]

It is quite likely that one of the Parkers attended the announcement of the Cooper Prize award, or that Bowman or Brodie introduced Gray to them in some other way. Gray would of course have been familiar with the fine biography of Astley Cooper the Parkers had produced, for he would surely have done his homework on the benefactor of his prize. There is always the possibility that Gray approached the Parkers directly himself. To have a book published was a step up for an ambitious young doctor, and Gray would have wanted to maximize the impact of work he had in hand: there was not much point in an unpublished manuscript languishing in a library, or in allowing the meat of it to be swallowed up and lost in a mere article. Henry Gray was quite an entrepreneurial young man, and it's likely that he would have learned from a number of sources that the Parkers were go-ahead publishers. Their shop window was a cornucopia of medical and other learning at that great intersection, Charing Cross, mid-way on his walk between St George's and the Royal Society.[76] Gray might well have seen the advertisements for the Parkers' growing medical list in *Fraser's Magazine,* or the pages of advertisements bound in at the back of their other medical books, like Bowman's *Physiology.*

Of all these possibilities, the most plausible seems to me to be the Bowman connection, but that is not to rule out some other means by which their introduction was brought about. When one tries to analyse why Gray went to the Parkers with his first book, the number of possible routes which come into play make it seem unlikely he would have gone anywhere else, although this is probably an illusion of hindsight.

Henry Gray was probably delighted when, in June 1854, a copy of this first book was laid in his hands. Such a moment is always one of curiosity and elation for author and publisher, one of the joys of the whole business.

The Parkers had done a fine job. Gray's *The Spleen* was nicely produced, its 380 pages were bound in dark brown cloth, blind-embossed with a geometric border on the front and back boards, and with rich chocolate-coloured endpapers. The whole thing was nicely set out, and the illustrations had reproduced well. The book felt much smaller and neater, yet somehow more significant than the prize essay itself (such is the transformation wrought upon manuscripts by print), which now seemed somewhat cumbersome. The new book was good to hold, substantial without being heavy. The spine was blocked very simply in gold:

THE
SPLEEN

———

GRAY

with 'PARKER & SON' neatly in small capitals much further down at the foot of the spine. The volume carried the same title as Gray's Astley Cooper Prize essay/dissertation, and the text was only marginally altered from the prize essay text in its beautiful script. With Gray having done all the work to create the essay, its transmutation into a published book had not been too onerous an achievement for him. It received good reviews. *The Lancet*, for example, referred to it as an admirable essay, commended the full and satisfactory manner in which the subject had been investigated, praised its beautiful woodcuts, and had no hesitation in recommending it.[77]

Of the fifty-six illustrations which accompanied Gray's prize essay on the spleen, a significant number were by Henry Vandyke Carter, the originals of which, as we have seen, do not seem to have survived.[78] One possible reason for the disappearance of the entire portfolio may be that Gray had requested it back from the Astley Cooper examiners, so as to be able to have the illustrations engraved for the book. It was probably still in his possession at his death. The Cooper archives preserve a record that the previous prize-winner, Thomas Wharton Jones, was granted exactly this facility, so it seems feasible that even though no record has survived of Gray's request, he nevertheless might have obtained the same concession.[79,80]

Gray would have a real feeling of éclat from the prize and the prize money, and the new book would have brought him a glow of pleasure, especially as he could give a copy to his mother, and other important figures in his life. Soon after its appearance, in August 1854, Gray was appointed Lecturer in Anatomy at St George's, and was confirmed again in the post of Curator of the Anatomy Museum, at Kinnerton Street. Gray was riding high.

In his biography of Thomas Huxley, Adrian Desmond has reproduced a wonderful sketch of Huxley, drawn by himself. He is strutting along, like a cock of the walk, nonchalant after the award of the Royal Medal at the Royal Society, in 1852.[81] Top hat, frock coat and waistcoat, silk neckcloth wound around a high winged collar, and a jaunty little stick. Huxley had a wicked sense of humour, and this is him, aware of his own absurdity, with his nose in the air and his top hat tipped back, right foot forward. After years dredging the oceans for jellyfish he was back on land at last, and getting honours. Gray must have felt pretty much the same after being awarded the enormous prestige, and the fat purse, of the Astley Cooper Prize in 1853, and seeing

The Spleen out in the bookshops, in 1854, after all that weighing and tabulating, all those post-mortems, all that messing with spleens and blood.

Of course we lack knowledge about Gray's own view of himself—whether or not he had a Huxleyesque sense of his own absurdity is beyond knowing, though no one seems to mention his sense of humour. There were contrasts between Gray and Huxley, no doubt, but there are some curious parallels too, between these close contemporaries in the scientific metropolis, not the least being the force of their individual sense of ambition, their devotion to dogged hard work, and their being showered with honours on their way up. Each was hungrily looking for permanent work, too, prestigious if possible, so as to establish households of their own.

Now, in 1855, the lives of these two anatomists collided. They must each have been well aware of the other's existence. Huxley had gained his FRS in 1851, a year ahead of Gray, and was well known in histology circles for having translated (with his friend George Busk) Albert Kölliker's *Manual of Human Histology*, for the Sydenham Society.[82] Scientific research received much better funding in Germany, and German histology was a serious force: every British microscopist and histologist knew it. Prestigious translations, especially for the Sydenham Society, were (like papers at the Royal Society or medal-winning) steps up, and earned money besides.[83] Being younger, Gray would have made less of an impression on Huxley, whose field of vision was far broader, but the Triennial and the Cooper Prizes may have brought him to Huxley's attention, even though Huxley was dismissive of prizes. And Gray was surely aware of Huxley's recent Royal Medal. The trophy they were both eyeing in 1855 was the Fullerian Chair in Physiology—a three-year lecturing post at the Royal Institution, with an annual salary of £100 a year.[84]

An impressive bundle of testimonials—no doubt from Brodie, Bowman, and a host of supportive St George's men, and more besides—was delivered to the Royal Institution in Albemarle Street on 10th May 1855, accompanied by a letter from Gray offering himself for the post.[85] Huxley did likewise. Gray later submitted another sheaf of testimonials in support of his application.[86] On paper, Gray looked excellent, personable, intelligent, well qualified and well-placed to get the post. And, after all, a previous winner of the Astley Cooper Prize, Thomas Wharton Jones, had walked into the job three years earlier. Gray hoped that he might do the same, but it was not to be. Huxley was the coming man: his brilliant lecture on 'Animal Individuality' at the Royal Institution earlier that year landed him the job.[87]

This was the first time Gray had tried for something and failed. He had been serious and assiduous in his cultivation of all the acquirements of the successful Victorian surgeon. All the way through his career, whatever he had set his mind on he had got, every exam, every prize he'd tried for: the Royal College's Triennial, and the Astley Cooper Prize, Fellow of the Royal Society at twenty-five years of age. He had been Demonstrator of Anatomy, Museum Curator, Post-Mortem Examiner, Hospital Governor at St George's. Then there was his Surgical post at the Dispensary, his Assistant Surgeoncies in private practice and Anatomy Lecturership at St George's, the *Spleen* book—all were his, and now this.

To Gray, rejection at the hands of the Royal Institution must have felt like a catastrophe, a terrible personal failure. And the whole thing was very public. All his supporters knew he had applied, and all would know he had failed. He was probably glad to retire back to the quiet atmosphere of his anatomical museum at Kinnerton Street, to hunker down, lick his wounds, and think of a new project.

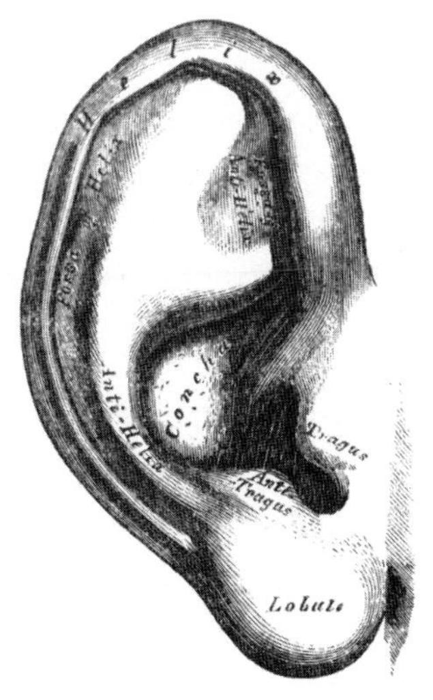

This accomplished self portrait looks to have been created after Henry Vandyke Carter's arrival in India. It reveals the essential quietness of his personality, his painstaking and skillful artistry, and his just pride in his MD.

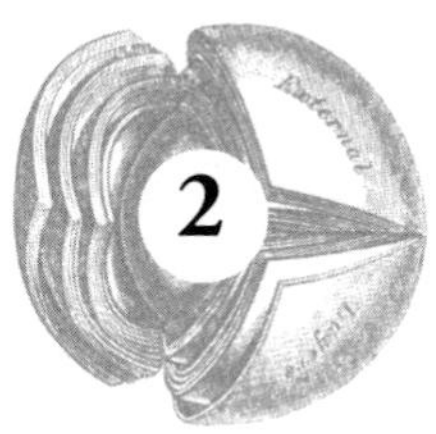

2

IMAGES

Dr Carter of Scarborough

There are a good many curious things about *Gray's Anatomy*, not the least of them being that although the book's famous selling point has always been its illustrations, the name of the man who drew them is hardly known. The book has been seen as Henry Gray's, and Gray's alone, and always referred to as if that really was the case. Gray was obviously a key force in the book's creation—the words are his, and the name on the spine is his—but above all else Carter's illustrations gave the book its distinctive visual appearance.[1]

We have seen that the two young men knew one another, and had worked together, before they became jointly involved in the work on this anatomy textbook. We have seen, too, that Henry Vandyke Carter was younger than Gray, and had come from what Londoners refer to as the 'Provinces'—meaning anywhere outside the metropolis. Carter had been born in Hull, but was christened and raised in the Yorkshire seaside spa town of Scarborough, where his parents had settled in 1831, soon after their first son was born.

Carter's father, Henry Barlow Carter, was a fine marine artist. A Londoner in origin, he was a man with a passion for the changeable moods of English coastal waters, and the dramatic beauty of traditional sea-going craft. An exhibition of his work at the Maritime Museum, Hull in 2006, revealed what a splendid painter he was, and the manifold influences upon him of contemporary artists such as JMW Turner and of his own artist friends and acquaintances, such as Samuel Prout, Clarkson Stanfield, Copley Fielding, and Peter De Wint, particularly in the use of watercolour, a medium he employed with great dexterity and charm.[2]

Young Carter's middle name perhaps reflects his parents' hopes for his artistic abilities. 'Vandyke' was the early Victorian spelling of the surname of

Sir Anthony Van Dyck, the seventeenth-century Flemish painter famous for his fine portraits of members of the English aristocracy and royal court. The more immediate reason for the choice of name may also have been that it was also the middle name of his father's friend, the Yorkshire artist Anthony Vandyke Copley Fielding, for whom their first child may have been named.[3] Despite his middle name, and the artistic skills he had learned at his father's knee, the boy did not choose to become primarily an artist: young Harry emerged from his years at the Old Grammar School in Hull having acquired the aspiration to become a medical man.

We know far more about Henry Vandyke Carter than about Henry Gray, because on his fourteenth birthday his grandmother gave him a diary, a gift he kept for the rest of his life. His diary-writing was almost stifled at its inception, however, when the headmaster at Hull Old Grammar School (JD Sollitt, Carter's uncle by marriage) discovered his diary, and humiliated his shy young nephew by reading out loud to the assembled school passages he found objectionable, then and there 'declaring his opinion of it, such being that of displeasure'. Young Carter had wrongly assumed that his privacy would be respected. Uncle Sollitt may have been anxious to demonstrate to his nephew, and to the school, that he would show no favours. During a reflective mood when he was much older, Carter commented upon his uncle's reprehensible behaviour, and expressed his own surprise that he had borne it 'so calmly at the time'.[4] Happily for us, despite this searing experience, Carter returned to his diary-keeping, and became an adept at using his own personalized shorthand to encrypt it. He annotated the little volume his grandmama had given him: 'The notes in shorthand will be understood by no one but myself.'

Carter kept up his diary-writing throughout his time at St George's, and for the first year or so of his professional life thereafter. Whether he persisted subsequently is not known. He preserved these youthful diaries until his death in 1897, and they were treasured by his family. In the late twentieth century, the Wellcome Library, in London, acquired them from a descendant.[5] It is a source of great regret that similar materials are lacking for Gray.[6]

There must have been other materials of Carter's we do not have, materials covering his medical experience, from apprenticeship and medical school right through his later career: these would have been notebooks, sketchbooks, work journals, and later laboratory and clinical notebooks. Their existence can be inferred because these surviving diaries mention virtually nothing of the daily details of Carter's student or working life, or his thoughts about it; because we have diary references to sketches, and because publications survive from his later life which mention clinical notebooks, and show that he executed prodigious amounts of work, distilled from his own pre-existing notes.

When, after studying at St George's, Carter was awarded an Anatomical Studentship at the Royal College of Surgeons under the comparative anatomist Richard Owen and the microscopist John Quekett, Carter was expected to keep a regular journal of work. Carter had to leave it behind when he left the College, and fortunately it still exists. As we shall see, this journal records a vast amount of work over the two years he was there. Carter probably kept such work-journals later in life, when he was doing important clinical and scientific work, and I have occasionally prayed that these materials, too, will one day come to light. This is not to say that his surviving diaries—intensely personal little volumes, full of small closely packed script—are of any lesser value. These modest diaries of Carter's are a gift, and a joy, allowing close observation and insight concerning the reserved young man who created the illustrations for *Gray's Anatomy.*

At least half, possibly more, of Carter's writings in these small volumes concern religious matters. His family was of a non-conformist turn of mind, and although Carter was christened in the principal Anglican church in Scarborough, St Mary's, the family did not worship there. The church they regularly attended belonged to a vibrant and demanding independent evangelical sect, with decided forms of worship and self-monitoring. Carter's mother Eliza was a deeply pious woman, and her influence upon her eldest son (who was aware that he took after her) was strong.[7]

Part of the reason Carter kept a diary, and was familiar with methods of writing in shorthand long before he arrived in London, was because it was regarded as normal at this period among religious non-conformists to keep spiritual diaries, and to record sermons swiftly in shorthand during services. Recording his own spiritual reflections, the gist of the sermons he heard every Sunday, and thinking about and commenting upon them afterwards was something Carter did regularly, right through his time at St George's Hospital. Often only a line or two in shorthand looks to have been enough to remind him of key thoughts or readings, or a preacher's chosen biblical text and angle of argument, but his religious musings and introspections occupy pages and pages of script.

Since a good proportion of the material in Carter's diaries was written *en clair*—not encrypted—much of what he wrote about his life can be accessed without decipherment. But he was usually careful to use his private shorthand when writing about others. The painful episode at school looks to have been related to something Carter had written concerning other people; he seems to have been determined never to permit his views in this respect to become common knowledge again. Apart from his often very lengthy religious perambulations, Carter's daily diary entries tend to be short memos, often abbreviated and cryptic, as if briefly reminding himself what he had done, or what he had thought: using the diary as an *aide-memoire* rather than as a personal chronicle. Longer and more revealing entries appear occasionally, but in general his records of his own doings tend to be short and concise.

Carter's weeks ran from Sunday to Sunday. He had a long-standing agreement with his mother that every Sunday was to be fully observed as the Sabbath, with walks in the fresh air to church and back (often twice), Bible reading, prayer, and no work or study, unless it be religious in character. Evenings on Sundays and Thursdays were to be devoted to mutual prayer and religious contemplation.

Sabbath observance was an intensely serious matter in Victorian England, and a cause of deep social contention. Religious moralists wanted everything but churches closed, while others fought hard for museums and art galleries as well as markets to be open for working people, for whom Sunday was the only day off work, six-day weeks being the working norm, and with long days, too. Carter's upbringing derived from the somewhat puritanical wing of an effective lobby that created the Victorian Sunday, the long-term effects of which lasted well into the twentieth century. A French visitor described a wet Sabbath in London: 'Shops closed, streets almost empty, the aspect of a vast and well-kept graveyard. The few people in this desert of squares and streets, hurrying beneath their umbrellas, look like unquiet ghosts; it is horrible. A thick yellow fog fills the air ... After an hour's walking one can understand suicide.'[8] Carter didn't contemplate suicide, but he was frequently low in spirits, especially in winter-time.

These prayerful arrangements may have reassured his mother that her boy would not go off the rails in London, known proverbially of course as a locus of all kinds of dire temptation and moral danger. Carter genuinely endeavoured to fulfil his promises, but it is clear from his diary that he suffered mostly alone under a terrible burden: although he strove for the comforts of a deep belief, he lacked faith. He described his predicament memorably in his diary: 'the face of God is hidden & my path dark.'[9] Carter fought hard for any crumb of religious comfort, and worried himself sometimes to near breakdown over the problem. At times he could barely contain his own concern, once confiding in his mother:

> but Oh! Mother, there is a something—a dark and almost impenetrable veil before my eyes—a shield before my heart—a hardened condition of my heart—which stifles if it allows of any slight conviction ... quenching the spirit ... and disheartening my soul[10]

Carter's religious musings circled round and around this heavy affliction, but he does not seem to have found a way out of the difficulty before the last diary ends. The simple comforts of faith eluded him. He searched continually for the slightest Providential sign. The damaging legacy of Calvinism—wanting to be chosen but doubting his worth—caused the young Carter to live a tormented existence stretched between the tenterhooks of hope and fear, battered by self-criticism and self-doubt, and sustained only by a dogged determination to exercise whatever God-given talents he might discover himself to possess.

In his diaries, Carter comes over as having an appealing determination about him, focused on a desire for faith, yet rooted in clear-eyed and often painfully honest critical self-knowledge. He knew himself to be reserved, and he found it hard to make friends. Carter had a very finely tuned sense of what was unworthy, paired with an inner ambition to strike high. It is not difficult to like him: he comes through as a deeply decent sort of person, trying to do the right thing, but often paralysed by his own self-doubt, hovering between smiting himself for the smallest sign of arrogance (we would probably regard it as confidence) and believing himself abjectly undeserving.

Despite his strong sense of his own unworthiness, Carter seems to have felt that he was destined for something. He had aspirations to do something memorable with his life, to find what he had been sent here for. As he grew up, his aspirations focused increasingly upon medical research. But in the 1850s, while he was training and trying to find his feet, there was no such thing as a *career* in medical research. A person had to become a doctor, and work at their research interests in and between the demands of medical practice.

Carter's uncle, John Dawson Sollitt, was an important figure in mid-nineteenth-century Hull. He was an active figure in the burgeoning educational life of the city, especially in the field of adult self-improvement focusing upon its public library, the mechanics' institution, and its various public societies. The education Sollitt provided for boys like Carter at the Old Grammar School was forward-looking, rich, and varied. Besides the three 'R's, the school's curriculum had a strong emphasis on science in its broadest sense: human and animal biology, botany, chemistry, geology, physics, astronomy, and particularly microscopy were only some of the subjects covered by the school's unusually extensive library, which no doubt, young Carter quarried.[11]

We do not know whether this rather advanced scientific training was what decided young Carter upon medicine as a career. The artist's life—as both his parents knew—was profoundly insecure, so it may have been their wish that their eldest son should have another means of earning a living. Carter himself may have acquired a partiality for scientific investigation at school. His uncle Sollitt was an enthusiast, and an adept, at the forefront of the scientific use of the microscope. The Hull historian Arthur Credland has shown that there were close associations in Hull between medicine, science, and the arts, and that Sollitt knew Edward Wallis, proprietor of the Hull School of Anatomy.[12] A scholarly cousin, Dr Henry Clark Barlow, was initially an architect, later a medical man and Dante scholar, and may have inspired ideas about medicine and scholarship.[13] There was perhaps also another significant influence: Carter's mother seems to have been unwell for a considerable period before she died, at a comparatively young age.

The family's finances were not sufficiently robust to allow young Carter to become a physician, which required a university education prior to a training in medicine. So he chose the next best thing, which was to train as an apothecary, and then qualify as a surgeon. This was both a more economical and a more traditional route to medical qualification for religious non-conformists, who as non-Anglicans were excluded from Oxford and Cambridge. Later, Carter would gain his doctorate from the newly founded University of London, which was open to all religious minorities, but at this stage such a qualification was perhaps beyond the horizon of his hopes. Initially, no doubt, too, the apothecary route was favoured because—after having been so long away in Hull—young Carter could live back at home with the family in Scarborough.

Apothecary-surgeons at this date were the general practitioners of the medical world, but the standard to which they educated their trainees was variable. Carter spent the first nine months as an apprentice working for a local general practitioner partnership, Travis and Dunn, surgeon-apothecaries of Newborough Street, Scarborough. But, although the elder Travis was interested in scientific pursuits, young Carter was far from happy. He chafed with annoyance at being treated as a dogsbody by the older apprentices, loathed their loose conversation, and felt he was learning nothing.[14,15]

Carter appealed to his parents to place him somewhere more worthwhile, and in 1848 they arranged for him to transfer his training to a London surgeon-apothecary, John Sawyer, whose premises were in Park Street, a fashionable district lying between London's Park Lane and Grosvenor Square. This time, young Carter landed on his feet. Sawyer was a good master to work for, intelligent and well educated, and supportive of the young man's aspirations.[16]

While Carter was living in as an apprentice with the Sawyer family, he also signed up to attend the medical school of his master's alma mater St George's Hospital, just a cut across the park at Hyde Park Corner. Carter completed all the required courses, in anatomy, materia medica, botany, physiology, chemistry, surgery, medicine, medical jurisprudence, and hospital practice in surgery and midwifery.[17] Indeed, he was something of a star pupil, winning awards in botany, anatomy, chemistry, and surgery, and both a junior and a senior scholarship. These last were important achievements which significantly eased the financial burden upon his parents.

Carter records very little about the atmosphere of his training in London. It would have been good to have a description of Sawyer himself, his shop, and the kinds of practice young Carter experienced while he was there. I imagine the shop itself with sanded floorboards, a shop-parlour with a rug; tall glazed cupboards to display and protect their contents and banks of polished mahogany drawers full of ingredients such as medicinal herbs and minerals. There might have been a few blue and white apothecary jars, wood and glass screens for privacy, and a heavy curtain separating the shop from the private business areas of the ground floor (including Sawyer's consulting room), the basement stores, and a closed door protecting the private domestic regions upstairs.

There would have been much bottle-washing and sweeping for Carter—spruce in clean shirt and apron—as well as grinding and blending ingredients, for all the decoctions, draughts, liniments and embrocations, linctuses, elixirs, ointments, infusions, powders, julaps and endless pill rolling the job required. The polluted air of Victorian London, full of soot and imbued with yellow fogs, would mean dusting and polishing too. Shine and efficiency were an important part of the image, so the shop window especially would require attention, with its great glass carboys and jorums of coloured liquids, and displays of pharmaceutical products to advertise the proprietor's capabilities. Sawyer may have had an assistant, but in the shop much work would surely have fallen to the apprentice: the delicacy of corking and sealing bottles, cutting and rolling pills, folding powders into papers, wrapping and tying up packets and parcels, looking after leeches. Like many religious non-conformists, Carter probably regarded everyone as equal in the sight of God, and the need to exhibit deference, in door-opening for visitors, the diplomacy and sociability of shop-talk and attendance, bowing and scraping, and attending better-off customers to their carriages, much of which to him probably smacked of hypocrisy, was irksome. A terse comment makes clear his impatience with all this: 'am determined to have as little as possible to do with shop and prescription work ... source of too much annoyance—Sawyer must see it.'[18]

Looking back on this time at a later stage in his life, Carter noted that Sawyer probably regarded him as 'steady... some odd notions, too much pride & self-esteem ... hardly practical enough and not very conciliating ... (to which his daughter would add bashful) ... some talent ...very philosophical, and decidedly industrious'.[19] While Carter cast a cold eye on the money-grubbing involved in general practice, he valued the therapeutic knowledge he had acquired, and gained real pleasure from being able to prescribe for his mother. The nature of her illness is not clear from the diary or the letters between himself and his sister Lily, but it turned out later to have been a slow form of breast cancer, which eventually killed her. She seems to have been in a lot of pain, and her loving son was pleased to be able to send his mother prescriptions for treatments she needed (though he recommends she has his prescriptions made up at a *different* Scarborough shop to the one he had worked in) and little packages of pills he lovingly made up for her himself. He discouraged her interest in the well-advertised proprietory panaceas purporting to heal every ailment, yet containing few active ingredients.

In the environs of Park Street, general practitioners of Sawyer's well-qualified sort would probably have been called into wealthy households only for minor complaints, below stairs for servants, and perhaps to stables for animals' health problems. Carter may have attended to the shop when his master went out on house calls, unless there was an assistant (or Mrs Sawyer?) to assume that duty, which may have allowed Carter to witness for himself some of the stark contrasts of London society, though probably not (in that wealthy district) its poorest extremes. This was the decade of the hungry eighteen-forties, when the Poor Law was at its harshest, governments collapsed across Europe, and Chartism was at its peak. At one point in 1848, Carter conveys a tremor of the middle-class fear he was probably surrounded by in Park Street on the day of the great Chartist demonstration on Kennington Common, and he personally witnessed a moment at the Surrey Chapel which caused him disquiet, when a woman preacher mentioned Chartism and was swiftly silenced by a plain-clothes policeman, whom Carter knew only as a churchman, Mr Sherman.[20] But in general the apprentice is almost as silent on wished-for matters as the hospital student is about the wards, the work, or the characteristics of his fellows.

In his book *The Medical Student*, the comic writer and contributor to mid-Victorian *Punch*, Albert Smith, ridicules the 'new man' who arrives at the medical school dressed like a provincial, calls everyone 'Sir', is deeply grateful to anyone who takes the trouble to explain things to him, and takes his hat off in the dissecting room. The sense that Carter or someone very like him was a model for Smith's innocent abroad is occasionally very strong as one reads, especially when Smith mentions that the new man 'thinks that beautiful design is shown in the mechanism and structure of the human body', avoids beer, regards other students as dissipated and irreligious, uses shorthand, and only eventually learns to smoke. Carter is not exactly the greenhorn Smith presents, having observed quite a bit of life before he got to St George's, but there is a very real sense in which he remained an innocent, and, to the more worldly of his fellow students, probably an object of mirth. Looking back upon his time at medical school, Carter wrote:

> at the Hospital was on the whole, p'raps esteemed a retiring, close fellow—some will say sulky & proud, serious, but hard working, and rather clever—fair memory, good comprehension—steady. Here again joined no set of men—too sober for the jolly fellows, & hardly 'fast' enough for the 'nobs' or well-dressed men—too serious for either—made the acquaintance of some of the 'nobs' but wanting 'material' it did not last—was always for bettering myself—associating with those above me (not always the most instructive society)—one of F[ather]'s precepts—where could learn or gain something: among the working men held a good standard; there being the usual 'emulation' amongst us. Amongst the Teachers, was noted for attention & enquiry: perhaps asked too many questions after lecture[s] and displayed too much zeal in detecting errors and deficiencies in their lectures, hence two [or] three (not the cleverest) are not well pleased with me to this day.[21]

In Albert Smith's book, the innocent is the only person to turn up at the Botany exam, and so gets awarded the prize. This is exactly what happened to Carter![22] One cannot fail to notice and enjoy the parallels, but even so, one's laughter cannot be anything but generous towards him. Smith was probably describing his own experience too, from a more rakish position of hindsight than Carter himself ever saw. On the day Carter heard he had won the Botany prize, he was confronted with patients dying of cholera within hours of entering the wards. One morning, thirty-two patients were found to be dead as he did the rounds.[23]

Despite his potential for social isolation, Carter seems to have loved medical school. He devoted himself to study, enjoyed dissection (which he found difficult at first, without a guide), and was determined to glean as much knowledge out of the experience as he could. He seems to have sorted out quite quickly who the time wasters were, and whose opinion he wanted to gain. When he wrote in his diary about the 'working men', Carter probably meant the men who, like himself, were devoted to working seriously, among whom Henry Gray would no doubt be counted.

At the beginning, and for quite a time afterwards, Gray probably seemed highly sophisticated to the younger man; he would have been categorized as a 'nob', meaning wealthy by comparison to Carter. But in June 1850 Carter noted in his diary that Gray was a snob, in the sense of being someone with a vulgar admiration for wealth and social position. What provoked this awareness is unexplained; but there it is, a short and shocking revelation: 'B&G snobs!'[24] The 'B' cannot be identified for certain, but there can be no doubt that the 'G' refers to Gray, with whom Carter was working daily at the time. The exclamation mark expresses his own amazement.

At the time, Carter was busily creating a variety of pieces of artwork for Gray's *Spleen* essay, so perhaps he was suddenly alert to something demeaning in his own social position, or something Gray said or did made Carter perceive him in an unwelcome new way. Carter was well aware of Gray's relative wealth and his insider status, but he had just discerned that Brodie's sophisticated young protégé had the mindset and social values of an upper servant. For someone of independent mind like Carter, such a discovery at first generated surprise and disbelief. It is a measure of Carter's situation and character that he continued working for Gray, and was still able to record admiration of him. The revelation probably became less remarkable over time. Gray is 'very clever and industrious', 'a capital worker', and 'a nice fellow' at various later stages. Aside from the social snobbery, Carter regarded Gray as a good model, not just for industry, but for the standard of his work —'what does, does well!'—and his decisiveness: 'think—do it'.[25]

These periods of intensive observational artwork for Henry Gray look to have been a summer arrangement, when the dissecting room and the school were not claiming their time. Carter mentions using Gray's fine microscope to make illustrations, so it looks as though much of the work was done at Gray's home in Wilton Street. The trouble with this sort of work was that while it helped Gray's career, and brought Carter a little extra money (of which more in a moment) it was occupying time that Carter really should have been devoting to his own studies. Carter might persuade himself that he was learning things, and he clearly needed the money. He had been so low in funds that he'd missed getting home to Scarborough the previous Christmas, and had spent bleak days in London shuffling about trying to keep warm and observing the skaters on the frozen Serpentine.

The largest tranche of work Carter undertook for Gray was to illustrate his competition essay for the Astley Cooper Prize, on which Gray was working between 1850 and 1852. Carter mentions having created paintings and drawings for Gray in this period and also undertaking wood engraving. How much money he actually earned from Gray isn't clear. It is not likely to have been very much. He often notes his hours of work, but sums of money received are hard to find.

At the very outset Carter seems to portray a scenario in which Gray had asked Carter about artwork, and that Carter had responded as if it were an honour to do it for him. He notes being miserable about his own boasting. Carter then writes the word 'shabby', and notes that another artist had charged a shilling an hour—as though Gray had invited Carter to name a price, and that Carter had discovered later to his own disappointment that he had asked far too little, and undersold his labour. Carter had known about the perils of working to commission from his artist father: here was his first real experience of it, and although he thought Gray had been shabby not to tell him what London medical artists' rates were, typically Carter's annoyance was mainly towards himself.[26] From that 'shabby', it appears that he and Gray set out on an uneasy note. On a later occasion Carter mentions having received £3 from Gray, but this was quite a while after the busiest periods of work, so it may be that he forgot to record other payments. He did count the hours he worked for Gray, so there was some kind of tally going on.[27]

London life bustled outside. Carter mentions seeing a Jack-in-the-Green on May Day, visits Greenwich Fair at Easter-time, swims in the Serpentine, occasionally goes to art galleries, and always to church. One lovely moment emerges in a letter to his sister Lily in May 1851, when he gives us a glimpse of the carnival excitement of patients and staff crowding round the windows in a ward on the topmost floor at St George's facing Constitution Hill for 'a capital view of the procession' accompanying the Queen and Prince Albert leaving the palace for the opening of the Great Exhibition. 'The crowds of people present and visible to us were vast.'[28] When Florence Nightingale later became a Governor at St George's, her first act was to demand that patients' beds be set further apart, and knowing this, one cannot help wondering how these gaggles of patients and staff got near the windows, and whether beds had to be moved aside for this exciting view to be seen.

A year later, after a tough examination on the eve of his twenty-first birthday in May 1852, Carter gained his Membership of the Royal College of Surgeons of England. That October, there was a further examination at the Hall of the City of London Company of Apothecaries, at Blackfriars. Under the gaze of the enormous portraits hung in the dark panelled rooms of this grand old hall, dating from just after the Great Fire of London, Carter finally obtained his apothecary's licence. Although he felt chastened with pity for a fellow St George's student who failed the exam, Carter himself was elated. He wrote to his sister the next day:

> My dear Lily,
> Your late very interesting letter would have been earlier answered had I not been engaged on a subject of great importance. I mean, working for the Examination at the Apothecaries Hall. This you and dear Mother will be glad to hear … I passed last night: and having worked only since I left you, and alone, I don't mind telling you, it is looked upon by the gentlemanly idlers at St George's as something of a feat, the Hall [exam] being a perfect 'bugbear' to such. I am heartily glad it is over, for I did feel somewhat nervous, and rather 'knocked up' with night-reading. The Examiners said they had very great pleasure in giving me their certificate.[29]

Carter always wrote delightful letters to his sister Lily. Even when he was in his lowest moods, his letters to her are warm and loving. He reported quite a celebration that evening with the Sawyer family for his double qualification. He signed himself, with not a little pleasure, 'H.V. Carter, M.R.C.S.E., L.A.C.' that is, Member of the Royal College of Surgeons of England, and Licentiate of the Apothecaries' Company.

Although now that he was doubly qualified his medical education was nominally over, Carter was evidently aiming higher. As a sort of reward to himself for qualifying, almost immediately afterwards, he went off to Paris, a medical Mecca for doctors who wanted to train themselves well in clinical and scientific medicine. He was thinking hard about how to approach his future career: whether he should remain a generalist, or if not, what specialism he might choose to focus upon, how he might navigate existing professional structures. Improving his French was another important consideration, and Carter worked hard at this while he was in Paris, attending plays with the script before him, so he could make out every word, and pick up good pronunciation, reading French books, and assiduously attending medical lectures and ward rounds. While away from home, Carter continued his Sabbath observance and attended the English church in Paris. He commented thoughtfully on the sermons, and continued to monitor his own moral progress. He was highly critical of lazy and dirty-minded students, and various 'fast' characters whose paths crossed his.[30]

Carter wrote home that in Paris he could live on £3 a month, which puts the six guineas the Apothecaries' certificate had cost him into extremely sharp perspective. He studied anatomy and surgery as well as midwifery, and attended ward rounds with some of the most celebrated doctors of the day. As he studied, his French improved apace. Carter was certainly lonely in Paris, and sometimes racked with religious doubt, but he grew to like it for its opportunities, and during his visits to the various medical facilities open to foreign students, he was fortunate enough to encounter one or two serious fellow doctors with whom he would remain in touch in after years.

In his personal habits Carter was abstemious, but in Paris he indulged a liking for the occasional smoke, sauntering over the bridges of Paris, looking at the moon, and watching passers-by. There is an extraordinary passage in his diary in which Carter described for his own interest the effect upon himself of smoking a pipe, which reveals what a careful and sensitive observer he was:

> pipe and stroll in quai. nearest open place and representative of Hyde Park. London, Bridge Scarbro'! effect of tobacco—mild—pleasant taste—warmth—stronger flavour. slightest nausea—then in 5 minutes a sort of general numbness or relaxation—hardly feel ground walking on—sensation diminished—thoughts expand … impression less attended to in fact a sort of intoxication follows—this lasts as long as smoke—increasing somewhat it might end in … prostration p'raps—feel no more afterwards. p'raps on contrary in a more genial mood. formerly had a slight headache wch. soon passed off. along with all this a creeping cool sensation over surface of body—p'raps increased by cold—occasionally tendency to increased nausea: these are undoubtedly the effects of the tobacco—never felt them after a cigar wch is more heating of stronger taste & effects much longer.[31]

This extract gives a good impression of Carter's abbreviated diary-writing style, but its length is uncharacteristic: most of his entries are shorter. He occasionally made notes so evocative that you can sense him sizing up the view as if to paint it—the light-effects of moonlight and mist on water, the river barges and their reflections. He had an artist's eye, of that there can be no doubt. Occasionally his reported visualizations took the form of anatomical structures, for which he dreamed up ideal drawings:

> design in head for making drawings of arteries &c. but want steadiness—would like to do it yet intend to leave Paris in a month! shew great want of decision in all that do—money goes fast.[32]

These ideas remained no more than idealizations in his mind's eye for now, but when he made the decision to extend his stay into January 1853, Carter did note in his diary that he was 'drawing', a sign that this was serious work. Sadly that is all he says. In the light of his later work on *Gray's Anatomy*, it would be intriguing to know what Carter produced in Paris actually looked like: such drawings might provide us with an indication of his evolving anatomical style, and the thought behind it. Sadly they seem to be lost.

The Paris diary reveals Carter asking himself a vital question, evidently (for him) both inescapable and unanswerable, vexing away at the intersection of his religious musings and his medical training: 'On what element of mind does the Holy Spirit act?'[33]

At this stage, Carter had no idea what he was going to do, or where he might practise, though he did know what he hoped to avoid. He was qualified as a general practitioner, and yet it seems he didn't yearn to be one. He had chafed under the burdens of apothecary shop-work, but was aware that running a successful shop was crucial to economic survival as a general practitioner. He had observed the economics at work, seen the problems caused by unpaid bills, the difficulties of being on-call 24 hours a day, and all the shifts to which general practitioners were driven to make a living. Carter lacked the capital to start up in business alone, and did not relish the idea of joining as an assistant. 'Have but little mind to face troubles of practitioner', he wrote at the beginning of a new diary in January 1853.[34]

Carter also felt stymied by himself:

> Have a love of science & the higher branches of the profession: too little confidence in self to strike high, and risk consequences: too little decision to follow one branch: too little energy & too little of other requisite qualifications—perhaps ought to be content with a lower station; yet & here seems the rub,

> my ambition is but just enough raised to cause inquiet, and I flatter myself understanding and mind are not wanting:—this is the poison—too little humility & trust in Providence: and former success only fans the flames.[35]

The suggestion of a post on offer in Scarborough provoked him to express his sense of being capable of higher things—though as yet he had no idea what they might be.

Carter had hoped that he would go far at the medical school by winning prizes, but as he had progressed through St George's he had begun to perceive that was not actually how the system worked. Cleverness and hard work were not enough. The trouble was, all the Hospital jobs were tied up, and without help such a career was almost impossible to attain. Help, at this time, invariably meant the patronage of an influential figure: it was something Carter not only lacked, but scorned. Obtaining professional progress by such means he found repugnant and unworthy. Like Richard Monkton Milnes, another high-minded contemporary, Carter wanted to work his way 'unpatronized or not at all'.[36]

Carter was of a newer school, one that believed profoundly in meritocracy—the best person for the job. This is exactly the ideal Gray had proclaimed when he used Lord Nelson's motto: *let him who merits bear the palm.* But Gray had Brodie behind him nevertheless.[37] The tide was turning against patronage in the metropolitan hospitals and their medical schools; but it had not yet changed at St George's. There were people ahead of Carter well qualified and waiting for whatever posts might become available, and with help at hand.

Of the possibilities for research in general practice, Carter seems to have had a poor opinion. There had been some general practitioners—such as James Parkinson (of the *Shaking Palsy*)—who had managed successfully to combine research and practice, but they were very rare birds, and Carter knew it. George Eliot's novel *Middlemarch* was not yet written, but the bright young enthusiast Dr Lydgate touched nerves when he appeared in fiction because he was the type for a mid-Victorian phenomenon—the forward-looking and idealistic young scientific doctor to whom existing professional structures were closed or uncongenial, trying to make an independent way in the world. A comment in one of Carter's letters expresses his uncertainty at this period. Carter mentioned to Lily that he was toying with ideas of going on a short scientific posting abroad. This would probably have been such a job as Thomas Huxley had filled on HMS Rattlesnake, as a ship's doctor/naturalist.[38] Carter was thinking about short-term opportunities available in Australia or India, but urged his sister:

> don't alarm [Mother] Lily, I may just as likely live in a country dell in Yorkshire.[39]

London was much more costly to live in than Paris had been. Carter's scholarships had ended, and he was no longer living-in at the Sawyers'. He found lodgings in a shabby area of Pimlico, to the south of Belgravia, a district we would nowadays call Victoria, after the railway station soon to be erected in the old canal basin nearby. Lack of money was a running theme throughout Carter's medical training period, as it was (and sadly is, again) for many poorer students. He had observed that those with too much money were often lazy, or were unfocused, dissipating their talent, and—as we've seen—he was careful to avoid such 'gentlemanly idlers'. Now, though, his scholarships ended, Carter was having to learn to live on very thin finances indeed.

In Paris he occasionally ate out, but towards the end of his stay there, mentioned snack-type meals which cannot have been very nutritious, such as a couple of bread rolls with a drink of hot chocolate. He was aware of being thin. In January 1853 he had found his sputum tinged with blood, a sign that he might have consumption (tuberculosis). He suffered alone with this news, and did not frighten Lily or his mother.[40] Back in London, now that he was qualified, he might have obtained occasional Saturday work as a locum for his old master, John Sawyer, or the odd artistic commission from staff or colleagues at St George's, but none of this was lucrative. His earnings were not only slender, but irregular, and hard to come by, always in arrears. It looks from his diaries as if his father was having a lean time, too, and was not able to give him much in the way of allowance, or, that Carter had chosen to use whatever parental funding he had to finance further postgraduate medical study, and had little left to live on. At this period, Carter was suffering money shortage particularly sharply: there is occasionally a slightly light-headed feeling in his writings that suggests that he was sometimes hungry.

An old medical school song performed at St George's wittily characterizes the predicament of the newly qualified medic, who keeps trying to find paid work, but who always returns in periods of unemployment, to the Hospital. The refrain is 'Back to the Corner again'.

> I'm here in a silk-lined frock coat
> With a very professional voice
> Lounging around with the surgeons
> As though I were here by choice.
> My knowledge of medicine's rusty
> My knowledge of surgery's nil,

So I'm watching the simplest dressing
And how to prescribe a pill.
A man who has squandered hundreds
Learning a noble trade,
Joyfully thinks how some day
He shall be thrice repaid.
Then his hopes he pins to a brass plate,
Labelled L.R.C.P.,
Ah, first year's man, that was my plan,
And now, well, as you see, I'm

Back to the Corner again, Sir,
Back to the Corner again,
Three years on the sea,
A year as GP,
And back to the Corner again.[41]

The song well describes Carter's predicament at more than one lull in his career in the next few years, and indeed Gray's position after his failure at the Royal Institution.

Carter had been pleased with his surgeon-apothecary qualifications, and had found medical life in Paris absorbing, but he saw clearly enough now that they did not qualify him for science. He looks to have decided even before he had left Paris that the ship's doctor idea was 'by no means essential' to his own advancement in life, and that what he really wanted to do was to stay in England, where he would be within reach of his unwell mother, and could work for a university degree in medicine.[42]

In the short term, he was aiming for a two-year special Anatomical Studentship at the Royal College of Surgeons, which came with a small annual salary of £100. This, he hoped, would allow him to scrape by, while he was studying for the university bachelor's degree in medicine (M.B.) that was now in his sights. The College's Hunterian Museum was famous across Europe, for its vast and growing collections of human anatomy and pathology, comparative anatomy, and fossils. The Anatomical Studentship promised all the interest of a serious beginning in a scientific institution. He must try to survive financially until the competitive examination in June. Back at the Corner again, a series of entries in February 1853 show Carter endeavouring to find or create a niche for himself, as he sought advice from various older doctors:

> 4th February 1853: 'To George's see Pitman ... Gray working and well placed as curator & fully alive to his advantages—envy him.'

> [5th February 1853]: 'Gray told me yesterday that Hawkins said his preparations &c 'were a credit to England'! high praise & very satisfactory to Gray. See Pollock who said my turn would come soon.'
>
> 14th February 1853: '… saw Owen who was kind and promised help'.
>
> 28th February 1853: 'Dr Crisp advised to advertise in his rough way'.[43]

At some cost to his already thin budget Carter took Dr Crisp's advice. On 3rd March he called at the Strand offices of the medical journal *The Lancet*, to place an advertisement, leaving a note for the Editor entreating that it be given a good spot. Carter proudly pasted the cutting into his diary: it indicates the kind of work he was pitching for:

> Medical Artist.—A young gentleman, M.R.C.S. and acquainted with Pathology, the Microscope, &c., is desirous of assisting gentlemen engaged in scientific research by making Drawings. Specimens will be furnished on address to H.V.C, No. 85, Upper Ebury-street, Pimlico.[44]

How much work this advertisement actually brought Carter isn't at all clear. There are occasional references in his diary thereafter to small sums of money deriving from work of this kind, and to money still owing. Most of his commissions appear to have come from people he knew from St George's, however, so the advertisement may not have brought him much work, beyond reminding colleagues and superiors that he was on hand, was serious about using his artistic talents, and wanted paying. No major projects ensued until 1855, when he was commissioned to create a substantial series of over forty display images on microscopy, and some anatomical diagrams for the Medical School, work that probably came his way through Gray.[45]

Carter managed to survive, studying hard, attending medical rounds at St George's, and working away in the dissecting room at Kinnerton Street in preparation for the demanding exam at the Royal College of Surgeons. In June, after two full days' gruelling examination, in mid-summer heat, dissecting the brain and front limbs of a small mammal, and of a large bird, he was rewarded with the College Anatomical Studentship.[46]

The Hunterian Museum at the Royal College of Surgeons in Lincoln's Inn Fields sustained a direct hit during the London Blitz, and now holds only a shadow of a remnant of what Carter was surrounded by during his time there.

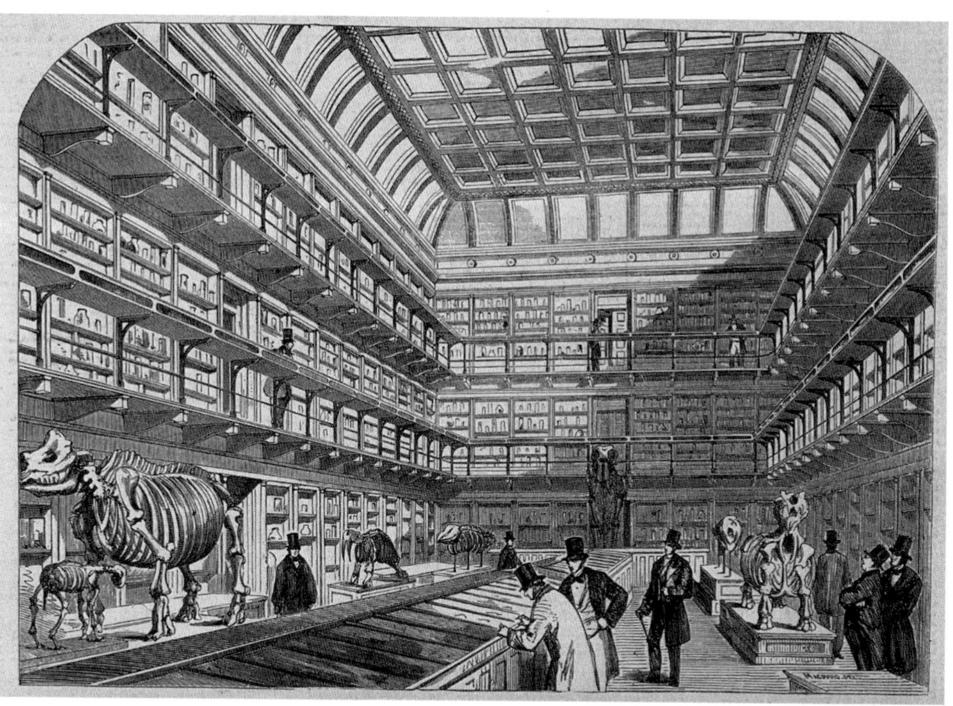

The Hunterian Museum at the Royal College of Surgeons, Lincoln's Inn Fields, London, when Carter was working there under Richard Owen and John Quekett. By Macquoid, from the *Illustrated London News*, 20th May 1854.

Pictures of the great multi-galleried museum as it was show it to have been an extraordinary repository for the study of comparative anatomy. Built around an extensive core collection from the eighteenth century surgeon John Hunter, by 1853 it contained large numbers of human specimens displayed alongside those of thousands of animals, from all corners of the natural world. The collection was rapidly expanding, receiving donations and accommodating purchases of curiosities—animal, human, and fossil—from anywhere and everywhere. It was overseen by two of the most remarkable scientists of the day: the famous comparative anatomist Richard Owen, and John Quekett, one of the most important microscopists of the Victorian era whose textbooks on microtechnique and histology were among the first in the modern era.

Carter inhabited this building for two years when the Hunterian Museum was at its peak, before Quekett died, before Owen resigned to establish the Natural History Museum at South Kensington, and before Darwin's theory of evolution. For the College, the point of Anatomical Studentships was to provide economical assistance in the curating process. The labour provided was 'various, useful and valuable to the College' and 'greater than could be obtained for the same money under other arrangements'.[47] Carter was worked hard. For the duration of the studentship, between June 1853 and June 1855, he worked diligently, dissecting and drawing everything he was asked to—which in that place at that time, was a considerable amount—from human specimens to those of exotic mammals, reptiles, fish, and birds, from all over the world. He was constantly on-call, so as to relieve Quekett and Owen from the responsibility of showing visitors round the Museum. There was also a lot of shifting of displays, and labelling, bottling, preserving, and drying of specimens. Between bouts of what turned out to be Quekett's fatal illness, Carter drew for him, wrote from his dictation, and helped him catalogue the College's histology specimens. Carter also dissected and drew for Owen, produced drawings and dissections for other figures in the College hierarchy, and created special display materials for use in College Lectures.[48]

One of the more unusual of Carter's tasks for Richard Owen was to work on the dissection and recording of a full-grown female anteater. The animal had been purchased by the Zoological Society, in September 1853 for the then enormous sum of £200, from some itinerant Germans who were discovered exhibiting it in the street somewhere near the British Museum. The creature generated enormous interest but survived less than a year in the Regent's Park Zoo. The anteater's corpse was donated to the Royal College as a specimen for the Museum, on condition that any morbid appearances discovered during post-mortem examination were recorded in drawings, and a written report concerning its anatomical interest and the causes of its demise were delivered to the Zoological Society. The work on this animal at the College, Owen thought, was a world first: 'no description of the soft parts is anywhere published'.[49] Within a week or two of its arrival at the College, Owen delivered a verbal report on the dissection to a meeting of the Zoological Society, accompanied by a number of 'accurate and beautiful' drawings, which Carter had made.[50]

Henry Vandyke Carter learned a great deal while he was at the Royal College. Besides all the obvious skills—like cataloguing, caring for specimens and museum displays, and relating socially to visitors—he was involved in scientific work with Quekett which he looks to have enjoyed greatly, finding specific gravities for the bones of a great variety of creatures, learning accurate titration methods, learning practical skills like shaving thin slices of bone for microscopic examination, and so on. Some of these were very new techniques: he was learning advanced scientific investigation in the natural sciences. The technical education he received under Quekett at the Royal College stood him in good stead in his own later researches, and when—later in his career—he had charge of a museum himself. Carter greatly valued the opportunity to attend the College's lectures, especially Quekett's, and in May 1854 noted: 'The Histological Lectures are now ended—doubtless to the regret of many, and amongst those, myself.'[51]

The variety of work in which Carter was involved can be suggested by the fact that on the 7th February 1855, Richard Owen laid before the Committee twenty specimens in spirits for the Museum's Physiological Series, including parts of a Great Anteater, Indian Rhinoceros, Sun Fish, Indian and Egyptian Crocodile, Seal, Fin-backed Whale, Sperm Whale, Frog, Wombat and Cuttlefish, with their appropriate descriptions for the Catalogue. These and others had been finished since Owen's report to the previous meeting a month earlier.[52]

Carter was learning fast: learning from his elders, and observing them. A revealing passage in his diary during this period (and an unusually long one, for Carter) reveals him pondering the scientific men he saw about him at the College, and the relative merits of generalism and specialism in scientific medicine. Carter is thinking aloud in such a mature way (he is still only twenty-two years of age here) that the passage is worth quoting:

> [18th March 1854]: Wish to record present tendency of mind as regards differences among men in social & mental rank. Mixing so much at the College with scientific subjects, and seeing not a little of men filling the grades of scientific standing in our Profession, mind naturally occupied with observation and involuntary induction thence. All ideas now concentrated on distinctions and the means of attaining them: seem to be always dreaming of writing or publishing some great fact or facts—always hovering about—watching the adventurous ones, and thinking of essaying to fly, self: very much disregarding however the preparatory hops and short flights, of those who now fly high.
>
> Every one seems to have confidence in himself above all that is around or external to him: willing perhaps to learn but still judging in his ignorance, which is allowed and admitted. Must not such judgment be imperfect? The whole resolves itself into this: the more one knows of a man, the more difficult it is to retain favourable opinion or implicit reliance on him. Of course there are exceptions yet the first fact mentioned, <u>partiality of knowledge</u> generally holds good: it is quite independent of correctness of knowledge. For false facts are far more abundant than false theories: (so says Cullen).
>
> Owen remarked in his lecture, today, that simply to <u>observe</u> requires a training: two persons are generally concerned in every fact—one discovers

> part: the other completes & corrects. .... Owen—about as perfect as any yet seen—yet not difficult to trace steps in his detections—they are not intuitive—Quekett—knowledge more partial, partly perhaps because his field more extensive and less worked yet, & not so easily, has good powers of perception, based on memory &c... [Carter names numbers of men he has observed] ... in fact the majority, wanting on most sides, but one: [Benjamin] Carpenter may be a partial exception. Ed Smith & Crisp—notorious instances [of partiality, or limited vision]...In fact pit these men against each other and every one shews a blank side: exceptions may be spared because of their extreme speciality, few caring to judge them of that ...
>
> A mans eminence consists then in a cultivation of his speciality to the highest pitch. Quandary of student who wants to learn everything—is he to exercise self-denial and deliberately choose one subject out of many that he is almost equally interested in, or is he to wait till circumstances seem to indicate, or offer to him, one path? ... or is he to go on, generally, and remain content with a fair knowledge of all, and so remain obscure? ... all departments of science are intimately connected and a mediocre knowledge of all will certainly advance him in any one ... Arguments may be made for either ... the general practitioner is perhaps most useful to the world; when well informed, he is inferior to neither, though obscured by either; being, after all, a receiver, rather than a creator: a giver at second-hand. I wait, gathering all I can, meanwhile.[53]

These questions were deeply significant for Carter, his desire to be a generalist tugging against his perception that eminence in the scientific world was associated with specialism. Typically, he left the matter floating, awaiting Providence to show him a path.

The workload at the RCS was onerous, the pay meagre, and relationships were sometimes strained. Quekett was repeatedly absent due to deteriorating health, and Owen was distracted by the strategic development of his own career. Carter was certainly not receiving a shilling an hour for his exquisite drawings, and although there were rare languors, at times the work expected of him invaded time and energy he needed for personal study.[54] He became aware that a measure of exploitation was involved in his post.

The fundamental problem was that world-class scientists like Owen and Quekett genuinely needed salaried assistants, for which the College would not pay.[55] The studentships had been promoted as a boon to those who won them, and competition for them was fierce, but in fact the direction of benefit was questionable. For Owen and Quekett the studentships were invaluable, as it relieved them from a great deal of drudgery, and from having to drop their work for visitors. But the pressure on Carter was eventually driven to a pitch where—quite uncharacteristically—he actually remonstrated. Relations were renegotiated satisfactorily. But by the end of his two years at the College, Carter had grown to perceive that the studentship had compelled him to bestow intellectual and artistic benefits gratis on superiors, and he had become bold enough to record his awareness of the fact: 'In the course of this work have made out some interesting and practical facts, which will form original points in Mr Hewett's Lectures', he wrote in his college work diary for May, aware that other eyes would see what he had written.[56]

Apart from such matters, probably the most significant problem with the studentship was the working accommodation, which was atrocious. The room in which Carter had to work was in the roof space, high up above the Museum. It was one of three that opened off a gloomy corridor, adjacent to the macerating room, where all the museum's animal specimens were soaked and injected for preservation. The atmosphere in there was ghastly: it was piled high with bottles and jars, hanging bladders of chemicals and waxes for preserving specimens, lead for sealing specimen jars, and tanks of preservative alcohol and chemicals. The rooms had ventilators and windows permanently open, but the design of the roof space was inadequate to clear the fumes from all the specimens and the chemicals used to preserve them, and the dust of decay and decomposition. Combined with the naturally rising vitiated air from within the building, 'certain odours' from the walls of boxes of dry bones (including lion, penguin, albatross, wolf) and specimens stacked within his room itself, the place made Carter ill. He mentions illness on several occasions in his diary during the period he was at the College. From reports of the working conditions, and the kinds of material he was sometimes exposed to, one cannot really wonder, especially when cholera and the fear of it was stalking the city outside.[57]

Outside the College, Carter was still attending lectures and ward-rounds when he could. In what was left of his personal time, he was studying for a University of London Bachelor of Medicine, a degree established as a radical alternative to the old exclusive Oxbridge qualification. Set to high standards, it had considerable status. In November 1853, Carter went through a personal crisis when he failed at his first attempt. He was devastated. 'The examination at the L. University has overthrown me—<u>I am rejected</u> … I burn to wipe out the stain.'[58] But Carter persevered, and a year later, in December 1854, he passed with honours.

By this time, if not before, it seems that Carter had become well and truly hooked on the idea of a life in research. The burgeoning field of microscopy was where he was happiest.

Today, we have far more effective microscopes, whose results may be demonstrated by specialized forms of photography. No such equipment existed in Carter's day: the person using the apparatus had not only to be a skilled operator of the available equipment and a meticulous observer, but in order to share their results, had also to be a good recorder of what could be perceived. Working under Richard Owen at the College had not been easy, but it had taught him that there was new knowledge to be discovered, and that with hard work, quite unprepossessing people like Owen himself, could find, record, and publish it. Carter's long experience with the microscope, and his artistic aptitude, gave him an enormous advantage: he was both a fine observer and a fine recorder of what was discernible through a Victorian microscope. Carter knew there were worlds to be discovered, and even in gross bodily anatomy nothing was to be taken entirely on trust: there were still secrets of the human organism to be found and verified. The relationship of eye, hand, and scientific exactitude was intense: Carter was in his element: drawing at the edge of what was known.

The difficulty looming for him, however, was that there were virtually no salaried jobs in microscopy at the time, and certainly no career structure. Having no desire to trouble his parents for further financial assistance, Carter needed to earn money to survive another couple of years, until he could get his full university doctorate in medicine. He was pondering the quandary of his longer-term future, and still no nearer a proper job, when the Royal College of Surgeons studentship came to an end in mid-June 1855.

Doubtless Carter was glad to vacate his fume-ridden room in Lincoln's Inn Fields, but the loss of the meagre regular income was a serious concern. During his last month at the College, Carter was again full of self-doubt: 'What manner of man am I?' he asked his diary. As the prospect of unemployment grew closer, the pitch of his fear rose once more: 'Money matters ... future uncertain'. News from the Crimea hardly helped his mood: 'This war is dreadful', he confided.[59]

Carter's mother may have sensed something amiss, because there is a charming letter from her with his papers in the archives from early July 1855, when Carter's brother Joe was staying with him in his lodgings at 33 Ebury Street. She writes:

My dearest Boys,
I owe you each a letter, this scrawl must be addressed to both, I trust I need not endeaver to impress on your minds the certainty of my dearest love, my continual and most solicitous anxiety for your welfare temperal & spiritual, daily I remember you, I have been particularly anxious about you in this severe weather. Pray have <u>good fires</u>, and everything comfortable, <u>do mind this</u> ....
With our united love I remain <u>my darlings</u> [underlined four times]
Your affectionate Mother
Eliza S Carter.[60]

Back at St George's again, Carter sought to earn some money as a medical artist, as before, and with Gray's help, managed to obtain a large project creating a large series of forty new display drawings of microscopic subjects and other diagrams specially for the Medical School. Other paid work was scanty, and only tiny sums were trickling in, irregularly. By August he was short of money, and overcome with anxiety: 'No adviser, no companion, and the cream of life passing away.'[61]

Two things however, showed the professional tide might at last be turning for him. Two older colleagues at St George's came forward to help: Prescott Hewett proposed Carter as a Member of the Pathological Society of London. Then, Athol Johnson asked Carter for some special illustrations, and other help on a paper he was writing for presentation at the Pathological Society. Johnson would later give Carter public credit for his efforts.[62, 63]

Carter was still worrying about the coming year, finishing the big series of drawings and diagrams (he still awaited payment for those already on display) and anxiously doubting his fitness for the doctorate, when Henry Gray came to him with a proposal: would Carter undertake the illustrations for a new manual of anatomy for students he had in mind?

ENTERPRISE

J W Parker & Son of West Strand

Before Piccadilly Circus got its neon lights, and became proverbially the busiest junction in all London, the place with that status was Charing Cross. The publishers of *Gray's Anatomy*, John W Parker and Son, had their shop and offices there, and indeed lived there, too, so it is to Charing Cross our narrative now turns.

Geographically placed at the pivot of access to Whitehall, St James's, and the City—between government, crown and financial centre—Charing Cross was London's hub. It was probably inevitable that the old settlement should have become a focus of pedestrian and wheeled traffic, where criminals were pilloried, and shops thrived. Historically all distances were measured from this spot. Dr Johnson found the full tide of existence at Charing Cross, and in the 1850s this was where George Augustus Sala met his readers to begin his tour of London, in his famous *Twice Around the Clock*.[1]

Before the clearance of Trafalgar Square and the arrival of the railway in the 1860s, the area had been clustered with the alleys and the higgledy-piggledy premises of an organic settlement, an area of cheek-by-jowl accommodation and opportunities for commerce, business, employment, and social intercourse. There were hotels, coaching inns, and classy tailors, warehouses for luxury goods—glassware, hats, and lace—food shops and cookshops, with accommodation above, below, and behind for living and working and making. In the back streets were innumerable small workshops, of assorted kinds, many of which were associated in a variety of ways with the printing industry. The Strand was the western extension of London's printing centre in Fleet Street.[2]

Nash, the Regency architect, had planned a grand square at Charing Cross, and a grandiose redevelopment scheme to push his stucco confections (like Carlton House Terrace) eastwards along the Strand. He did not live to see Trafalgar Square, and not all his plans ever came to pass. The demolition of the innumerable ancient small shops and houses behind St Martin's-in-the-Fields, however, did result in the dignified 'Pepperpots' of the West Strand development: a long terraced front, clad in stucco, with two round towers as corner features, standing close to each other like a domestic cruet. West Strand had shops at street level (like Nash's Regent Street) but of a different kind from those they had replaced: higher ceilings, larger display windows, regular dimensions, and higher rents.

Although the outer shell of the Nash building still stands, the shop John W Parker occupied between the 1830s and the 1860s has vanished. The entire triangular Pepperpots block was gutted in the 1970s, and somehow, number 445 was lost in the subsequent renumbering.[3] But if you look at the windows of the upper levels of the stucco terrace, and count along from the westernmost pepperpot towers at Adelaide Street, the sixth and seventh windows are those which provided light to Mr. Parker's upstairs front sitting room. His shop, and the side entrance door which gave access to the upper storeys, lay directly beneath.

There were two Mr Parkers. Rather confusingly, father and son had the same first names, John William, and both used both names. For reasons of clarity I shall refer to them by age: senior and junior, or older and younger. The senior Mr Parker was himself the son of an officer.[4] As a young boy he had been apprenticed to one of the most go-ahead printers in London, William Clowes, and had then established himself as an independent copperplate printer near Leicester Square. Soon, probably through becoming involved with the publication of a well-known musical magazine, *The Harmonicon*, Parker again became more closely involved with Clowes, and was eventually persuaded to become manager of his thriving press.[5]

Clowes had originally occupied premises in Northumberland Court, a back street on the south side of the Strand diagonally opposite the site of the new Pepperpots. There he had established his own typefoundry and his busy hand printing-presses. Clowes's steam-presses are said to have been the first to be used for book-printing in London, but the noise of his works prompted the aristocrat who owned the old mansion of Northumberland House next door to sue Clowes for nuisance in 1824. The court found the printer entitled to follow his trade, so the wealthy lord offered Clowes a large sum to move away instead. Clowes took the money and relocated to the works of the printing-machine entrepreneur Applegath, just south of the new Waterloo Bridge. He went on to become one of the most prolific and important printers in London.[6]

In 1830 Parker was living in Stamford Street near Clowes's works, with a house and stables, and his own printing workshop at the back. His fortunes had risen from an insurance value of £400 in 1817 to £3,000, so he was doing very well indeed.[7] Parker may have been led to believe that he might be taken into partnership as Clowes had no sons, but when Clowes's daughters both married other printers, Parker decided to try his fortune elsewhere. By May 1833, he had established himself back at Charing Cross, no longer describing himself as a copperplate printer, but as a printer and publisher.[8] Parker's home and printing press was now in St Martin's Lane, but his retail premises were more respectably situated in the new Pepperpot building, at 445 West Strand. Being clever and entrepreneurial himself, and perhaps with help from Clowes, Parker swiftly found work printing books and pamphlets for the Society for Promoting Christian Knowledge, and other work on his own account.

The 1820s and 1830s was an era of political turmoil, with the agitation for an extension of the franchise and for social and political reform. It was in this period that the mass reading public emerged as a force to be reckoned with. The government tried its utmost to control news and book publishing, imprisoning and bankrupting many individuals who published or sold 'unstamped' news sheets and other radical literature, but eventually the cultural pressure became so great that the government was forced to back down, and reduce its 'taxes on knowledge'.

Of course there was much concern on the part of the 'haves', about the new reading matter now becoming available to the 'have-nots', and a great many people became involved in a variety of ways in generating emollient, uncontroversial, or 'improving' literature. The *Penny Magazine* was a product of this era, issued by Charles Knight for the utilitarian Society for the Diffusion of Useful Knowledge, whose works were printed by Clowes. The Society for Promoting Christian Knowledge and other religious organizations created a mass of cheap 'educative' material in a religious vein. Parker published a great deal of this material for the SPCK. He was far from being a radical, but he was an interesting publisher insofar as he seems to have chosen not to publish the more mawkish and sentimental of the religious tracts of the day, preferring to focus on educative works on a wide range of subjects—especially on nature and the wonders of creation—for an intelligent reading public.

The politics of knowledge-distribution was by no means invisible in this period, and educative works were mass-produced to keep the lid on a vibrant radical culture that, for at least fifty years after the French Revolution, was suppressed because it was perceived as dangerous by many in power. Parker's output contributed to a flood of educative literature, generating managed texts for teaching, legitimating the status quo.[9] Though his SPCK books did guide the reader in an establishment and a religious path, Parker's output tended

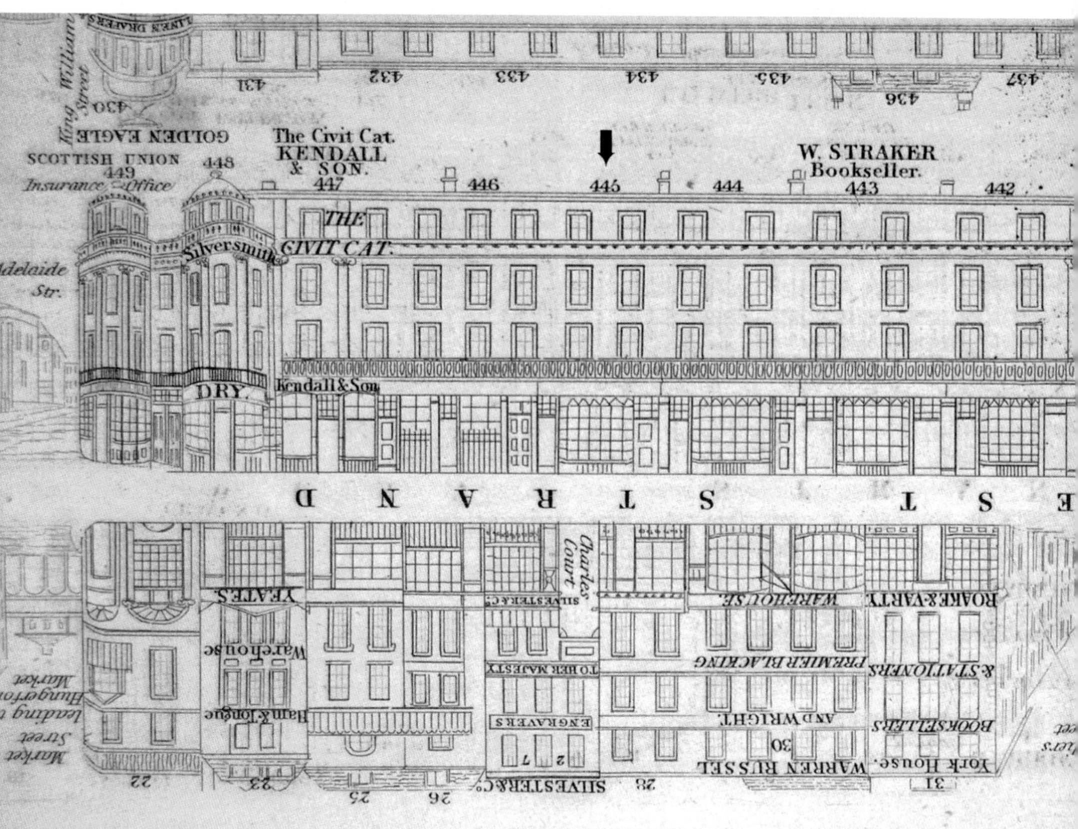

The 'Pepperpot' building at West Strand, from Tallis's *Street Views of London*. The Parkers' premises at 445 Strand occupied the shop below the sixth and seventh windows counting east from the western pepperpots at the corner of Adelaide Street. The shell of this building survives, but the inner structure has been gutted, and the Victorian numbering is now lost.

not to patronize too much, and many of his books were well-illustrated and beautifully produced for their price.

Several books from the 1830s and early 1840s period suggest Parker had family members helping him behind the scenes. There are two Parker authors on his early book lists: Anne and Elizabeth. Sometimes their works first appeared under initials, only later being attributed by name. EP's delightful little book of *Popular Poems,* declares Parker the publisher on the title page at 445 Strand, and printer of St Martin's Lane at the back—a sort of printed paper embrace.[10] Anne Parker was the author of *Fables and Moral Maxims,* advertised as being illustrated 'with 100 cuts', showing Parker's love of illustration.[11] Both women may have helped with other works for the SPCK and for Parker himself which appeared anonymously. They are mostly smallish books, on parenting and other domestic matters, with titles such as *Baby Ballads, Easy Poetry for Children, Pretty Lessons for Good Children, Rhymes for my Children* by a Mother, and *The Mother's Book: Education and Treatment of Children.* Another more substantial work suggests that the entrepreneurial Parker was ahead of his time in capitalizing on the domestic knowledge of female relatives: the *Family Hand-book; or Practical Information in Domestic Economy including Cookery, Household Management and all other Subjects Connected with the Healthy Comfort, and Expenditure of a Family: with a Collection of Choice Receipts and Valuable Hints* appeared anonymously in 1838, twenty years ahead of Mrs Beeton.

We can glimpse Parker's busy efficiency at work from a hurried note written to the wife of the famous abolitionist Thomas Clarkson, a prominent Parker author on the subject of slavery.[12] It is an extraordinary palimpsest of a letter, created in January 1839, when the Clarkson family's agent in London, CH Sheppard, reported back to Mrs Clarkson at Playford Hall near Ipswich, on various errands he had undertaken for the family, concerning among other things delivering a turkey, and recent communications with Lord Brougham. Parker senior drew a line under Sheppard's beginning, and wrote:

> The corrections on the plate shall be carefully made. I contemplate printing the work in small Pica type—a clear and distinct page—and making a handsome volume of about 700 pages. But the exact size of the page—and the number of pages, cannot be determined on, till I have seen the new matter. The sooner, too,

> that I can have the new matter, & begin printing, the better, or the book will be too late to have the fair benefit of the publishing season. John W Parker.

Then follows an explanation by the agent:

> I afterwards went to Parker the publisher, who had just come in from Cambridge—and folding my paper down—procured from his pen what he would communicate. He also showed me a volume—and if the type and style are a fair specimen—the book will make a very handsome appearance. It was large 8° (octavo). But he cannot, as he says, safely print, till he knows the quantity of new matter that is to be prefixed. I am to see the plate when the correction is engraved.

The book concerned was Thomas Clarkson's *History of the Rise, Progress and Accomplishment of the Abolition of the African Slave Trade,* which did indeed appear later that year, with 615 pages, printed by Parker on his own press in St Martin's Lane. The plate discussed in the letter could have been the portrait frontispiece of Clarkson or one of the book's foldout illustrations, among which was the extraordinary—and now famous—diagram of hundreds of African slaves chained to the decks of a slaving ship.

Back in the late 1820s, after a request for help by Cambridge University, William Clowes had deputed Parker to advise the University concerning the management of its in-house Press, which was in need of modernization. In 1836 Parker took over the management of the entire operation, being officially declared University Printer. Over the next seventeen years, he transformed the antediluvian regime of the printing shop there, and the distribution of Cambridge books. As soon as he took over the job, Parker banned the use of child labour at Cambridge University Press. He introduced printing by steam, better management, and recruited new staff. The University had wanted a new broom, and had found it.[13]

He had also to re-educate his academic employers. A letter survives in Parker's hand to Dr William Whewell, the University's Vice Chancellor, which demonstrates Parker's businesslike manner and his polite refusal to defer to ancient university rank, or ancient university practice. Parker is respectful, but not subservient. He speaks as the defender of the Press, standing on his own dignity as a bookseller and a skilled practitioner of the art of printing, the aristocracy of manual trades.

Rev. Sir Dec 7/42

I venture to call your attention to certain customs which prevail in respect to business done at this Press, whereby much loss of the time of persons employed is occasioned, and gentlemen having the business done are frequently put to unnecessary expense.

1. As to sending for persons, whose time is somewhat costly, to receive directions, though frequently on business of trifling amount, and requiring such persons to take proofs, when printed. It frequently happens that calls for these purposes are made over & over again before the parties can be met with.

All the real requirements of most cases of this kind, would be fully answered, if a few lines of written directions were enclosed with the MS to be printed, to the acting Printer, who might be directed to send proofs under cover, and these could be returned for Press, with directions as to the number to be printed, paper, mode of delivery of copies &c &c. In cases of Examination Papers, or other private matters, all possible secrecy would be ensured by the packets being marked 'Private'.

2. As to commencing works which are but in part ready for the press, and feeding the press with copy in small and irregular supplies. Much costly time is wasted, and much valuable material neutralized, by this practice. If gentlemen were charged with all the waste of time and property thus caused they would be greatly surprized at the amount—when it is not charged, a serious loss is inflicted on the Press. Generally speaking, there is very little practical advantage in beginning to print before a work is sufficiently prepared to keep it going at the rate agreed upon—but when this cannot be done, it is highly desirable for the Press, that the work should be suspended, in order that copy may accumulate, and in the meanwhile both hands and material would be disposable for other business.

3. As to the necessity of giving precise directions, in cases of Examination Papers, and other business comprehended under the general term of job-work. For want of such directions, it sometimes happens that papers extend to a greater number of pages than have been contemplated by the employer. The expense caused by an error of this kind may be large in proportion to the amount of the job, but is incurred before the error can be known, and the fair adjustment of it may be difficult. Occasionally, when part of a second page has been occupied, a question has been started as to whether the matter should not have gone into one, or whether the whole, or only the printed portion of the second page was chargeable. A still greater difficulty has been when, exceeding two pages (though making a portion of a third) a job has been charged as four pages. Now, by the universal custom of the Printing trade, both man and master are justified in the same charge for a short page as for a whole one, of the work of which it forms a part, and whenever a job exceeds two pages it is chargeable as four. This custom has all the force of a law in the printing trade, but whenever it is

desirable and possible to avoid it by compression of the matter, the printer is always ready to do so, provided he receives previous directions, and a discretion is left him as to making the pages larger.

Trusting you will excuse my troubling you with these technical details,
I remain, Revd. Sir ,
Your very obedient servant,
John W Parker.
To the Reverend
The Vice Chancellor.[14]

Whewell must have come to respect Parker for what he was, as the two men remained in good relations for many years after this letter. Parker later published a number of Whewell's most important books under his own imprint, and Whewell sent other academic authors to him.[15]

Until then, the University Printer's job at Cambridge had been a full-time local printing post, but one reason Parker had been appointed was because of his knowledge of the London bookselling scene. He was the key figure for London sales and marketing: effectively printer and publisher for Cambridge University Press. He had every intention of persisting with his own publishing, too. For many years he travelled to Cambridge regularly two days a week, leaving his London business in the hands of a loyal assistant, Mr William Butler Bourn, and later also his sons, whom he took into partnership.

Parker appreciated that the University's old monopoly on the printing of Bibles was a thing of the past, but he instituted a vigorous publishing programme, and tried to reorient the Press in terms both of output and marketing, especially in the publishing of textbooks, theology and philology. Many works were inexpensively mass-produced. A prominent project the Press undertook during his time was a series of fifty-five volumes of Protestant Reformation texts, intended to restate the Anglican case during the religious controversy generated by the high church Victorian Oxford Movement, with its leanings towards Rome.[16]

The Press produced fine books, too. Robert Pashley's remarkable *Travels in Crete*, and *Salopia Antiqua* by CH Hartshorne, were both travelogues blended with anthropology and archaeology, and both beautifully illustrated in a grand format, with wide margins and wood engravings of great delicacy and interest interspersed through the text. During Parker's time the Press issued some fine bibles: one was exhibited at the Great Exhibition of 1851, printed at the behest of King William IV.[17]

But over time, it became clear that the University was unable to embrace change to the extent Parker had hoped. The modern historian of the Press at Cambridge, David McKitterick, has said that Parker's employers, the Syndics of the Press, were 'disinclined to compete, or even to consider it'.[18] Trade was looked down upon by those at high table, and competition regarded as unworthy, even in the University's own bookselling arm. Reinvestment was not a priority either: the University took for itself the profits Parker generated for the Press, which probably limited what Parker hoped he could do there. If his efforts began to tail off, and competing publishers in Cambridge began to take advantage of his part-time commitment, it was understandable; he had brought as much change as the place would allow at the time. Parker (who had become a widower) had remarried in 1848, and had started a new family, so he had more to keep him at home in Charing Cross. Towards the end of his association with Cambridge, he travelled there only every other week. Eventually, in 1852, Parker walked away from the job.[19]

His association with Cambridge over, Parker senior returned full-time to the West Strand business, which he had gradually built up during the same period. The frustration of the experience at Cambridge may have influenced his involvement in two quite political (for him) projects. The first was an interesting and far-sighted tract, entitled *Words by a Working Man about Education, in a Letter to Lord John Russell*, which argued for an Education Act to legislate for the public funding of schools for the children of working families. The second was an important agitation within the publishing industry that year: the 'Bookselling Question'. This was an attempt to break the monopoly of publishers and the cartel called the Booksellers' Association, which fixed book pricing. Parker and his son John William, and others in the campaign, wanted booksellers to be free to sell at whatever level of profit suited them: booksellers purchased their books, the argument went, so why could they not sell their own property at whatever price they chose? John Chapman, the radical publisher of the *Westminster Review*, instigated the campaign, and in 1852 the Parkers (having resigned from the Booksellers' Association) joined him and the publisher Bentley to mobilize opinion.[20]

A large number of writers and others, many of them eminent—Dickens, Cruikshank, Carlyle, JS Mill, Tennyson, Whewell, and Charles Babbage—supported the campaign. The hope was that if books were cheaper, sales would rise, and a wide range of reading matter would be democratized. Charles

Darwin wrote to say that 'booksellers, like other dealers, ought to settle, each for himself, the retail price', and John Stuart Mill spoke for many when he commented 'I think that there is no case in which a combination to keep up prices is more injurious than in the case of books.' The parallel with the fixed price on corn, and the abolition of the Corn Laws, was much in mind. The agitation was successful when the Booksellers' Association was declared an illegal conspiracy, but most publishers, and many members of the Bookseller's Association, nevertheless preferred the old status quo.[21]

The Bookselling Question may have drawn Gray's attention to the Parkers. The peak of the agitation was in 1852, and the agreement for his spleen book was made in 1853, soon after Gray was awarded the Astley Cooper Prize. Authors and others were certainly hoping for a drop in the price of books, a wider readership, and a rise in authors' takings. The issue received much attention in the press, and Gray may well have been aware of the Parkers' principled stand in leaving the booksellers' cartel.

In London, Parker had moved away from printing on his own account to employing a variety of independent printers for the mass of work his publishing was generating. The list was growing, and authors were buzzing with recommendation. Even Thomas Carlyle commended him. Writing to Lady Bulwer Lytton, in 1851, Carlyle described Parker senior as 'a quick clever little fellow ... of honorable correct habits in business', and in 1856 told another writer: 'You could not have a better publisher than Parker.'[22]

Parker had taken his son John William into partnership in 1843. The younger Parker was not a trained printer. He was a bookish, thoughtful sort of a fellow, with a polymathic interest in the modern sciences, philosophy, and the arts, including the art of photography. His mother and his elder brother had died, and he himself was not robust. His mother's love of poetry, and his father's Anglican independence of mind and benevolent nature, had re-emerged in him.

As a boy, John William Parker junior had attended King's College School on the Strand, and then studied at King's College itself, where he was taught by the Professor of English Literature, Frederick Denison Maurice, and became actively involved with him in that branch of Victorian Anglican thought known as Christian Socialism. Young Parker brought friends home from College, and 445 Strand became a hub for a group of interesting figures who would later write for him when the Parkers acquired *Fraser's Magazine*.[23]

Parker senior had a long involvement with periodical publishing, from the early success under Clowes of the music journal *The Harmonicon*, through the SPCK's pious journal, the *Saturday Magazine*, and the monthly *Magazine of Popular Science and Journal of the Useful Arts*. The partnership also became the publisher and financial backers of a penny weekly newspaper *Politics for the People*, edited by FD Maurice and Parker junior during the turbulent summer of 1848, the high watermark of Chartism, and the year of revolutions across Europe. A number of Maurice's colleagues, and friends from the Cambridge Apostles—including Charles Kingsley, James Anthony Froude, and future Archbishops Richard Whately and Richard Chenevix Trench—were among its anonymous and pseudonymous contributors. The group rejected any idea of physical force to bring about political change, but felt serious matters such as parliamentary representation required redress, and that 'whatever a great number of our countrymen wish for deserves earnest consideration.'[24]

Poetry did not have a high profile in the columns of *Politics for the People*, but its final issue featured a sonnet that seems emblematic of its authors' hopes, and perhaps too, those of its publishers:

> Not all who seem to fail, have failed indeed;
> Not all who fail have therefore worked in vain:
> For all our acts to many issues lead;
> And out of earnest purpose, pure and plain,
> Enforced by honest toil of hand or brain,
> The Lord will fashion, in his own good time,
> (Be this the labourer's proudly-humble creed),
> Such ends as, to his wisdom, fitliest chime
> With his vast Love's eternal harmonies.
> There is no failure for the good and wise:
> What though thy seed should fall by the way-side
> And the birds snatch it; yet the birds are fed;
> Or they may bear it far across the tide
> To give rich harvests after thou art dead.[25]

The paper lasted for only seventeen issues, from May to July 1848, but in the long-term, the grouping that had coalesced around FD Maurice that summer became influential within Anglicanism. They remained friends, and continued to meet regularly long after the editorial work was given up. Maurice and Trench were later to fulfil the melancholy duty of conducting young Parker's funeral service.

Portrait photograph of John William Parker the elder, printer and founder of the publishing house.

Portrait photograph of John William Parker the younger, publisher and Editor of *Fraser's Magazine*.

The spirit of the sonnet chimed with the Parkers' own motto on their trademark emblem, which often appeared on their title pages: *PER ASPERA AD ARDUA TENDO*—I strive, through difficulties, to lofty things.[26] It also appeared in the collected annual volumes of the monthly *Fraser's Magazine*, which they had acquired in 1847. The younger Parker was installed as the journal's new Editor. He guided the journal in a fresh direction towards liberal Christianity, retaining some of the older contributors (such as Thomas Carlyle) and adding new ones from the Maurice circle, as well as important independent outsiders, including John Stuart Mill, Alfred Tennyson, and even that stalwart of John Chapman's *Westminster Review*, George Henry Lewes.

As he had done with *Politics for the People*, young Parker regularly entertained his writers and other friends at 445 Strand, which almost certainly had its own inner sanctum on the ground floor behind the shop, as did many London shops at this period. Meals were probably brought in from a neighbouring chop-house, and served upstairs. References to these dinners appear occasionally in the diaries of participants: largely male gatherings, full of earnest and interesting talk, and good humour. Arthur Munby, for example, mentioned having met JA Froude and the poet Arthur Hugh Clough at Parker's; Henry Kingsley met Thackeray.[27] Entries in the memoirs of Sir Frederick Pollock give an inkling of the heterodox company at these events:

> 17th June 1858. 'Dine with JW Parker at his rooms above the publishing shop in the Strand, at which he gives frequent men's dinners. Met Theodore Martin, WG Clark, Alfred Tennyson. Brought home proofs of 'Hanworth', coming out in Fraser's Magazine.'
>
> 13th April 1859: 'Dine with JW Parker. Sir George Cornewall Lewis [Home Secretary], [Richard] Owen, [Arthur] Helps, Lord Bury, etc.'
>
> 23rd May 1860. 'Dined with JW Parker, Theodore Martin, [AKH] Boyd, [Henry] Buckle, Arthur Helps, Canon Robertson, [Samuel] Smiles.'[28]

Bookshop sociability was an old London tradition, long predating the upper-class salons of the eighteenth century, and literary meals were not a rare phenomenon in the printing district of Fleet Street and the Strand.[29] They offered opportunities for general chat, as well as briefing and debriefing about the journal's content, brain-storming for new ideas, immediate and long-term directions for the journal, suggestions of names to attract for reviews or articles, as well as socializing and gossip. Such evenings no doubt helped the Editor, bookish John W Parker junior, keep his finger on the pulse of things, and also allowed for important decisions and introductions to be made. Parker was instrumental, for example, in bringing together JA Froude and Francis Palgrave, keeper of the Records at Rolls House, the influential figure behind the subsequent establishment of the Public Records Office. Palgrave was seeking a historian to examine the unknown contents of a great cache of newly discovered records from the Reformation. Froude's first novel *Nemesis of Faith* had been publicly burnt at Oxford, and he was working as a jobbing journalist. The result of the meeting was Froude's famous multi-volume *History of England*, and its author's eventual appointment as Regius Professor of History at Oxford. Many years later in 1879, Froude, writing to John Skelton, recollected these gatherings affectionately as days of golden memories.[30]

Quite a number of Parker and Son's book authors also wrote for *Fraser's Magazine*, and the chronologies of their involvement would make an interesting study. Among contributors during the period the Parkers published the journal (1847–63) were such eminent Victorian figures as JA Froude, Henry Kingsley, GH Lewes, Arthur Helps, George Hogarth, AKH Boyd, Walter Savage Landor, Thomas Love Peacock, WF Ainsworth, John Forster, Leigh Hunt, Thomas Carlyle, Arthur Hugh Clough, George Eliot, George Meredith, John Stuart Mill, Coventry Patmore, Herbert Spencer, and Frances Power Cobbe.[31]

In the mid-1850s, at the time Gray and Carter were discussing the possibility of a manual of anatomy, the Parkers' list was well-established, extensive, and impressive. The partnership still had a considerable publishing output for the Society for the Promotion of Christian Knowledge. A sheaf of advertisements for their SPCK books has been found bound in at the back of a contemporary bestseller, *Conversations of a Father with his Children.* Set in small type, listed from algebra and ancient history through reptiles and shells to wild animals, it occupies eight pages, and contains 130 works, priced from threepence to three guineas (£3. 3 shillings). Their backlist still featured quite a preponderance of theological and educational works, as well as mathematics and classics, largely associated with Cambridge. Cambridge authors, like Whewell, had continued their association with the former University Printer after his return to Charing Cross, and a growing number of new authors came from King's College, London, along the Strand, and via *Fraser's*.

The Parkers' output covered a wide sweep of predominantly non-fiction subject-areas, including translations from Theodor Mommsen's *Italy*, and other travel books like WG Clark's *Gazpacho* and *Peloponnesus*, CH Cottrell on *Siberia*, works on the Near East and ethnography, and a magnificent book on *Siam* by Queen Victoria's plenipotentiary to China, John Bowring, after a rare diplomatic visit. His account of the journey ran to two volumes, with tipped-in illustrations, colour printing—including detailing in gold—and fold-out facsimile documents and maps. Newer parts of the world favoured by emigration became a feature, such as New Zealand, the United States, as well as South America and Africa. Travellers' commentaries often contemplate social conditions at home as well as abroad. Slavery, in James Stirling's *Letters from the Slave States*, and the reform of British colonial governance, were important themes.

They did publish novels, but not many: the early work of Charles Kingsley (*Yeast*) and Charlotte Yonge's *The Lances of Lynwood* were among them, as was George Whyte-Melville's *Digby Grand*. The partnership's output included an impressive amount of music (mainly that of the entrepreneurial choral master and pioneer of mass music-teaching, John Hullah) and of poetry, particularly new collections of old English balladry, and an extensive range of anthologized poetry from Chaucer to Dryden and Cowper. There was also a smattering of modern poetry, by Charles Kingsley, George Meredith, Coventry Patmore, and Frederick Tennyson.

There were biographies and works of history, including HT Buckle's best-selling *History of Civilization*, the early volumes of JA Froude's *The History of England*, as well as Arthur Helps's *Spanish Conquest*. Philosophy and its history had grown to be an important area, too, with significant works on the history of ideas by William Whewell and George Henry Lewes, as well a string of works by John Stuart Mill: *Logic*, *Political Economy*, *Liberty*, and *Utilitarianism*. The more one looks at the Parkers, the more distinguished they seem to be. They are not famous names in the history of publishing like John Murray or Smith, Elder or Macmillan's, but to have come from small beginnings to a list of this nature bespeaks considerable sagacity and acuity.

JS Mill gives us a glimpse of the Parkers as publishers. In two letters to his beloved Harriet Taylor, written three days apart in March 1849, he describes his state of mind concerning the second edition of *Political Economy*, the printing of which was underway. Mill is at first apprehensive, even suspicious, clearly fearful that his book might become 'a disagreeable object'; then reassured. Something otherwise intangible of the atmosphere and the hands-on practicality of the business comes through in his words. The Parker mentioned here is, I think, the elder, but this is based on a hunch that the younger worked with Mill in an editor–author relationship, while the elder oversaw the printing and production:

> 14.3.1849: 'What a nuisance it is having anything to do with printers. Though I had no reason to be particularly pleased with Harrison, I was alarmed at finding that Parker had gone to another [printer]. & accordingly, though the general type of the first edition is exactly copied, yet a thing so important as the type of the heading at the top of the page cannot be got right—you know what difficulty we had before—& now the headings, & everything else which is in that type, they first gave much too close & then much too wide, & say they have not got the exact thing, unless they have the types cast on purpose. Both the things they have produced seem to me detestable & the worst is that as Parker is sole owner of this edition I suppose I have no voice in the matter at all except as a point of courtesy. I shall see Parker today & tell him that I should have much preferred waiting till another season rather than having either of these types—but I suppose it is too late now to do any good—& perhaps Parker dragged out the time in useless delays before on purpose that all troublesome changes might be avoided by hurry now. It is as disagreeable as a thing of this sort can possibly be...'

> 17.3.1849: 'The bargain with Parker is a good one & that it is so is entirely your doing ... the difficulty with the printer is surmounted—both he and Parker were disposed to be accommodating & he was to have the very same type from the very same foundry today—in the meantime there has been no time lost, as they have been printing very fast without the headings, & will no doubt keep their engagement as to time.'[32]

In the event, the second edition of Mill's *Political Economy* appeared later the same year, 1849, printed by Parker's favourite local firm, Savill and Edwards.

The Parkers were also publishing a wide range of books on the study of language, and languages. The works of Richard Chenevix Trench promoted the entire field of English philology, and were an influence towards the later development of the Oxford English Dictionary.[33] There were books on Latin, Greek, Arabic and cuneiform, and they also issued a cluster of works on German, a modern language whose profile was rising in the mid-century, with the growth of interest in German (Protestant) theological and scientific output, and—as railways developed—increasing opportunities for travel. As we have seen, travel—geographical and temporal—was another important feature.

Churchmen had a high profile on the Parkers' list, probably because of the SPCK connection, but also because many naturalists were churchmen, and many Cambridge men were destined for the Anglican clerisy. Church reform was a key theme, in books by churchmen Richard Whately and Julius Hare. There were sermons and other biblical and moral commentary by among others, Charles Kingsley, CJ Ellicott, and Alexander Bain, as well as a fine book on churchyards, *God's Acre*, by Elizabeth Stone.

The Parkers also had a militaristic arm. They published Joseph Allen's *Navy List*, *Historical Records of the British Army*, and Admiral Smyth's nautical guide to the Mediterranean. Both John Smeaton on *Lighthouses* and WS Gilly's famous *Shipwrecks of the Royal Navy* were Parker books, while the Crimean War seems to have prompted them to publish a cluster of works on military tactics, skirmishing, on musketry, and on temporary and permanent fortifications.[34] The Parkers were also quick to enter the field after the Indian Mutiny, with a memoir by Major Hodson 'of Hodson's Horse', *Twelve Years of a Soldier's Life in India.* Such books in the shop window would surely have attracted passers-by, aficionados of the local barracks by Trafalgar Square, and top brass visiting Whitehall, St James's, Horseguards, and the Admiralty, all of which lay close by.

Science had a broad presence on the Parker's list, ranging from natural history—such as William Broderip's beautiful *Leaves from the Notebook of a Naturalist*, and Edward Stanley's *Birds*—to geology, including Adam Sedgwick's *Palaeozoic Rocks and Fossils*, as well as works in the fields of chemistry, astronomy, and geography.

Pathbreaking works on sanitary reform also featured. Florence Nightingale's writings on the sanitary shortcomings of the Army in the Crimea, and her famous *Notes on Hospitals*, appeared under the Parkers' imprint. They also published the work of the architect Henry Roberts on model housing for the working classes, influential upon Prince Albert. Social reform was a strong feature, including education for the working classes and for women, favourite topics for FD Maurice, who went on to establish the Working Men's College, and Queen's College London, for women. The need to end transportation, and to reform national policy on prisons was a particular focus, especially in the work of Matthew Davenport Hill. The Parkers also published informed commentary on the 1851 Census, and the journals and other output of the Statistical Society, as well as the Transactions of the National Association for the Promotion of Social Science.

Arising no doubt in part from their involvement with *Fraser's Magazine*, a Parker speciality was the literary genre of the essay. Their output in the essay form ranged from reprints of the essays of Francis Bacon (a great success, as were a series of secondary works about Bacon from a number of modern authors) to the charming essays of Arthur Helps (later Clerk of the Privy Council, biographer of Prince Albert, and personal friend of Queen Victoria) like *Companions of my Solitude*, and the Rev. David Badham's amiable book on 'fish tattle', originally a series in *Fraser's*, regarding fish 'ichthyophagously' rather than 'ichthyologically' (not to study, but to *eat*). The Parkers published an impressive series of more serious essay collections, too, with contributions from eminent scholars from both Oxford and Cambridge. As we shall see, the publishing house would become notorious in 1860, after the publication of a book entitled *Essays and Reviews*, whose clerical authors were accused of heresy.[35]

Essays and Reviews and Froude's *History of England* were almost certainly book ideas spawned during one of the younger Parker's literary dinners. The idea of expanding Parker and Son's medical list was probably also his, and may have met his father's approbation increasingly after his printing days at Cambridge were over. Father and son shared a keen awareness of the value of the shop's geographical position, in a marketing sense, and a desire to maximize it, especially in terms of what audience lay relatively conveniently nearby. Sales from the shop were profitable: the bookseller's cut on their own books (the proportion of the cover price that usually went to other booksellers) accrued to the firm itself.

The Parkers' shop at 445 Strand was positioned between the Royal College of Surgeons at the south side of Lincoln's Inn Fields, and the Royal College of Physicians, on the St James's side of Trafalgar Square, Pall Mall East. The Medical Society of London was still in Bolt Court, off Fleet Street, while *The Lancet*'s offices were just along the Strand, as were those of the Medical Directory. Four major London teaching hospitals were within an easy walk. The new Charing Cross Hospital—which had opened soon after West Strand was completed—stood right around the corner, just behind the eastern Pepperpots, sharing a site with the Western Ophthalmic eye hospital. King's College Hospital was also close by, just north of St Mary-le-Strand church; while St George's was a stroll away, west, across St James's Park. The Westminster Hospital—ten minutes down Whitehall—made a fourth. Other successful general publishers, such as Churchill, Highley, Longman, Murray, Renshaw, and Walton, had good medical lists, why not Parker and Son?

Perhaps the best way to capture the nature of the Parkers' medical list in the 1850s is to consider the books they advertised at the back of *Gray's Anatomy* when it made its first appearance in 1858. As one reads down their list (shown overleaf), the breadth of medical interest is evident, and the care with which books were chosen.

The Parkers started off, understandably enough, with Henry Gray's book *The Spleen.* Listing the author first was a courtesy, and to place him in such high company as the authors that follow served to promote his status still further. Moreover, *The Spleen* was moving so slowly off the shelves that it was virtually dead stock: the Parkers must surely have hoped that the appearance of Gray's latest work would help shift some more. Albert Kölliker's new *Manual of Human Microscopic Anatomy,* next, was a major scientific work translated from the German, and was still in production at the time, so this was an advance announcement of its arrival. Histology then was like genetic or stem-cell research now: *hot,* and Kölliker was its biggest name.

The rest of the list offers an interesting clutch of books, judiciously chosen and placed to represent a balance of scientific and general interest, as well as authorial eminence, in a variety of branches of medicine, which *Gray's*—with both anatomy and surgery in its title—would help complete. It starts with the Parkers' lead title in physic, the great textbook of medicine *The Principles and Practice of Physic* by Thomas Watson, a grand figure in the medical world. Thomas Mayo's book on *Medical Evidence* followed probably because he was President of the Royal College of Physicians, and because the issue had a strong appeal for doctors. Another of his books, *The Philosophy of Living,* rounds off the textbook list, before the featured medical specialisms. Between Mayo's volumes are listed a series of big textbooks, starting with Todd and Bowman (of which more in a moment), followed by works on epidemic diseases, John Tomes on dentistry, and WA Miller's medical *Elements of Chemistry.*

This is not an extensive medical list. Churchill's, for example, was far larger, but he had been building it up for years. For a publisher only recently entering the field, it was a creditable list, covering physic and the care of acute and chronic diseases, but also dentistry, mental illness, physiology and microscopy, chemistry, medical law, and even personal well-being.

Towards its end, the Parkers' list opens right out to present a small group of popular books, situating medicine in the wider context of other sciences. These were important books on scientific ideas, logic and philosophic thought, rounded off with Buckle's bestselling *History of Civilization.* The list as a whole is reflective—intended to invite young doctors to ponder the many branches of their own discipline, and to think largely about the breadth and potentialities of the medical calling, and its place in the cosmos. The list was a modest cornucopia.[36]

Many Parker books lying in libraries today look nothing like they did when they were placed on display in the window at 445 West Strand. Numbers of them have lost their original bindings, having been rebound by their first owners soon after purchase, or more recently (and often unsympathetically) by libraries, as Victorian binding materials have finally given way. It is often only by luck or by neglect that an original Parker binding survives. This is particularly true of well-used Parker books such as *Gray's Anatomy*, but it is also the case with numbers of others.

Those that remain in the covers they were given at the time of their publication are often discoloured or faded, brittle, or in poor condition. Few have survived anywhere in their pristine clarity of outer colouring, almost all original shades have attained at exposed places an indeterminate faded greyish brown, whatever the colour of the flank of the boards might have been. The clean endpaper is rarely to be seen, the plain whiteness of original pages has darkened and yellowed with age. Page edges are darker still, ingrained with 150 years of dust. Flyleaves, and sometimes title pages or owners' signatures have been torn or cut out, they have been inscribed, annotated, and used. They are often weak at their hinges, and browned spines flake or peel or split at top and bottom, or flap from one side. Rarely, one can yet find one uncut, which is somehow even more pitiable for a book than being damaged by overuse.

What I have to say here about how Parker books looked when they were displayed in the shop is based on prolonged searches in a variety of libraries, in numerous bookshops, and through the ether on the Internet, all peppered with a glint of something which emerges when one attempts to examine these books with their original appearance in mind. One has to try to look at them as though they were zappy and new, and for this, imagination is called for.

Overleaf: JWP'S medical list in 1858. Advertisement inserted at the back of the first edition of *Gray's Anatomy*.

BOOKS PUBLISHED BY JOHN W. PARKER AND SON.

The Structure and Functions of the Human Spleen.
By Henry Gray, F.R.S.
Lecturer on Anatomy at St George's Hospital.
With 64 illustrations. 15s.

Manual of Human Microscopic Anatomy.
By Albert Kolliker.
With Numerous Illustrations. Octavo [no price given]

Lectures on The Principles and Practice of Physic.
By Thomas Watson, M.D.
Fellow of the Royal College of Physicians.
Fourth Edition, revised and enlarged. Two Volumes, Octavo. 34s.

Medical Evidence in Cases of Lunacy.
By Thomas Mayo, M.D., F.R.S.
President of the Royal College of Physicians. 3s. 6d.

The Physiological Anatomy and Physiology of Man.
By Robert Bentley Todd, M.D., F.R.S.
AND
William Bowman, F.R.S.
With numerous Original Illustrations. Complete in Two Volumes. £2.

Diseases of The Kidney, their Pathology,
Diagnosis, and Treatment.
By George Johnson, M.D.
Fellow of the Royal College of Physicians, Physician to King's College Hospital.
Octavo, with Illustrations. 14s.

On Epidemic Diarrhœa and Cholera.
By the same Author. 7s. 6d.

Lectures on Dental Physiology and Surgery.
By J. Tomes, F.R.S.
With 100 Illustrations. Octavo. 12s.

Use & Management of Artificial Teeth.
By the same Author. 3s. 6d.

Elements of Chemistry.
By William Allen Miller, M.D., F.R.S.
Professor of Chemistry, King's College, London.
Complete in Three Parts. With numerous Illustrations. £2. 6s. 6d.

The Philosophy of Living.
By Herbert Mayo MD
5s.

Management of the Organs of Digestion.
By the same. 6s. 6d.

Familiar Views of Lunacy and Lunatic Life.

WITH HINTS ON THE PERSONAL CARE AND MANAGEMENT OF THOSE AFFLICTED WITH TEMPORARY OR PERMANENT DERANGEMENT.

By the late Medical Superintendent of an Asylum for the Insane.

3s. 6d.

German Mineral Waters, and their Employment for the Cure of certain Chronic Diseases.

By S. Sutro, M.D.

Senior Physician to the German Hospital.

7s. 6d.

Spasm, Languor and Palsy.

By J.A. Wilson, M.D.

Late Physician to St. George's Hospital.

7s.

Gout, Chronic Rheumatism and Inflammation of the Joints.

By R.B. Todd, M.D., F.R.S.

Physician to King's College Hospital.

7s. 6d.

History of Scientific Ideas.

FIRST PART OF THE 'PHILOSOPHY OF THE INDUCTIVE SCIENCES.'

Third Edition.

By William Whewell. D.D., F.R.S.

Two volumes, Small Octavo. 14s.

History of the Inductive Sciences.

By William Whewell, D.D., F.R.S.

Third and Cheaper Edition, with Additions. Three Volumes, 24s.

The Senses and the Intellect.

By Alexander Bain, M.A.

Examiner in Logic, Moral Philosophy, &c., in the University of London.

Octavo, 15s.

A System of Logic.

By John Stuart Mill.

Fourth Edition, revised. Two Volumes, 25s.

The Biographical History of Philosophy,

FROM ITS ORIGIN IN GREECE DOWN TO THE PRESENT DAY.

By George Henry Lewes.

LIBRARY EDITION. Octavo, much enlarged and thoroughly revised. 16s.

History of Civilization in England.

By Henry Thomas Buckle.

The First Volume. Second Edition, with an Analytical Table of Contents.

Octavo, 21s.

Parker books were not uniform in appearance. Not only did book fashions change between the 1830s and the 1860s, the period Parker and Son were in business, but their own taste shifted in that time. In addition, there were periods of cotton shortage which affected the manufacture of the book-cloth which was the dominant material for covering books in London in the era the Parkers were actively producing them. Designed for the mid-to-low end of the publishing market, and mass-produced in the specialized workshops found within a stone's throw of the Strand, Fleet Street, and Paternoster Row, Parker books were typical London trade books of their time. They were in no way unusual for their era, except perhaps in their happy mix of high printing standards and affordability.

In the early years when Parker senior was apprenticed and setting himself up in business, it was still quite commonplace for books to be sold without bindings. The folded sheets forming the pages (quires, gatherings, or signatures) were sewn together at the back, and sold in paper wrappers, without boards or leather covers. The idea was to make it easier for purchasers to have them hand-bound to suit their own tastes and pockets. This method of publication also made books cheaper for those who could not afford binding, and many were happy to leave them in their paper wrappers, even though it left their contents more vulnerable. This is also how periodicals like *Punch*, *Fraser's*, and other magazines continued to be published, and it is still sometimes possible to find their printed paper wrappers surviving safe within more substantial bindings.

From the mid-1830s on, mass-produced publishers' *casings* came into their ascendancy. This was the new outer skin in which most Victorian books after about 1840 made their appearance. Casings were made in one piece, an outer sheet of paper or treated cloth was folded over at the corners and edges of two book-size boards which served to cover the front and back of the book, and between them, a long narrow fillet of paper or card was laid down to create the hollow half-tube which appears behind the spine when a book is opened. Another narrow fillet was usually glued to the sewn gatherings of the book itself, helping hold the gatherings and sewing firm, and to prevent the gathered backs of the book's pages from becoming stuck to the spine visible on the shelf.

These 'hollowback' casings were mass-produced by hand, in labour-intensive workshops, in whatever numbers a publisher's edition demanded. They were hand-pressed in embossing machines, with patterns and sometimes with gilding, to produce the final casing, to which the ready-sewn book would be glued with tapes or strings left hanging loose from the sewing of the gatherings, and a sort of loosely woven bandage called 'mull', which bridged the gap at the crucial hinge between book and casing. The endpapers came last. In Parker books they were usually very plain: printed on one side only in a single colour with a matt surface, on a paper so thin that you can usually discern the weave of the mull through it, as well as its imprint opposite. The endpapers covered a multitude of sins, concealing the paper/cloth foldovers from the outer cover as well as the inside surface of the boards, the tapes and/or the mull, holding everything neatly in place together, with the help of deftly applied natural glue.[37]

At the outset casings were fairly plain, made in a limited range of sizes, and were titled to suit with paste-on labels on the spine. As time went by, edition binders created special casings for publishers' entire editions, with bespoke titles and publishers' names embossed on the spine, and ornamental designs for the front boards, blind, gilt or (later) printed with coloured inks. The design of book casings became more ornate as the century progressed, but the Parkers ceased publishing before designs became really florid.

There were other ways of binding books, some of which were so expensive as to make eyes water, but this cheap, mass-produced manner of covering a book suited Parker senior, and his son. We know from his work at Cambridge that Parker senior knew perfectly well how to produce a book fit for a king, so it can be inferred that if the books which came out of 445 Strand were not printed on hand-laid paper and bound in hand-tooled leather, then this is how the Parkers wanted to pitch their wares. They made cheap, mass-produced trade books for an audience they wanted to reach.

The average cost of Parker medical books was low, as you can see from the prices on their list. Many of these books would have been sold as paperbacks, themselves each carrying advertisements too. This far from grand manner of production was chosen for Todd and Bowman's textbook, *Physiological Anatomy and Physiology of Man*, much more akin in outward appearance to the old paper-wrapped volumes of the past, but inside, it was closer to the SPCK books, with their generosity of modest illustration.

'Todd and Bowman' was an unusual book, not least because collaborative authorship between physicians and surgeons was uncommon. Robert Bentley Todd was a physician, while William Bowman (whom we have met before at the Royal Society) was a surgeon. Both men worked at King's College, London. Physiology was a rapidly expanding field at the time, and Todd and Bowman were writing while the knowledge-base was shifting. Part-publication was evidently a convenient method of publishing on the hoof, for making the latest findings in this important subject-area available in a cheap format for information-hungry medical students.

Illustrations in Victorian books and periodicals were generally printed by engravings on wood. Steel engraving and lithography were expensive, and were not easily combined with letterpress text. For clarity of line, comparative cheapness, and for use in combination with text in the same page set-up on steam presses, wood engraving was the illustrative method of choice. The illustrations in Todd and Bowman were modest, generally smallish but informative wood engravings, typical of the technical textbook work of their time, pale, simple, not in the least flamboyant, and unsigned.

While it is common knowledge that Victorian novels were issued in parts, that scientific works were too, tends not to be so. Todd and Bowman began appearing in 1843, and continued to be issued in a cheap paperback binding right through the 1840s and into the mid-1850s, the entire period Gray and Carter were training at St George's. The final part of the series was not completed until 1856, and was in production at the Parkers at the same time as *Gray's Anatomy* was being created.[38]

We usually know which printers the Parkers used, because they identified themselves in the books they printed. But we cannot so easily identify the Parkers' bookbinders. Fine bookbindings are commonly signed or embossed somewhere discreetly with their maker's mark, but the common casings produced by mid-Victorian London edition binders were rarely signed or embossed with any form of identification. Just occasionally you might find a bookbinder's ticket (a tiny paper sticker, bearing name or trademark) or the remnant of one, glued at the very back of the book, usually on the fixed endpaper at the inner edge of the back board.

Surviving Parker business ledgers are uninformative on this question. The *costs* of bookbinding are itemized, but no bookbinders are named. We have, I think, to assume one of two things: either the Parkers always used the same firm to bind their books, and so did not need to name them in the ledgers; or, the printers subcontracted out this final element of book-making, and charged the publisher directly but separately. Because of the variety of casings and binding styles so far found on Parker books, the last seems the more likely.

Few records survive from this period, and the network of relationships between publishers, printers, and bookbinders may never be fully mapped out. But although little is yet known about the identification of casing and binding styles of London edition bookbinders, scholars of bibliography are beginning to tease out the history of book production in the workshops of London, and within the next decade or so, much more will be known. Only twice, so far, have I found a binder's ticket in a Parker book, and in each case they identified the binder as one of the most successful and prolific bookbinding firms in London: Burn and Co, of Hatton Garden.[39]

We can infer from the nature of the dealings we have seen between Parker and Son and their authors, like Clarkson, Whewell, and Mill, that a great deal of the Parkers' business was done informally. Agreements were reached swiftly, and put into action. There was a lot of face-to-face dealing, and matters could be sorted out promptly. This is probably also how bindings were decided upon. Generally, the premises of publisher, printer, and bookbinder were close by, within five or ten minutes' walk of one another. Things could be seen and agreed quickly between competent parties, especially when they had long-standing relationships. Just as the intellectual evenings at 445 spawned friendships and ideas, Parker senior probably had his own friends within the printing fraternity, whose business was done amicably.

Whether the bookbinders were commissioned by Parker and Son, or by their printer, the Parkers must have had an input, because there is a house style. Just as for a modern publisher, within the Parkers' list at any one time choices were made about differing characteristics of size, shape, and appearance for what was felt to be appropriate to content, and of course what was economical at the time. So, for example, a military book on a new rifle musket issued in 1855 was a martial red, with a cloth-covered paper cover, blind embossed and boldly lettered in gold, with an austere plainish interior, technical illustrations as separate lithographed plates bound in at the back, and a leaf of advertisements for other books on military topics published by Parker. It was light in weight, and gave the impression of being easy to carry in a knapsack.

The books the Parkers created for the SPCK were generally smallish and attractive both in binding and internal layout, which makes them feel as if they were designed with a domestic and predominantly female audience in mind, the sort of thing which could be read out in schoolrooms, Sunday schools, or loaned from village libraries organized by vicars and their wives, and perhaps also distributed as school prizes. Their embossed cloth casings were sometimes delicately ornamented with foliage or scrolls, and frequently a favourite Parker yellow used for the endpapers. The typefaces were well chosen and nicely spaced for unconfident readers, without seeming pedestrian. That the SPCK and the Parkers were interested in literacy education is shown by advertisements in Parker books published under the auspices of the SPCK, for reading books and posters on the phonic principle. Parker's SPCK books were often beautifully illustrated with wood-engravings of modest size, offering enticing views of the world to intrigue and entertain the reader. Children, and even older readers, would probably have loved these books for their beauty alone, and this is just as the SPCK would have intended, as their business was to convey a Christian message without too heavy a hand.

Theological works tended to be larger in all dimensions, looked plainer, and felt more substantial. During periods of cotton shortage, such as during the Crimean War, when the embossed book-cloths were too expensive to use or in short supply, Parker theological books were cloth-covered only along the spine, with about an inch or so overlapping at front and back on to the boards, and the boards themselves were papered in a plainish dull grey, which was probably thought suitable for a vicarage. At other times, they were clothed in dark cloth, often navy, black, or brown-black, and uniformly embossed, often with a morocco leather grain laid at a diagonal. Scientific textbooks (like Gray's *Spleen*) were invariably in dark green or mid to dark brown cloth, embossed in geometric or straight line 'picture frame' designs, over morocco graining, and with darkish endpapers in shades of chocolate or coffee brown. They tended to be set in legible type of a decent size, and gave an impression of condensed but comfortable mental application. History books were similar, large, serious, dark, but with ornamentation at their corners.

Colourful bindings were largely confined to novels and poetry, which were slighter in size, and usually in cloth with a finer weave or more finely embossed detail. Invariably colours also spread inside to the endpapers: a primrose yellow was not uncommon, occasionally a fine bold Titian red, or rarely, a warm lilac. One novel, *Hanworth*, mentioned above in Baron Pollock's diary entries, had a mock leather casing of embossed rich turquoise book-cloth, and golden-tan endpapers. In only one instance so far found, a book of poetry had endpapers of a surprisingly pleasant shade of pink, which colour has deteriorated unevenly over a century and a half on a quiet shelf, to a curious pewterish grey.[40]

One of the small benefits of books in poor condition is that one can see their structure more easily than if their bindings were intact. Someone in the Rare Books Room of the British Library once told me that a scholar somewhere has written a paper in praise of broken bindings. Try as I may, I have not yet been able to track this article down, but I imagine that I and its author would get along very happily, because of the curious joy I have experienced in examining Parker books in poor condition. It's a bit like archaeology, trying to understand the layers of effort which went into the business of making the books which were sold at 445 Strand.

The colours of casings and endpapers, book sizes, typefaces, and choices of printer were all, probably, conscious choices made by one or other of the Parkers themselves. But the nature of the fillets which were inserted in the hollowbacks, between the casings of their books and the sewn gatherings of pages, were probably not. They were a part of the bookbinding trade no one seems to talk of much, which involved the reuse of waste paper. The paper fillets would probably have been cut by child or female labour, and glued dexterously in place over the gauze mull over the sewing; and between the boards, by the casing maker.

Hidden for years, they become visible only when the books fall apart, and they are fascinating. Although some are plainish and brownish and uninteresting, I have found newspaper advertisements, usually fragments of long classified columns, cut in such a way as to preclude identification, sometimes printed on grey, pinkish, or creamy-white newsprint, containing truncated gobbets of shipping news, stock prices, auctions and deaths, advertisements for books or monster clothing sales, tailors and liquidations (one assumes) of stock, or bits of advertising puffing other books.

Occasionally, the fillets are long narrow extracts of beautiful handwritten manuscript in copperplate, cut in such a way as to obliterate anything researchable, filleted in such a way as to be so fragmentary that it might make Sherlock Holmes weep. Part-names, or part-words emerge, so well disguised as to make one wonder if this was not a Victorian form of shredding, devised by the Circumlocution Office itself. Sometimes these things have faced each other silently in the quiet of the hollowback, a newspaper face to face with the brown ink of elegant handwriting, for 150 years. In the back of Froude's *History of England* I found an account that looked to be dated 12th May 1795, for a half-year's interest on a sum of £5560.9.8. And, in the brittle back of the British Library's only copy of Henry Gray's book on the spleen, lies a fillet from a vibrant cartoon. A kind bookseller has identified it for me as the front wrapper of *Diogenes*, a mid-Victorian rival to *Punch*, which went out of business in 1855.[41] Neither Gray nor the Parkers, probably, had any idea that their sombre binding, with its masculine straight-line design, and serious endpapers, had a defunct cartoon hidden away beneath the fabric of the book.

One of the most appealing aspects of the books produced by the Parkers for the SPCK is in the delicacy and charm of their illustrations. These were cheap books, mainly for children or for self-educators, and the illustrations are designed and placed in such a way as to be embedded in the printed text. It is sometimes possible to discover the identity of their wood-engravers, but many left no mark or monogram to disclose their authorship. Among those so far identified as engravers for the Parkers are the Dalziels, Mathew Sears, Ebenezer Landells, and Edward Whimper, all of whom represent the cream of the jobbing engraving trade in London in the mid-nineteenth century.

I am looking at a typical example of a Parker SPCK book on the desk before me. It's small, by which I mean that if you held out your hand, the foot of the book would just fit across the palm above your thumb, and its head would fall no higher than the tip of your tallest middle finger. It is covered in a pale grass-green cloth, with no author credited on spine or title page. The browned spine, about ¾ inch wide, is embossed vertically in gold: 'ELEMENTS OF BOTANY'. The rest of the casing is blind-embossed all over in a very fine-grain vertical stripe, and over that, with a sort of picture-frame motif on each of the boards, and a diamond-shaped curlicued crest (which reminds one of a Jacobean knot-garden) embossed at centre-front and centre-back. Inside, the book has Parker yellow endpapers, through which the contours of the two tapes from the sewn spine, and the woven inch of mull, are visible. So too, is their impression on the facing endpaper. In this copy, Clara Forsaith has signed her name on the front free endpaper, and dated the deed: 15th May 1852. Her writing is spidery, and leans to the right, and she evidently liked a fine nib for her pen. Her ink has browned.

The title page states simply: 'THE ELEMENTS OF BOTANY: FOR FAMILIES AND SCHOOLS. PUBLISHED UNDER THE DIRECTION OF THE COMMITTEE OF GENERAL LITERATURE AND EDUCATION APPOINTED BY THE SOCIETY FOR PROMOTING CHRISTIAN KNOWLEDGE. *THE SEVENTH EDITION*. LONDON: JOHN W. PARKER, WEST STRAND. M.DCCC.XLVIII.' Facing it is a full-page wood engraving (unsigned) of a verdant place, full of growing plants, captioned: 'PLANTS PRODUCED FROM SEEDS HAVING TWO, OR VERTICILLATE, SEED-LEAVES.'

The colophon on the reverse of the title page shows the book to have been printed by Harrison and Co, Printers, St Martin's Lane. This was a printer who used Parker's old printing premises, and with whom Parker had been in partnership for some of his working life, so perhaps also his presses, too. The book's *Preface* stresses the value of the study of botany, in agriculture, food, medicine, and so on, states that botany may be a feminine pursuit, and ends by quoting the Bible, concerning the planting of trees. There is a good analytical table of contents, and a glossary of terms, and then we're into the text, which takes us on a guided tour of plant anatomy, from roots and stems, through leaves, buds, hairs, flowers, bracts, calyxes, and corollas, and all the other detailed anatomical terms used to analyze and describe a plant, including a simple elucidation of the Linnaean classification system. The book ends with an interesting tabulated listing of the 'Principal plants, or their produce, useful to Man, with their botanical names' using vernacular English names, old scientific names (genus and species, natural order), and their (newer) Linnaean classification. The project is modest, but global, and sacred.

This charming and informative little book has 139 pages, and over 60 illustrations. Most are very small simple outline-engravings of the parts of plants under discussion, lying in with the text. There are three full-page wood-engraved plates, whose artists and engravers so far remain unidentified.[42] These full-size plates are not tipped in, but have been printed with the regular pages, and they have been allowed to remain blank on the reverse to prevent letterpress showing through.

This modest volume has no bookbinder's ticket at the back inside cover, but between the main text and the back endpapers, there is a bound-in list of Parker's other educational books, eight pages long. No one examining this book when it was published in 1848 would have been able to predict that a decade later the very same publisher would produce *Gray's Anatomy*. Yet curiously, looking at it now, with that knowledge in mind, it is possible to perceive certain characteristics which do in fact reveal interesting parallels:

- the good organization of the book, designed so that the reader can make their way around it without difficulty,
- the plainness and modesty of its self-presentation,
- the good value (much content for modest price: it cost only two shillings),
- the sheer generosity of the contents, and the fact that it does not patronize the reader, rather, that it stretches the reader's ideas as to what constitutes the discipline of botany, and the self-discipline of studying botany, while also providing concisely the means to do so,
- the integration of image and words, by the use of simple but accurate wood-engravings,
- leaving blanks on the reverse of printed wood engravings in a two-shilling book bespeaks an uncommon respect for illustration, and for an illustrative technique cheap by comparison with copper or steel engraving.

There is one image in this book to which I should like to draw attention. A diagram appears on page 63, concerning the normal arrangement of the different parts of a flower, 'supposing it to be regular and perfect.' It shows the calyx and the corolla of a perfect flower schematically, with the letters of the words CALYX and COROLLA set in circular motion, neatly placed on the structures they seek to reveal. You have to turn the book slightly, or set your eyes to a slight curve, to read it at first; but once you have understood the labels, you grasp why they are there.

The interweaving of word and image is closely similar to the process used by Henry Vandyke Carter for the anatomical illustrations he created for *Gray's Anatomy*. Among the reasons his style of anatomical drawing was appreciated at 445 Strand was its clarity, its simplicity, its suitability for wood engraving, and perhaps because Carter's process of naming structures integrally with typography chimed with this example of Parker's.

When we look at other illustrated Parker books, a family resemblance reveals itself. Bishop Edward Stanley's beautiful little book on *Birds*, for example, is small, and full of information. Though rather fatter than *Botany*, it is neatly bound and put together, well-organized within, and beautifully

The corolla & calyx of a flower: diagram with intrinsic labelling from *The Elements of Botany*, published by JW Parker for the Society for the Promotion of Christian Knowledge (1853). Actual size. From the author's collection.

illustrated. Once again, the pages printed with wood-engraved plates have been left blank on the reverse. This uncommon respect in Parker books towards wood engraving as an art form carries through to larger works, too. For example, in 1855 they published *The Lances of Lynwood*, a romance in mediaeval costume by Charlotte Yonge, probably for the Christmas market. Bound in an ornamented rich dark royal blue by Burn & Co, it featured full-page illustrations by Jemima Blackburn, wood-engraved in black line by the Dalziels, in a contemporary Pre-Raphaelite style. The novel is not particularly memorable in itself, but the illustrations are unusual, and it seems that the Parkers thought so too, as each was individually protected with a tipped-in sheet of tissue.[43]

One reason *Gray's Anatomy* did well in 1858 was because by the standards of the day, its iconography was unusual. Although printing illustrations on wood was understood as a revival of a much older and cruder art form than engraving on copper or steel, in the 1850s the mediaevalism of the woodcut (now called wood engraving) had undergone a transformation, and had acquired something arty and rather self-consciously modern about it.[44] Queen Victoria and Prince Albert dressed up in what passed for mediaeval costumes, architecture was suffused with mediaeval stylistic touches, the Great Exhibition had an entire Mediaeval Court. The Mediaeval was associated both with knights and ladies, and with democracy, with Magna Carta, and with Caxton. The Pre-Raphaelites had made fashionable the sinuous black outline of the German Romantics. To us, wood-engraved book illustrations seem old-fashioned; to mid-Victorians, they looked modern and stylish.

Almost every Parker book so far examined has a colophon naming the printer, either on the reverse of the title page, or on the final printed page at the very back of the book, just before the endpapers. When Parker gave up printing for himself (which looks to have been at some time in the late 1830s or the early 1840s) his favourite regular printer became Savill and Edwards. They were local, only a minute's walk away in Chandos Street, just north of the Strand, and relationships must have been congenial. The Savill colophon is the one most frequently seen: probably seventy per cent of Parker output was printed on Savill's presses. The rest went to a variety of other printers, particularly to Harrison of St Martin's Lane, and Robson & Levey of New Fetter Lane.

Savill and Edwards's premises in Chandos Street were destroyed by a catastrophic fire in 1853, which also destroyed premises on either side.[45] Further study may reveal the impact of this fire upon Savill's customers, but one of its most immediate effects was that until Savills could rebuild, retool, and return to business, publishers like Parker had to go elsewhere to get work printed.

One of the printers who picked up work from Parker at this time was John Wertheimer, whose premises were on the other side of the City, in Finsbury Circus. Wertheimer was a linguist, and was well known for fine and complex printing, including (like Parker) of Bibles. How contact was made is not known, but Parker was a printer-publisher, and printers had their own bush telegraph. Few Parker books exhibit a Wertheimer colophon. The distances involved may have been uncongenial to Parker senior, who as we have seen liked to do business face-to-face. But Wertheimer's willingness to accommodate Parker in this emergency later proved significant, because it was to him that the Parkers returned in 1856 to print the completed two-volume text of Todd and Bowman, and a year later for the complex job of printing *Gray's Anatomy*.[46]

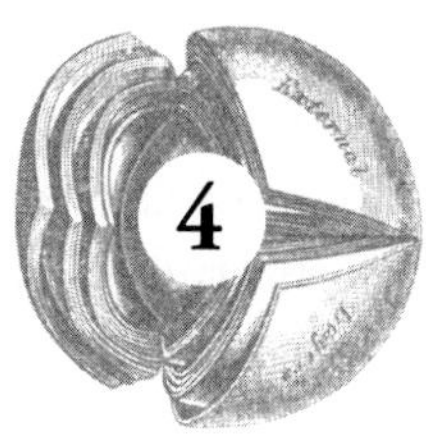

IDEA

Person or Persons Unknown

Considerable accord must have existed at the outset between Gray and his publishers for the project to develop as it did. Surviving Parker ledgers show that the publication agreement for *Gray's Anatomy* conformed to the Parkers' regular arrangements for works by commission. This was not a 'half-profits' agreement. Gray was commissioned to write the book by the Parkers, and did not put up financial capital towards its publication. John W Parker and Son shouldered the entire risk.

Father and son must really have wanted to publish an anatomy text, or they would not have invested so heavily in the idea. A new anatomy textbook on such a scale was an extremely expensive project, involving the long-term investment of considerable sums of money, so it is not likely that the Parkers would have wanted to commit to it without each having a strong inclination in its favour.

Parker senior had an interest in human anatomy that long predated this particular book. Back in 1837, when he was still the University Printer at Cambridge, and concurrently publishing in London for the Society for the Promotion of Christian Knowledge, Parker had produced a little book for the child and self-education market, entitled *The House I Live In.*[1] It was a small, slender volume, illustrated with simple but accurate anatomical wood engravings. The book was originally written by an American, Dr Alcott, but Parker had commissioned a London doctor, Thomas Calvert Girtin, MRCS, to edit and adapt it for an English audience. *The House I Live In* played on the idea of the human body as a building, a habitation for the soul. This was a long-standing Christian idea which cast God in the role of the Great Architect. The anatomical information this little book contained was simply phrased to soothe sensitive sensibilities, focusing mainly on the skeleton, and carrying the analogy of the house and the body to a short book length.

The cheaper end of the contemporary commercial book and magazine market for popular medical advice and information still tended towards a mysterious-seeming amalgam of medical obfuscation, herbal and folk remedies, and the advertising of quack and proprietary medicines. The subject of the human body in the early Victorian era was clouded with euphemism at one extreme, and salaciousness at the other. *The House I Live In* kept a clear distance from both, and from misinformation. The book cut through the fear and mystery surrounding the human body and its medical terminology, simply by explaining it. Unusually, it spoke straightforwardly about the need

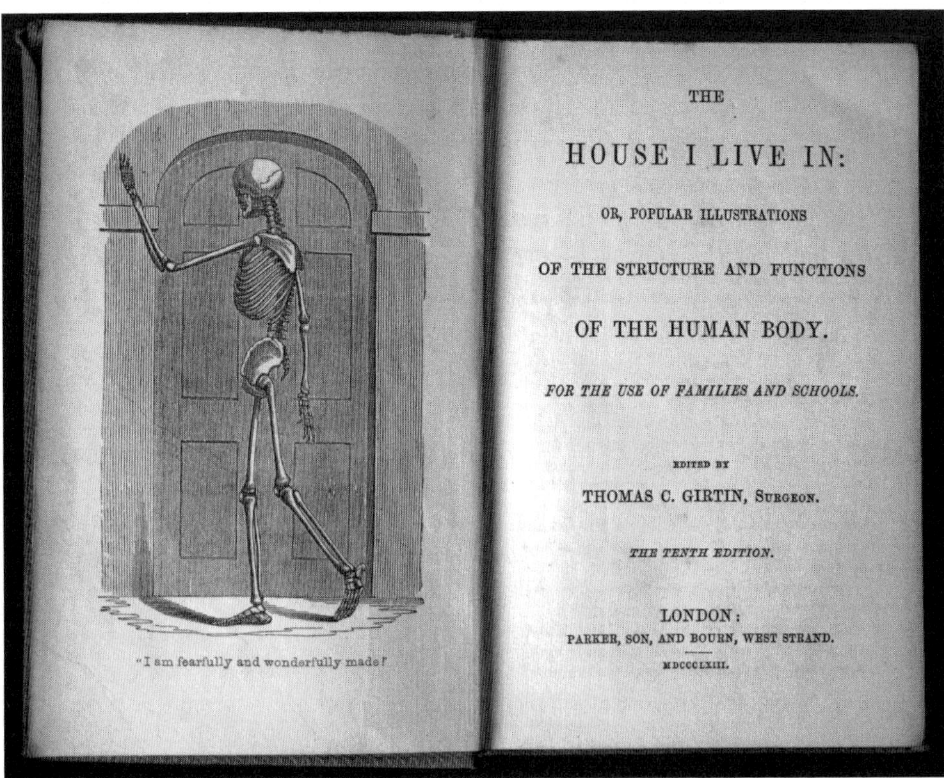

"I am fearfully and wonderfully made!"

THE

HOUSE I LIVE IN:

OR, POPULAR ILLUSTRATIONS

OF THE STRUCTURE AND FUNCTIONS

OF THE HUMAN BODY.

FOR THE USE OF FAMILIES AND SCHOOLS.

EDITED BY

THOMAS C. GIRTIN, SURGEON.

THE TENTH EDITION.

LONDON:

PARKER, SON, AND BOURN, WEST STRAND.

MDCCCLXIII.

The frontispiece and titlepage from the tenth edition of *The House I Live In*, one of JW Parker's best-selling books. Author's collection.

for human dissection. Communicating a sense of wonder in the complexity and beauty of the human frame, the book conveyed at the same time an essential innocence in exploring knowledge of bodily workings, treating the body as a wonderful mechanism.

The English editor understood and appreciated the simple frankness of its American author's original style, and left the little book's natural theology intact. The biblical caption for the frontispiece encapsulated the book's message: 'I am fearfully and wonderfully made.' There need be no embarrassment about the body, which (after all) is the image of God: we should understand our bodies, and should be able to speak about their functions without embarrassment or prurience, and prevent illness and premature death by observing simple rules of diet, cleanliness, and sobriety.

The House I Live In remains a rather charming book, even today. It was enormously successful, running into at least ten editions in Parker's publishing lifetime, between the 1830s and the 1860s. It managed to dispel ignorance without in any way transgressing decency, appealing to the natural sense of curiosity about the human body amongst the Victorian reading public. It did not cover sexual matters. It was the sort of book a Victorian mother could safely read to her children, an intelligent child could find and read alone, a schoolteacher might happily read aloud to a class. The book sold well after its first appearance, and steadily thereafter. A publisher's satisfaction in producing a book of such intelligent simplicity and charm would have been matched by the pleasure of its success. What had proved a delightful thing to produce, was also a delight to others.

To the younger Parker, this modest book would have been childhood reading, something he knew intimately, and had grown up with: part of the furniture of home. The sophistications of university might have made it an amusing thing to contemplate at a distance of years, the kind of thing undergraduates might chortle at. Yet young Parker also knew that its innocence was genuine, its educative function benign, and that the little book's popularity had helped provide the wherewithal to allow him to study, so his smile probably never became one of ridicule.

By the time young Parker became a publisher himself, he would have grown to appreciate the apparent artlessness of his father's little book in a new and different way. Personal enthusiasm for a volume was no foolproof recipe for sales, young Parker had found: Gray's *Spleen* had proved as much. Other publishers might choose to risk publishing prize medical essays, perhaps on more popular subjects.[2] *The Spleen* might have been a prize-winner, but it was not a best-seller: its subject was not of wide interest, and despite good reviews it had not sold. That particular attempt at expanding the partnership's medical list had been quite a disappointment. Publishing a book that sold well was an art he now knew had to be learned. *The House I Live In* was still selling.[3] By 1855, Young Parker would have known that far greater care must be taken to suit the market.

No one knows whose idea it was to create *Anatomy Descriptive and Surgical.* Pondering how the idea emerged initially throws up a number of possible scenarios. The most straightforward is that Parker senior suggested the idea to his son as a way of rectifying *Spleen*'s commercial losses.

The Parkers were textbook publishers: they would have shared an awareness of the output of close competitors—like Churchill, Highley, Longman, Murray, Renshaw, and Walton—all of whose lists included medical textbooks with long-term regular sales. The Parkers' medical list included some excellent books, several of which were doing nicely. An anatomy textbook would involve high outlay, would be quite an investment, but it would round off the list. And who better to create it than the prize winning young anatomist, Mr Gray FRS, who was already a Parker author?

This scenario renders the gestation of Gray's book a hard-nosed publisher's decision, and it may well be that the development of *Gray's Anatomy* was indeed primarily a commercial matter. After all, this is how most commissioned books are begun.

A further possibility, however, is that the idea emerged in conversation between young Parker and Gray, most likely after some enquiry from the author about his earnings from *Spleen.* Both young men would have been disappointed by the book's commercial failure. There were still over eight hundred copies sitting in the storeroom, unsold. Most of this dead stock would eventually be wasted (recycled), probably to offset bookbinding costs by being sliced up for reuse as bookbinding fillets.[4]

For each of these young men, the upset was personal as well as economic. Market failure is a blow to an author's ego. For Gray it would have tarnished the pleasure of being in print: he had expected to do well out of the publication of his prize essay. If he was disgruntled that no more money had materialized from all the effort of seeing the thing through the press, it would be understandable. For young Parker, there was the difficult job of explaining the hard economics to a hopeful author. Moreover, the book's failure had cast doubt on his own judgement. Back in 1853, he had perhaps to persuade his father that Gray's prize-winning *Spleen* would be a sell-out, and had discovered with time that Parker senior was far wiser about the likely market than his college-educated son had imagined. *The House I Live In* had never won a prize, but had been a far greater long-term commercial success.

To have good textbooks on the Parker list from big names such as Todd and Bowman, Watson, and Miller was not simply luck. Young Parker had made it his business to get the best people. He was well-disposed towards medicine, and well-informed about it. His friend Froude had badly wanted to train as a doctor, and some of his enthusiasm had probably rubbed off. Maurice had been chaplain at Guy's Hospital, and Kingsley's brother was full of good advice. Might Parker junior carry forward his father's labour with an anatomy book for adults? Conversations among his medical friends could well have alighted at some point on the relative merits and demerits of current anatomy books.[5] An anatomy textbook could well have been a project that had been thought about and planned for a long time. The investment costs would be high, so the opportunity might have been awaited and hoped for, as part of a longer-term plan. The publishing house already had books on physic, physiology, dentistry, and chemistry. It would have been a logical development to think about anatomy and surgery.

During the possible conversation with Gray, the thread of Parker's comments may have taken something like the following path: 'I'm sorry the Spleen book hasn't sold. ... Surely you can write something more saleable? ... We all want you to earn some money ... Textbooks are good earners .... You're an anatomist—play to your strengths—why not a textbook of anatomy?'

It is equally plausible that the idea had originated with Gray himself. After all, he had been using existing textbooks to teach anatomy to young medical students at Kinnerton Street for some time now. As curator of the medical school pathology museum, Gray would have been thinking too about how to re-organize the collection, how to make the fascination of its contents more accessible to students, how to structure existing knowledge, perhaps how to ease his workload, to free himself up for more research or private practice. New large-scale diagrams and micrographs Carter had created for the medical school were on display at Kinnerton Street, and they were impressive, equal to anything at the Royal College Lectures.

Gray's natural inclination would have been towards physiology or pathology, but the Parkers were already publishing a first-rate textbook in physiology, and the Sydenham Society had recently published a work on pathology by Kölliker translated by TH Huxley.[6] Especially after his failure at the Royal Institution, it was important to Gray to be up with those big names—Kölliker, Huxley, Todd and Bowman—on the spine of a big book. There may have been a moment when the idea came to Gray that he could capitalize on the field he knew best, that anatomy would answer.

Many teachers of pre-computer vintage may be familiar with an occurrence not infrequent in old-fashioned classrooms: when, in the act of wiping from the blackboard some fine example of chalk-drawn teaching left there by a previous lecturer, they momentarily experience a wave of pathos at the laying waste of human effort, and the evanescence of the teaching endeavour. Gray wanted his name on a major textbook, but was keenly aware that his own illustrative talents were weak: so far as I can discover, he never signed an illustration. He admired Carter's meticulous dissections, his eye, his grasp of significant detail, his graphic style—after all, why else should he have employed him to do so much for *The Spleen*? Carter saw things. His microscope illustrations for Gray had been superb, and they had cost very little. At Kinnerton Street, Gray was additionally familiar with Carter's display work for the medical school, and with his blackboard skills. Carter was immensely clever with pencil and chalk, and had an intelligent microscopic eye. He was also diligent, poor, and deferential. He would serve admirably. The idea could indeed have been Gray's own. As a team, the two of them would be formidable.

There is yet another possibility to consider, which is that the idea for the book originated from a woman. Ellen, the elder Parker's second wife, was the daughter of a doctor, and an experienced anatomical artist herself. She had worked alongside one of the finest draughtsman–engravers of the day, George Scharf. Her intimate knowledge of the anatomy of fossil creatures derived from her own work, illustrating specimens from the collection for which her late father, Dr Gideon Mantell, had been famous.[7] Her careful anatomical drawings had been used in her father's lectures, as displays in his well-known fossil museum, and as illustrations in his books. We do not know precisely which volumes filled the bookcases in the Parker family home, but among them were probably some in which Ellen's work might be found. She had an alert and informed scientific mind, and a graphic eye.

Ellen would have been a highly valuable co-worker alongside her husband and stepson in the process of considering topics for *Fraser's Magazine*, or new opportunities for book publishing. Had Ellen seen Carter's work for Gray's *Spleen*, she would have appreciated its quality immediately. She might well have asked the simple question: if they were keen to expand their medical list, might not a manual of anatomy be a good idea? It was an obvious question. If Ellen voiced it, she would have met swift endorsement from her husband, and his son.

It is quite possible that Parker senior had been looking at anatomy books for years, and that his interest had long predated *The House I Live In*. Printers were—indeed are—known for their love of their craft, and for their great breadth of bookish knowledge. Anyone with a love of print, and especially of copperplate printing (Parker's original calling) would have had an interest in the spectacular anatomical atlases printed in England in the previous century. William Cheselden's great book on the bones of the skeleton, *Osteographia,* (1733) is a prime example of these works, with its large images of all the bones of the human body, including life-size thighbones, shown in what we might think of as near-photographic detail. The book was poised between old and new worlds when it appeared, with animated skeletons in praying and dance-of-death-like poses, but with the proclaimed use of visual technology to draw them. Cheselden demonstrated the accuracy of his images by showing a camera obscura in use on his title page.[8]

An even finer copperplate anatomical work was *The Human Gravid Uterus* of 1774, William Hunter's great atlas of the pregnant womb.[9] The famous anatomy school Hunter had founded in Windmill Street was still trading round the corner from Parker senior's printing shop in Castle Street, Leicester Square, in the 1820s. Tales about the Hunter brothers, both of whom had worked close by only a generation previously in the era of the grave-robbers, were likely still current in the neighbourhood.

William Hunter, author of the *Gravid Uterus*, had taken twenty-five years to acquire the fourteen fresh bodies on which his great work was based, all of them belonging to women who had died during pregnancy or childbirth. The book's enormous illustrations, by Jan van Rymsdyk, show the structures of gestation from full-grown death in the womb, right back to the bubble of a five weeks' conceptus. The perfect dissections of these tragic bodies are recorded in the illustrations, revealing knotted or torn umbilical cords, ill-positioned placentas, ruptured wombs, infants lodged in breach and other impossible positions, stifled where they lay. These astonishing images, protected by heavy bindings, are poised between hope and hopelessness, death and life. They embody the most extraordinary juxtaposition of meat and love.[10]

A work such as this one by Hunter was a collectors' piece, prohibitively costly, and seldom available for sale. Yet Parker senior may nevertheless have had the opportunity to catch sight of one at some time in his working life. He had friends with access to medical and scientific libraries: some kind professional might have taken him just across Trafalgar Square to see the Library at the College of Physicians. Indeed—once he was financially established—he could obtain access himself to some of the best private lending libraries of London. There were also the numerous second-hand bookshops and bookstalls along and around the Strand, along Booksellers' Row in Holywell Street, Fleet Street, easterly along Paternoster Row and Cheapside, and scattered in the warren of alleys to the north of Charing Cross, famous for its bookshops even today. The vicinity of Parker's printing office and shop was swarming with booklovers, and an honorary member of their club, such as he, would have encountered many kindred spirits happy to show him their prizes, without having to make a purchase.[11]

Had Parker ever had the opportunity to handle a copy of William Hunter's book, he would certainly have savoured its handsome rounded type and beautifully spaced typesetting, its hand-laid paper and gold-tooled binding, and its astonishingly well-observed illustrations. The volume had been produced by the revered English printer John Baskerville, type-designer, book-designer, industrial innovator and artist of the book. A doctors' classic, the book was also a masterpiece of printing. Created by a surgeon and a printer–publisher, it was a great work indeed, examining the great mysteries of Creation and of the human condition.

We do not know for certain that Parker had the opportunity of examining this grand volume, or if he was ever able to compare it with a work by Hunter's teacher, William Smellie, whose *Sett of Anatomical Tables*, had been published in London twenty years earlier, in 1754.[12] Both books had been illustrated by the same artist, but in very different styles.[13] Whereas Hunter's book was intended for rich subscribers, Smellie's was not.

Smellie was a rare bird: like Hunter, committed to ameliorating the calamity of death in childbirth, but unlike him, committed to publishing his knowledge for practitioners lower down the social scale. He had caused something of a scandal in mid-eighteenth-century London, by teaching female anatomy to large mixed classes of men and women at his midwifery school. His *Sett of Anatomical Tables* used an inexpensive printing method of the day—etched mezzotint—to show ways to assist in cases of difficult birth, including turning the baby in the womb, and the correct use of the obstetric forceps of which Smellie was an early champion. Not for him the shameful professional secrecy of their inventors, the Chamberlen family of Essex, royal obstetricians, who had preferred to bury their instruments for over a century rather than share their knowledge. Their forceps were not rediscovered until 1813, by which time people like Smellie had pioneered their own alternative designs.[14]

Smellie intentionally chose mezzotint for his illustrations, because it was a faster and more economical method of illustrative printing. Characteristically, he also used cheaper format and production methods, so that his book would be affordable to his own students. His work continued in circulation well into the nineteenth century, especially after Elizabeth Cox, a printer–publisher working in St Thomas's Street (between Guy's and St Thomas's Hospitals) had them re-engraved to smaller dimensions, and reissued the whole work as a pamphlet.[15]

Parker's *The House I Live in* was an exceedingly modest nod towards such anatomical works. Although his little book was aimed at an even cheaper market, in an unassuming way it nevertheless partook of their beauty and their educative and ameliorative intentions.

So, a number of possible conversations between Gray and the Parker family in various permutations have been postulated, the chronology of which we cannot ascertain. But, once a mutual interest had been expressed, matters might rapidly have passed into a discussion of what was wanted, what was to be avoided, what was practicable: what *kind* of book they might create together.

It may well be that one conversation evolved into another: that from a chat in the shop, Gray was asked to stay to supper, and the whole group sat together to brainstorm about the book; or, the family meal at which the idea arose prompted a discussion with Gray. Either way, we can easily imagine an initial burst of enthusiasm, the collective talking through of the issues, and an agreement to separate to think more on the matter before any firm decisions were made.

All we have to go by is that note in Carter's private diary, on 25th November 1855:

> Little to record. Gray made proposal to assist by drawings in bringing out a Manual for students: a good idea but did not come to any plan … too exacting, for would not be a simple artist.[16]

This conversation looks to have taken place after an initial discussion involving the Parkers, by which time plans were already underway. The project is a 'Manual for students', and Gray sounds to have the confidence of an author who already has his publishers' backing. Perhaps, while reading Carter's words, we are vicariously witnessing the first direct result of a meeting with the Parkers. After all, both they and Gray knew that without a good illustrator, the book could not succeed, and Gray had one of the best young anatomical illustrators in London in his pocket.

But Gray would appear not to have received the ready assent he'd expected: the younger man was uncertain, doubtful, and not about to yield in haste to Gray's enthusiasm for an idea on this scale. He was not the pushover Gray had anticipated. The next year was probably already planned ahead in Gray's mind, but Carter had plans of his own, too. He demurred, had reservations, wanted time to think.

Gray would have grasped the situation. Getting Carter on board was crucial to the scheme's success. His illustrator was busy studying for his MD, was low, wanted persuading: he'd come round to the idea eventually. Gray would have to *court* him. He wasn't asking that Carter do any work *now*: all he'd wanted to ask was the loan of some of Carter's drawings, just to show the publisher.

Carter always found it difficult to demur, so he might have agreed to such a preliminary salvo. Gray would disappear off with the portfolio, leaving Carter to his thoughts. Gray could square things with the Parkers, and get down to using the Christmas break to plan the shape of the new book. Carter might be persuaded if money was forthcoming. If not, another illustrator could be briefed about what was wanted.

While Carter was battling his winter mood, Gray probably had a second meeting with the Parkers. He might have invited Parker junior to see the dissecting room at Kinnerton Street and to sup at Wilton Street, but on balance it seems more likely that the meeting would have been at 445 Strand. The autumn teaching term extended right up until Christmas, so this was perhaps a prearranged engagement for which time would have been set aside by both sides, after dusk left the dissecting room deserted.

Because we lack diaries or letters either from Gray or from the Parkers, we cannot be certain who represented the publishers: it may simply have been Parker junior alone. Later on, we shall see that the project was regarded as his, so it is quite possible that there were only the two of them. Both parties would have amassed a few books for discussion. The key point is that Gray and someone from the publishing house would have gone over these existing volumes together at some early period in the planning of the new book, to agree primarily how their own new work should compare, how it might look, how it should be priced. The intention was to achieve some mutually agreed clarity about what it should be, and what it should not.

A number of important topics would have been up for discussion: the new book's shape and size, its external and internal appearance, the text length, typesize and readability, spacing, size of the illustrations. It was important that a process was undergone which included looking over—and criticizing—other authors' and publishers' efforts, so as to clarify what was really wanted, and in what way this new book could win its audience—what its unique selling points would be. Without real examples, neither participant could be sure that they were agreeing about the same sorts of things.

The works Gray would have chosen were probably not large or showy; but they are likely to have been well-thumbed. These were books in regular use. Publisher and author were already agreed from their previous conversation that the new work the Parkers wanted to publish was a student text, not a collector's piece. So Gray probably brought only three or four books to the meeting, each of which he'd have known well. Their strengths and weaknesses would have been at his fingertips.

Having looked extensively at the books available to tutors of anatomy in the 1850s, I have chosen what I regard as the most likely candidates for the discussion Gray and young Parker had planned. I shall present them in order of size, just as they might have been lifted each in turn from the pile as it stood on the table in Parker's comfortable sitting room above the shop, or on Gray's desk at home or at Kinnerton Street—diminishing upwards in size.

The smallest volume at the top of the pile would be the first of the volumes to be considered, so it is most likely to have been Robert Knox's squat pocket-sized *Manual of Human Anatomy, Descriptive, Practical and General,* which had been published in 1853 by Renshaw, an enterprising publisher whose shop was just along the Strand from the Parkers. Dr Knox's *Manual* was diminutive (only 6½ inches by 4), but it contained a great deal of good material, and all well presented. The illustrations were very small, but Knox's descriptions were clear. The author was an experienced teacher of a previous generation, a disciple of the great French anatomist Xavier Bichat, and a survivor from the bodysnatching era. Knox was the very man to whom Burke and Hare had sold their murdered corpses back in the 1820s. Driven from Edinburgh, he had found a safe haven in London.

His *Manual* was competent and comprehensive within its small compass, rooted in Knox's wide knowledge of both anatomy and medical students. It was also cheap to buy: only 12 shillings and sixpence. The trouble with it was its size. It was so small that the letterpress text looked congested. The illustrations were rarely over a third of a page in size, and were mostly littered with tiny numbers and letters, with small and close-set explanations in footnotes at the bottom of the page. Probably for reasons of economy, the smallest boxwood blocks had been used—which had put a terrible strain on both the engravers and the illustrator, because it constrained not only what could be shown and labelled, but cut. A vertebra of the human spine, for example, was less than ¾ of an inch in height. The engravers were valiant in what they had done, but one had to go in very close indeed to see what detail they might have tried to put there, or to which miniscule near-illegible letter the footnote referred.

The next-sized volume was probably either *The Anatomist's Vade Mecum* or *The Dissector's Manual,* both by Erasmus Wilson. The *Vade Mecum* was an older book, into its sixth edition in 1854, published by John Churchill of New Burlington Street, an important publisher with a large medical list. Wilson's more recent book was a best-seller for Longman. Although they were issued by different publishers, they were really two volumes of the same work. The first covered human anatomy by system, bones, nerves, muscles, et cetera; the second demonstrated the relationships of the systems within the regions

---

A typical page from Erasmus Wilson's *Anatomy.* The wide margins and good typeface of this US edition did not make up for the paltry size of the illustrations, the same as in Wilson's *Anatomist's Vade Mecum,* a smaller book. Actual size. From the author's collection.

with the fourth and fifth metacarpal bones; by the two lateral articulating surfaces, with the magnum and cuneiforme; and by the flattened angle of its apex, with the semilunare.

Attachments.—To *two* muscles—the adductor minimi digiti, and flexor brevis minimi digiti; and to the annular ligament.

Developement.—The bones of the carpus are each developed by a single centre.

The number of articulations which each bone of the carpus presents with surrounding bones, may be expressed in figures, which will materially facilitate their recollection; the number for the first row is 5531, and for the second 4475.

METACARPUS.—The bones of the metacarpus are five in number. They are long bones, divisible into a head, shaft, and base.

The *head* is rounded at the extremity, and flattened at each side, for the insertion of strong ligaments; the *shaft* is prismoid, and marked deeply on each side, for the attachment of the interossei muscles; and the *base* is irregularly quadrilateral and rough, for the insertion of tendons and ligaments. The *base* presents three articular surfaces, one at each side, for the adjoining metacarpal bones; and one at the extremity for the carpus.

Fig. 38.

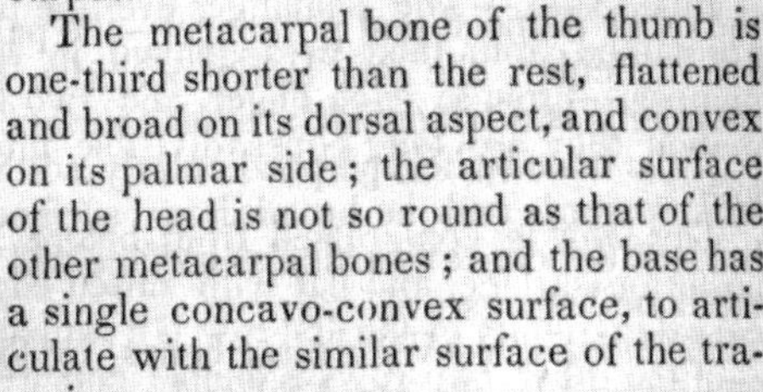

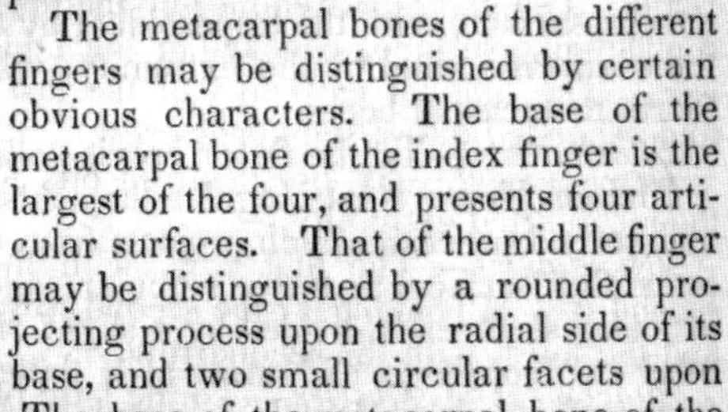

The metacarpal bone of the thumb is one-third shorter than the rest, flattened and broad on its dorsal aspect, and convex on its palmar side; the articular surface of the head is not so round as that of the other metacarpal bones; and the base has a single concavo-convex surface, to articulate with the similar surface of the trapezium.

The metacarpal bones of the different fingers may be distinguished by certain obvious characters. The base of the metacarpal bone of the index finger is the largest of the four, and presents four articular surfaces. That of the middle finger may be distinguished by a rounded projecting process upon the radial side of its base, and two small circular facets upon its ulnar lateral surface. The base of the metacarpal bone of the ring-finger is small and square, and has two small circular facets to correspond with those of the middle metacarpal. The metacarpal bone of the little finger has only one lateral articular surface.

Fig. 38. The hand viewed upon its anterior or palmar aspect. 1. The scaphoid bone, 2. The semilunare. 3. The cuneiforme. 4. The pisiforme. 5. The trapezium. 6. The groove in the trapezium that lodges the tendon of the flexor carpi radialis. 7. The trapezoides. 8. The os magnum. 9. The unciforme. 10, 10. The five metacarpal bones. 11, 11. The first row of phalanges. 12, 12. The second row. 13, 13. The third row, or ungual phalanges. 14. The first phalanx of the thumb. 15. The second and last phalanx of the thumb.

of the body, chest, abdomen, head, and so on. These are different ways of looking at anatomy, and teaching it, even today. Both works were widely used because the author was a good anatomist, and had a clear view of what needed to be known by a medical student. They were also affordable: each as cheap as Knox's—only 12 shillings and sixpence—but together, they were double the price.

Wilson's *Vade Mecum* was barely taller than Knox's book, his *Dissector's Manual* an inch taller, but all three weighed about the same. Wilson's illustrations suffered from similar problems to Knox's: they were far too small. The entire base of the skull in Wilson was only 2½ inches in height; the leg from the heel to above the knee, less than 4 inches, all spotted with tiny reference numbers, and footnoted in small close type that hurt the eyes. There was one excellent innovation in Wilson, though, that everybody liked, which was his scheme of showing the relations of the arteries in a simple and orderly diagram, stacked around the main vessel named centrally in a box. Learning to name these nerves and vessels was a bugbear for all medical students, but Wilson's diagrams meant that if you could memorize the central name, you were better able to recollect the initial letters of its relations.

Next in the pile would have been a volume a quarter of an inch taller again, and a little fatter. This was a volume of Quain, the work used by every student of anatomy and anatomist for the last quarter of a century, new and second-hand. Its full title was *Elements of Anatomy*, and it was widely regarded as the standard work. When it had started out in 1828, the book had been a one-volume work, small and fattish, with no illustrations at all. Its author, Jones Quain, had subsequently joined forces with Erasmus Wilson to produce an array of elaborate large plates, sold separately, either in plain black and white, or (for guineas more) hand coloured, issued in parts over several years. But most students couldn't afford the full set (at 20 guineas) or couldn't afford or manage to assemble it: plates got lost, nerves but no blood vessels, bones but no viscera—even bound copies were often incomplete, and then there was the difficulty of finding convenient shelf-room for the fat little book and the large plates.[17]

A typical page from Quain's *Elements of Anatomy*. Quain had an informative prose style, and margin summaries, which were useful for finding one's place. But the rest of the page design was poor, and the images were execrable, made worse by proxy labelling which referred the inquisitive reader to footnotes, inches away. Actual size. Compare with Carter's versions on pages 116 and 167.

sinus, the sacculus, the three ampullæ of the semicircular canals, and the cochlea, receiving one each. of the nerves.

Fig. 243.*

A

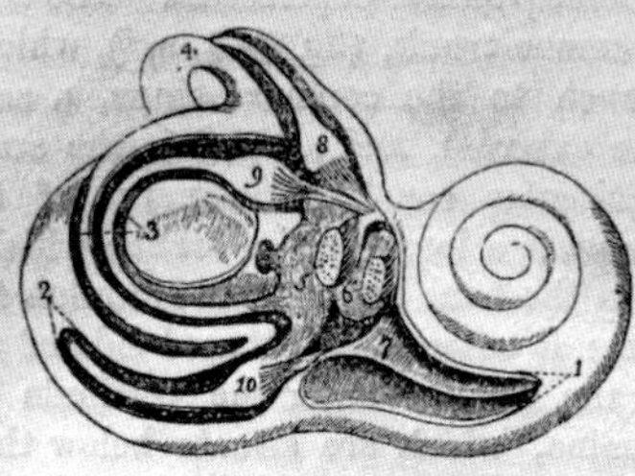

B

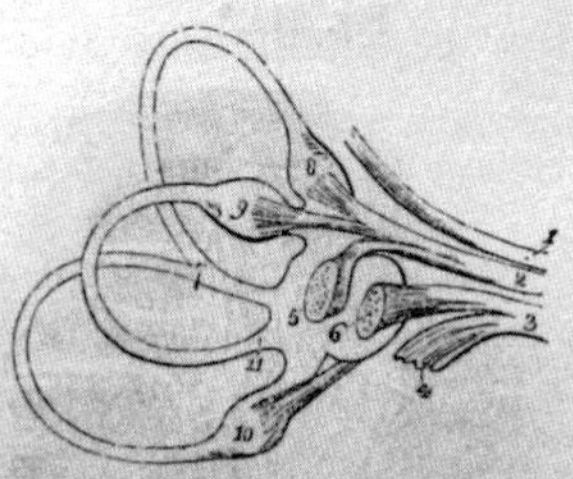

Primary Division of the Nerve.—The nerves for the supply of the *common sinus*, and of the *superior* and the *external* Nerves of vestibule and semi-

* A. Membranous labyrinth displayed in position by cutting away part of the osseous labyrinth. (After Breschet.)—Spiral canal of the cochlea. 2. Posterior vertical semicircular canal opened to show the membranous canal within. 3. External or horizontal semicircular canal; the whole of the membranous canal seen. 4. Superior vertical semicircular canal. 5. Common sinus. 6. Saccule. 7. Lamina spiralis. 8. Membranous ampulla of anterior semicircular canal. 9. Ampulla of the external, and 10, ampulla of posterior canal.

B. Membranous labyrinth detached. (Breschet.)—1. Portio dura of seventh nerve. 2. Anterior portion of the portio mollis, giving branches to the anterior and the external ampulla and to the utricle. 3. Posterior portion of auditory nerve, giving branches to saccule and cochlea. 4. Nerve to cochlea. 5. Common sinus. 6. Saccule. 7. Common end of the superior vertical and posterior vertical semicircular canals. 8, 9, and 10, as in fig. A. 11. Undilated end of external or horizontal canal.

The most recent edition of *Quain*, the fifth edition, had been issued in 1848, in two volumes—a weighty double-decker—each volume bigger than Knox. It was edited from University College, London, by Quain's successors, William Sharpey and Richard Quain, and published by Walton and Maberly of Upper Gower Street. This latest Quain was illustrated, but not generously: after pages of letterpress you might find a couple of smallish woodcuts on the Knox plan: the cochlear of the ear, for example, less than an inch in height, set amidst close type, dotted with nearly illegible miniscule numbers, or pierced with a sheaf of arrows with tiny numbers or letters referring to a key in a footnote close-set in even smaller type. Often too, the explanatory footnotes in Quain flowed round on to the following page.

Quain's illustrations did feel like the afterthought they were, but the work had important merits. Above all it was well written. Anatomy can be deadly dull, and although Quain looked painfully dreary, if you got down to reading it you would find that Quain managed to describe and explain the subject in a way that made it interesting and memorable. His volumes also had margin-summaries, so you could find your place quickly and revise more easily. But *two* fat volumes—enough to make anyone groan! And each volume cost twenty shillings, £2. 0s. 0d. for the two![18]

Finally, the tallest of Gray's probable sample volumes, and the most modern: *Human Osteology*, by Luther Holden of St Bartholomew's Hospital Medical School. Its first edition had just been published earlier that year, 1855. It was elegant in its dimensions, 9½ by 5½ inches, and although it weighed more than a single volume of Quain, the taller format, and the beautiful manner in which its pages were laid out, made it seem much less heavy going. The publisher, John Churchill again, could rightly be proud of this fine volume.

Holden hadn't tried to cover the whole of anatomy, his focus was simply the bones of the skeleton. His text was the fruit of years of teaching, simply and engagingly written. The letterpress text was in a fine clear typeface, generously well spaced and agreeably laid out. The illustrations were clarity itself: elegantly spare full-page line drawings, lithographed, individually bound-in in appropriate places. The bones of the human hand were small life-size—seven inches tall—and nearly three times the length of Knox's miniature version. The human spine was presented on a fold-out sheet which opened down to thirteen inches in length. It wasn't near life-size, but three times the size of Quain's. There was a small fold-out skeleton at the back, too.

The book had two original features: Holden had indicated the sites of the muscle-attachments upon the bones themselves using dotted lines. The most important novelty was that he did not favour the use of proxy labels: Holden's anatomical labelling was given in full, on or right beside the bones, so you could grasp immediately what it was you were looking at. The main drawback to Holden was that his book covered only the skeleton. It was brilliant for examining and understanding bones, but it featured nothing at all about dissection. Medical students had to learn all the rest from other sources.

Gray might also have brought along a couple of little pocket guides just to show Parker that even in the 1850s, unbelievably, publishers were still producing works on human anatomy without any illustrations at all, and that they continued to sell.[19] He might also have thought to bring Joseph Maclise's *Surgical Anatomy*, (Churchill, 1851) with its handsome large hand-coloured plates, or even a continental example, like Samuel Soemmering's spectacular osteological atlas, even though it was a decade old.[20] But these last two are doubtful: such costly things might serve for inspiration, but for a student manual they had no real relevance. By this stage Gray must have been aware that the Parkers' inclination and strength was towards economical books illustrated in black and white, using wood engravings set together with the letterpress.

By way of demonstrating the sort of quality and style the publishing house was accustomed to producing, for his part, Parker was likely to have assembled a few of Parker and Son's existing volumes. Gray and he had probably already had a conversation pretty much like this back at the time *The Spleen* was being planned; but this new book was much larger in scope, a completely different project. Quite a few good things had been produced in the interim, especially on the scientific side. The first few books Parker would probably have reached for were Todd and Bowman's *Physiology*, Watson's *Physic*, and Miller's *Elements of Chemistry*.[21]

Todd and Bowman's work on human physiology—which we have glimpsed at before—had been published in parts, and had developed with the field. It had been intended originally as a four-part work, sold cheaply in an inexpensive binding for a medical student audience, at seven shillings each part. But scientific understanding of human physiology had increased so swiftly in the previous twelve years, that the fourth issue had to be split into two, and the final (fifth) part was then being prepared for the press. The whole set would be re-published in two volumes, to appear in 1856–7. As the work had evolved, the wood-engraved illustrations, which had been smallish to begin with, had been given more space and grown larger in its successive parts, and the recent ones displayed more useful detail. For the final part-issue,

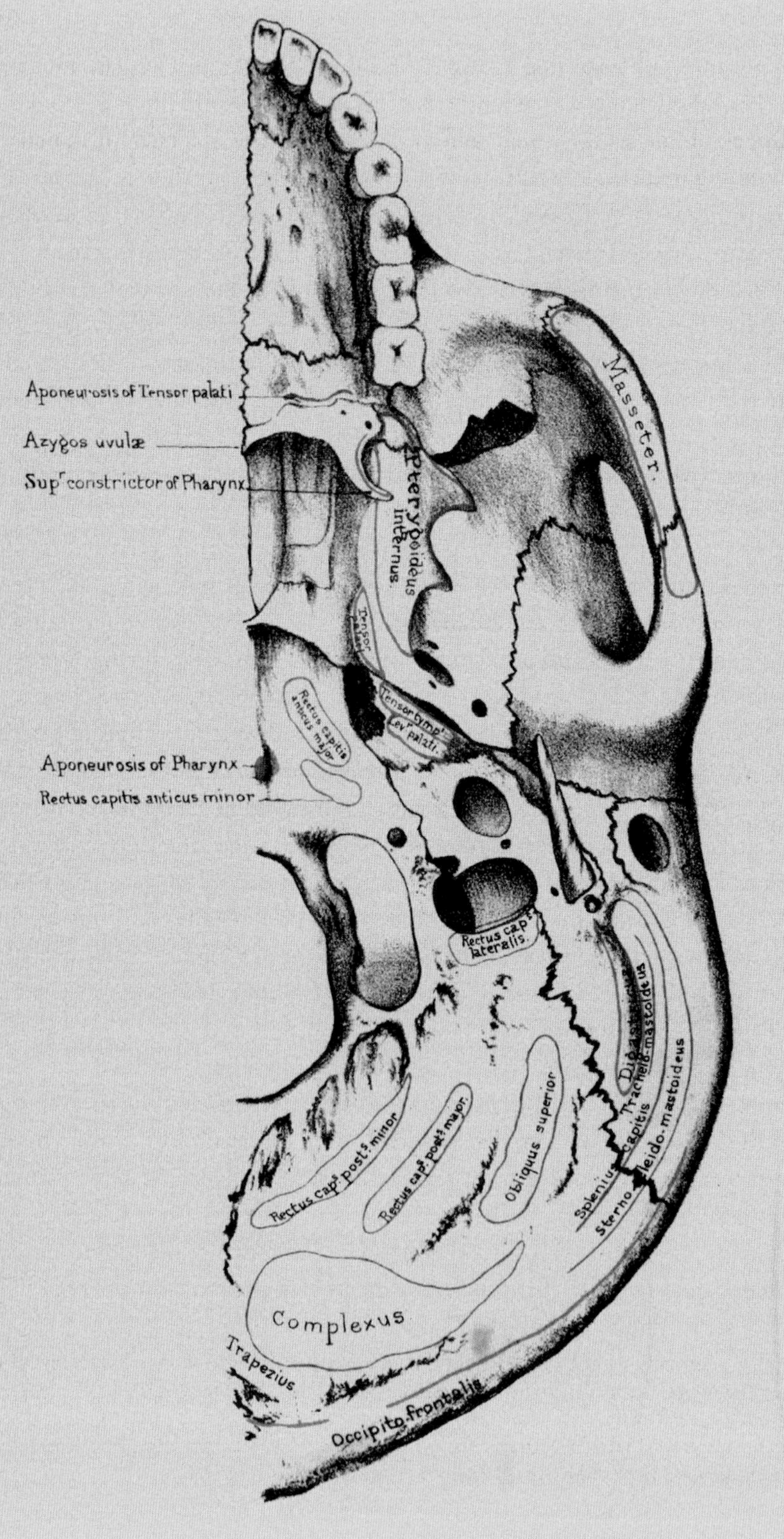

Aponeurosis of Tensor palati
Azygos uvulæ
Supr constrictor of Pharynx
Aponeurosis of Pharynx
Rectus capitis anticus minor
Masseter.
Pterygoideus internus.
Tensor palati
Tensor tympi
Levr palati.
Rectus capitis anticus major
Rectus capt lateralis.
Digastricus
Trachelo-mastoideus
Splenius capitis
Sterno-cleido-mastoideus
Obliquus superior
Rectus capt post. major.
Rectus capt post. minor
Complexus
Trapezius
Occipito-frontalis

the authors had requested to use three illustrations from Gray's monograph on *The Spleen*, including two fine examples by Carter.[22] WA Miller's book was a three-volume work, which sold at 46 shillings and sixpence. It contained everything required by a university student of chemistry. Although it was only sparsely illustrated, the wood engravings—mostly showing apparatus set up for experimental investigations—were of a good size, and neatly executed in black line. Thomas Watson's *Physic* (its full title was *Lectures on the Principles and Practice of Physic delivered at King's College, London*) was a substantial two-volume work, which sold for 34 shillings. It was not illustrated, but was brim-full of the wisdom and experience of a long-established and wise physician.

Gray would already have known these textbooks well—they were in daily use at St George's. Another well-known Parker book, John Tomes's *Dental Physiology*, perhaps might also have been singled out for discussion because of the variety of its illustrations, which were good for such a reasonably priced book—it sold for only 12 shillings.[23]

Parker would have been primarily concerned about how much investment would be required to make this venture run, whether Gray was really up to it, whether the large outlay could be recouped by the cover price in one edition, and whether it would continue to sell over time. The big questions for Gray were interrelated: the number of pages, the number of words he would have to compose, and the number and size of the illustrations. He had doubtless already made some estimates of his hopes in these respects: he probably wanted his book to be more impressive than the competition, particularly by comparison with Huxley's recent edition of Kölliker's textbook.[24] Gray wanted his name on the spine. Behind these presentational questions was his desire to know what kind of financial deal was going to be on the table: how much he himself might benefit from this big project, how much he could afford to offer Carter. The important thing was that Parker and he could jointly thrash out the project's viability, and the feasibility of each other's expectations. Both men were hoping for a commercial success *and* something to be proud of.

A typical illustration from Luther Holden's *Human Osteology*, which would have impressed contemporaries by the standard of book production and lithographic quality, by the size of its illustrations, and by the clarity and simplicity of its graphic style. The use of intrinsic labelling, which Carter developed in *Anatomy Descriptive and Surgical*, was pioneered by Holden from blackboard teaching techniques. Actual size. From the author's collection.

Of course there was no recording equipment in the room when Gray and Parker were discussing their plans—how one wishes to be able to have been there! But, because the end result of their efforts survives in its finished form as a volume, we can infer what sorts of things were decided, either at this early stage or at some other finalizing meeting soon thereafter.

The dimensions of the book were probably a primary consideration—because the book's size really determined the value of the illustrations, and much else besides. Gray wanted good informative illustrations, and so did the Parkers. Both parties knew, even at this early stage, that they did not want the button-sized pictures of the existing student pocket books of anatomy. To bring to the market one more small fat anatomy book would not have been envisaged by either party. Nor were there any plans to create an outsized one, which would be costly, unwieldy in the dissecting room, and difficult to curl up with to read, revise, shelve, or sell.

Printing in colour was still at an early stage of development for wood engraving. The results were so disappointing that the costs and complications involved could not yet be justified. Knox's publisher had tried it for the blood vessels in the *Manual*, but the blue and red were so dull that there wasn't much to tell between them. Holden had red and blue, and brighter, too, but the colour was applied to each plate by hand, which was very costly. So colour was possible, but not for this book.

Lithography, or tipped-in or bound-in plates of copper or steel engravings, would also have been ruled out on grounds of cost. The Holden osteology book was splendid, but in a book like that every plate was specially printed on special paper, left blank on the reverse, and by a specialist printer. Lithographs for books were cut to size from larger sheets—if you think about the Holden fold-outs and how they were made to fit, all by hand, how they were assembled at the binder's in their correct position in with the rest of the book's letterpress (from a different printer), and how such high-quality illustrations must be held carefully in position during the binding process, you can imagine the extra labour involved, and therefore the extra costs. Fine book, but no wonder it had a price tag of 16 shillings.

The same went for leather or half-leather binding, silk-braid bands at the gatherings to conceal the top and bottom ends of the sewing, or glued-in ribbons for bookmarks. Marbled endpapers were all very well, but (like gilt or marbled page-edges) those were in the fancy book league, with prices to suit. A bit of spattering on the edges might be managed, but that's all, and even that would depend on how the costings worked out nearer the time. This was a working book: it wouldn't be having gilt embossed panels on the cover, or a fancy frontispiece with its own individual leaf of tissue paper.

To cover the whole of human anatomy takes a lot of pages. Holden's book covered only the bones. Wilson's books were 626 and 698 pages each. Quain, with its two fat volumes, had 1363 pages in all. All these works had congested pages, condensed footnotes, and mean-looking illustrations, so small you could hardly see the structures they were talking about without a magnifying glass.

You couldn't prop any one of these books up, and see what you needed from where you were dissecting. The ideal would be a clear typeface, with decent spacing between the lines, and large illustrations—pages you could read if you had to, from say, a couple of feet away. Holden was good for that, but imitation of his pages—with their widely spaced type and small letterpress area—would seriously limit how much you could say in the text, unless you wanted to spread into two volumes, which would put you above the Quain in the cost league. And Holden's well-spaced text had the drawback that text didn't necessarily appear beside the appropriate illustration—it was often left facing the blank reverse of the plates. If the planned new book was to be a single volume work, with good type and large illustrations, it would have to be fairly tall, and fairly fat—even Knox was almost 700 pages.

Gray's book was to be a textbook, on the lines of Todd and Bowman, or Miller's *Chemistry*, in a single volume. Illustrations larger than in either of those Parker books would mean that this volume would have to be taller and wider than they, so royal octavo: that's about 9½ inches by 6, twice the page area of Knox's *Manual*. If it were to have about the same number of pages as Knox, 700 pages or so, that should be room enough, plus proper contents lists and index. It would need a good commercial binding. Round about 750 pages or so: Parker's calculations would be based on that.

Casing—bookcloth, blind-embossed in a plain way, brownish, serious. For the endpapers—again, plain colours: for textbooks Parker and Son generally liked shades of brown. Title page, contents, list of illustrations ... dedication page, perhaps? And the title? Gray's course at St George's was listed in *The Lancet* as 'Anatomy, Descriptive and Surgical', so that would do for now as a working title. Contents: anatomy and dissection, a few common operations perhaps, but to keep focused (and avoid duplication with other Parker books) not much physiology needed. A bit of microscopy to keep it looking up-to-date. Index.

If Gray had managed to borrow Carter's drawings, and if he hadn't done so before, this was perhaps the stage at which he brought them to the table. Parker already appreciated that larger illustrations were what was wanted, and would be greatly preferable to the mean little things in Knox and Quain. Students would much prefer more expansive and informative images if the book could be made to cover the whole of human anatomy. He knew the detailed clarity of Carter's work from *The Spleen*, and understood the young artist's dislike of ruining them with proxy labels. Had Parker chanced to see Carter's teaching drawings now, especially perhaps the Paris drawings of arteries, he would have grasped the value of their appeal immediately.

Blackboard teaching in anatomy is a skill that has probably died out in the UK since the advent of whiteboards, overhead projectors, slides, and computer graphics. But doctors who trained even as recently as the 1970s can still recall the extraordinary freehand blackboard drawings made by some anatomical teachers, using coloured chalks. Carter would not have been at all unusual in teaching in such a way: the blackboard diagram was a general norm in the teaching of anatomy, part of a tradition he helped pass on.

If Carter's drawings for his blackboard work were akin in quality to his eventual illustrations for *Gray's*, we can see exactly why Gray and Parker would have been enthusiastic in their favour. The eye could grasp what it needed at once. The anatomical names appearing directly on the appropriate structures meant that there would be none of those nasty little digits dotted all over like numerical smallpox. Furthermore, there would be no need for footnotes to explain everything. The students would love these illustrations. Like Holden's bones, you could see straight away what you were looking at, none of that to-ing and fro-ing with the eye. In books like Knox and Quain, sometimes a third of a page was lost in footnoting the illustration. Carter's illustrations could afford to be larger, because the elimination of explanatory footnotes would allow a substantial saving of space.

Because of the costs involved, it was important to agree how many illustrations to calculate for. Looking at the competition, it seemed that it would have to be between three hundred and four hundred. Gray had probably made a rough estimate of how many figures each section might require, adding up to somewhere in the region of 350–400. The cutting of them would be a special job: accuracy was paramount, and there'd be a lot of lettering to cut. It would need a skilled engraver. Might be a year's work to make them—350 or so, yes, easily! But, there would be quite a saving in *printers'* time: printers loathed the bother of composing footnotes—bespoke fiddly explanatory text created afresh in variable heights to fit at the bottom of every illustrated page, and in small type. These images made all the apparatus of proxy labels and footnotes look outmoded.

The concept of illustrations of this kind would have transformed Parker's idea of the book: from a good standard anatomy textbook, to something visually striking. For something of that sort, you'd need goodish paper—to prevent show-through from the printer's ink—and a good firm of black-line facsimile engravers.[25] The best thing about these unusual anatomical illustrations from an engraver's point of view was that their large size would make them easier to cut than the run of anatomical or other scientific work, though the lettering would be fiddly. The engraver would have to be a good letter-cutter, and someone who would honour Carter's dexterous use of outline and bold contrast. The Parkers were well-known as publishers of illustrated books, and they worked regularly with a number of competent engravers, so finding an appropriate wood-engraving firm should not have presented them with too much difficulty. Parker would confer with his father, who probably had someone already in mind.[26]

Once the engraving was done, you'd need a good printer, accurate and not troubled by scientific terms and odd bits of Greek and Latin. This is the sort of job that would have been bread and butter to Parker senior back in his Cambridge days. The timetable must be sorted out, too. Where were we? November 1855. The thing won't be ready till at least the end of 1856, even if an illustration was to be engraved every working day. The realistic plan would be to have it ready by mid- to late 1857, proofed and ready to go to the printers two years hence, so it could be printed and bound, and ready before the start of the new medical school year in October 1858. Gray and his artist had better start weaving.

So, just to recapitulate: the book was to be in appearance somewhere between Knox and Holden—a one-volume work probably in royal octavo—covering the whole of human anatomy, and with bold good-size illustrations. It would be brownish, in bookcloth, usual Parker endpapers and titlepage, dedication, contents, list of illustrations, 750 pages or thereabouts of content, and an index. And as for likely sales, the print run would have to be carefully considered. Pricing was a serious matter: this book would be costly to produce.

After such a positive preliminary meeting, Parker and his father, no doubt, would have had to sit down and do some serious calculations to see if a formula like this would work economically. The sums would not have been easy, because the constraints on price were tight, and the print run would have to be decided. The Parkers' previous textbooks in the medical sciences were not exactly inexpensive. Watson on Physic, in two volumes sold at 34 shillings. Todd and Bowman, also two volumes, was planned to be 28, but had ended up at 38 shillings. Miller, in three volumes, cost 46 shillings and sixpence. The volume discussed with Gray would not be cheap to produce, and it would have to compete with the best anatomy works already available in the student market. The crucial point was that when it reached the bookshops, it must be cheaper than Quain, at 40 shillings.[27]

As they separated to consider their estimates and calculations, neither Gray nor Parker could have had any notion how long this venture would run. Nothing was set in stone at this early stage, but things were shaping up. It was beginning to feel as though this book would really happen.

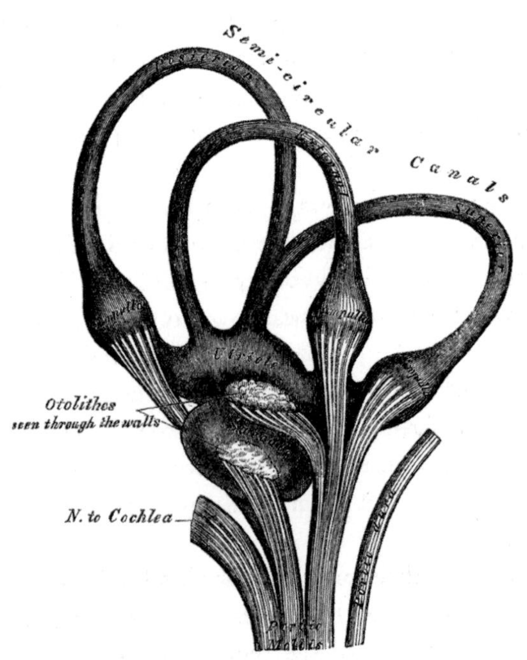

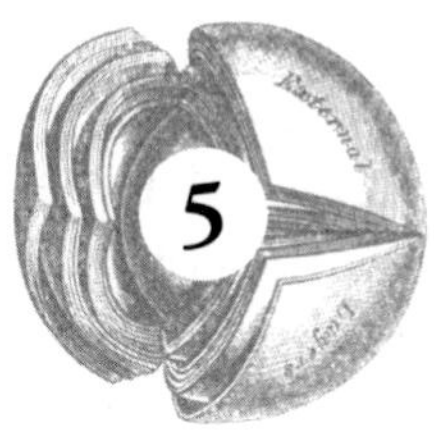

RAW MATERIAL

The Friendless Poor of London

The Victorian writer Ellen Barlee's book *Friendless and Helpless* features the story of a period in the life of a Portsmouth woman whose husband, after a period of unemployment, was offered work in Liverpool. He wrote home to ask his wife to pack up the family and join him. He would send money to the railway office at Paddington Station to cover the family's journey north. His wife used their last money to get herself and the children to London. Arriving at Paddington, she discovered to her dismay that the promised letter had not arrived.

There was no money for a telegraph, even had she known where to send it. She must find somewhere for the family to sleep for the night in a place she knew nothing of, and with no money to pay for it. The first thing this mother did was to sell her own shawl, to pay for a night's lodging, and the following night was able to find a cellar in which the family huddled together. But within a matter of days—the letter still not having arrived—they were evicted, starving, cold, distraught, and now homeless. Their only prospect seemed to be the workhouse, where the family would have been broken up completely. Rather than face such a terrible prospect, the mother resorted to begging on the street.

At this point, however, their fortune changed. A kindly person gave her the address of a homeless refuge, and they were taken in, hungry, cold through, and she now ill. The master of the refuge questioned the sick mother, uncovered the whole story, and wrote immediately to the man's employer. By locating the husband's whereabouts, everything was sorted out. The father's letter had somehow miscarried and had been returned to him as a dead letter, but until the refuge master contacted his employer, the father had been at his wits' end to know what had happened to his family, their previous home being deserted. The mother and her children were enabled to travel north together in good heart.

Communication being what it was in Victorian times, the simple miscarrying of a letter could have near-disastrous consequences. This was a real story, one among a myriad of other human predicaments the details of which will never be known because they were never recorded. We know of this family's crisis only because a charitable woman heard their story when visiting the refuge, and later wrote it up as a telling individual case history in a book she was then composing about the importance of emergency help for the homeless.[1] Ellen Barlee was evidently attracted to the story because it demonstrated the importance of charitable giving towards the support of such places: it was a real-life narrative in which attentive charitable kindness had wrought a genuinely happy ending. There was a further reason, too. This mother's story was perceived by Ellen Barlee as an exemplary case, because it served to illustrate a remarkable thing about the Victorian metropolis: the breathtaking speed with which a person—or indeed, an entire family—could innocently fall through social nets, and pitch into utter destitution.

The predicament in which this family quickly became mired, and which the charitable visitor recorded and passed down to us, shall have to stand for that of all those individuals, less fortunate than they, whose faces and body parts appear in *Gray's Anatomy*. Each one of them was an individual human being whose story we do not know. Some had died at St George's Hospital, others elsewhere: most likely in a workhouse in London, or in its hinterland. In each instance, no one had come forward to claim their corpse for burial. If anyone existed to make such a claim, they did not manage to do so in time to prevent the dead person's body from being consigned to dissection at Kinnerton Street.

The law governing the provision of corpses for the study of human anatomy at the time Gray and Carter were working at Kinnerton Street was called the Anatomy Act. Prior to its enactment in 1832, the only legal source of corpses for dissection in anatomy schools had been the bodies of murderers fresh from the gallows. For generations that source had proved insufficient, and teachers and pupils of anatomy and surgery had resorted to grave-robbery. By the 1820s, grave-robbing had become a well organized professional business, and the prices bodysnatchers could demand were high. So high did prices rise, in fact, that people were being murdered for the price their corpses would raise. The Anatomy Act was designed to put an end to this despicable trade. It also helped make the Victorian workhouse the hated institution it was, as the Act decreed that the bodies of those dying in institutions without anyone able to claim them for burial could be sent for dissection.

The law allowed individuals in institutions to 'opt out' of its provisions, by recording their wishes concerning dissection before witnesses. But of course, most witnesses would have been other workhouse inmates, powerless to protect their fellows' bodies from being transferred to dissection rooms, whatever the dead person's wishes might be. In the only known case in which a register of such wishes was officially kept, the number of people recording their desire to be decently buried was so high that *no one* went for dissection from that parish. The Anatomy Inspector felt impelled to visit the clergyman concerned to prevent such a form of registration for the future.[2]

The exponential growth of burial insurance in the early nineteenth century, and its cultural importance in poor communities until well into the second half of the *twentieth* century reflect the very long-term impact of the anatomy legislation, and the shifts to which huge numbers of ordinary people went to avoid burial (read dismemberment) as a pauper. As the literary scholar Tim Marshall has shown, the distress generated by unhappy knowledge of the pauper's fate impacted upon the imaginative literature of the Victorian era like a prolonged shudder.[3]

If a patient at St George's Hospital was fortunate enough to have next of kin at a settled address, the institution should have informed them of the fact of death. But, as in other such institutions, to expect such messages promptly if at all from hospital or workhouse staff was to be over-optimistic. If they were sent to next of kin at all, messages went only haphazardly, and were not to be depended upon. Hospitals and workhouses preferred to leave it to families and friends to keep themselves informed of a patient's survival or demise, by visiting regularly. For the Victorian poor, hospitals were places of fear, and visiting held enormous importance. A dead person's body could legally count as 'unclaimed' after forty-eight hours, so if for some unforeseen reason visiting had been irregular or temporarily suspended at a time when a death occurred, even patients with family living locally might end up being dissected.

St George's Hospital had a very large catchment area, and also accepted casualties, attempted suicides, and victims of other life-threatening accidents, who, if they were comatose at the time of admission, might not have had the opportunity to convey next-of-kin information, and might even remain unidentified at death. The bodies of those found dead out of doors were generally taken to workhouse mortuaries, and could end up in the dissection room at Kinnerton Street from there. We may never know what real-life illnesses and accidents, what separations or miscommunications, or what casualties and mischances of other kinds brought about such a denouement for these people's lives.

Very few patient records survive from St George's in Gray's era. The official protocol followed in cases where patients on the wards were known to be friendless prior to death, or if their 'unclaimed' status became evident after death, is not known. Mostly, such things were not written down in so many words. If the system at St George's was similar to that in operation at other teaching hospitals in Victorian London, their corpses would be removed from the mortuary to the dissecting room fairly swiftly, and, in the appropriate column in the patients' register where the name of the relative/friend/undertaker removing the body should have been entered, the name of the Hospital's own contract undertaker was inscribed. Staff who kept this great register would be privy to the real meaning of this particular kind of entry, so if relatives or friends turned up late to make enquiries, they would in all likelihood be sent off to see the undertaker, there to be informed that their relative had already been buried at the Hospital's expense.

This was what occurred in the case of Mr William Gillard when his wife died three weeks after giving birth at a different hospital: the Queen Adelaide Lying-in, in 1840. Mr Gillard was not informed within forty-eight hours of the deaths of his wife and newborn child. Three relatives, with the family's undertaker, went to collect the bodies of Mrs Gillard and her child as soon as they knew. They were given the mother's body, but were sent away with the information that the child was already buried, which was repeated on several subsequent occasions when further requests were made for the child's body. Mr Gillard eventually demanded to know the identity of the churchman that had read the burial service over his child, upon which the hospital's Matron and Secretary finally caved in, and gave him a note to take to the Windmill

Street Anatomy School. It read: 'The child that was sent to you some time ago is now claimed and I will Thank you to give it to the bearer'.[4]

We know about this case only because Mr Gillard was literate, and because his letters of complaint have survived in Home Office files. There were probably many more such instances in which poor people were fooled, fobbed off, or frightened into silence. It is clear from the way in which the hospital's note was eventually written and received at the school, that the recovery of bodies in this manner was not rare, and was likely always dealt with informally like this. The entire arrangement was informal. No official records existed at either end of the transaction. Prior to Mr Gillard's correspondence, nothing at all had been known about the matter at the Anatomy Inspectorate, or at the Home Office.

The delay in informing Mr Gillard had been a dodge designed to make claiming more difficult, as most poor families could not raise money quickly for a funeral, and a delay at the hospital would allow them even less time. The further delay, supported by the lies and misrepresentation inflicted on Mr Gillard's relatives and his undertaker, looks also to have been deliberate, designed to silence enquiry, and to allow the anatomy school more time to use the body even were it to be reclaimed. The child's body had indeed been partially dissected by the time Mr Gillard was at last told where he might find it. Mr Gillard never received an apology. The Home Office eventually sent him a reply, only after he had written a further letter of remonstrance, to the effect that it had been deemed inadvisable to take the matter any further. This final brush-off, for a man whose wife had just died, had taken a further four months. Collusion against the hospitalized poor and their relatives went to the highest level of the Act's administration.[5]

In the mid-1850s, all the burial grounds of London were closed for sanitary reasons, and the great public cemeteries were opened on the periphery of London. Reaching them involved very long journeys. Extramural burial probably provided added opportunities for corpse procurement.[6] The costs of verifying the truth of an officially asserted burial would have been prohibitively high, especially if exhumation was asked for. Hospital patients and their kin were poor. If any dispute ensued, poor families might be threatened with liability to reimburse burial expenses to the hospital, plus the costs of exhumation. Individual enquirers were probably silenced by such deceptions.[7]

Although the Anatomy Act had instituted an Inspectorate, the nature of the control exercised was in reality little more than paper shuffling. The profession had been involved in illegality for so long that the Act was difficult to enforce, especially as there was no form of compulsion by which it could become persuasive, and no penalties by which to punish contravention. The legislation was operated largely upon goodwill and silence. Since the Inspectors were usually hand-picked anatomists themselves, the system worked in the profession's interests, and official silence was largely maintained. The Inspectors' chief anxiety was to keep comment out of the public press, and to keep their main suppliers—workhouse masters and undertakers—sweet. Their main effort seems to have been to maintain secrecy concerning the operation of the entire unpopular process of procuring the bodies of the poor.

Memories of the grave-robbing and murder which had bedevilled the study of anatomy were still fresh among the public and the profession alike, especially in a metropolis full of anatomy schools and graveyards, large workhouses, and larger accumulations of poor. Memories are long, and bodysnatching was the stuff of stories told over pots of ale in the public house, or round the fire. The medical profession had made a good appearance of putting the old days behind it, and was becoming more respectable. The old private anatomy

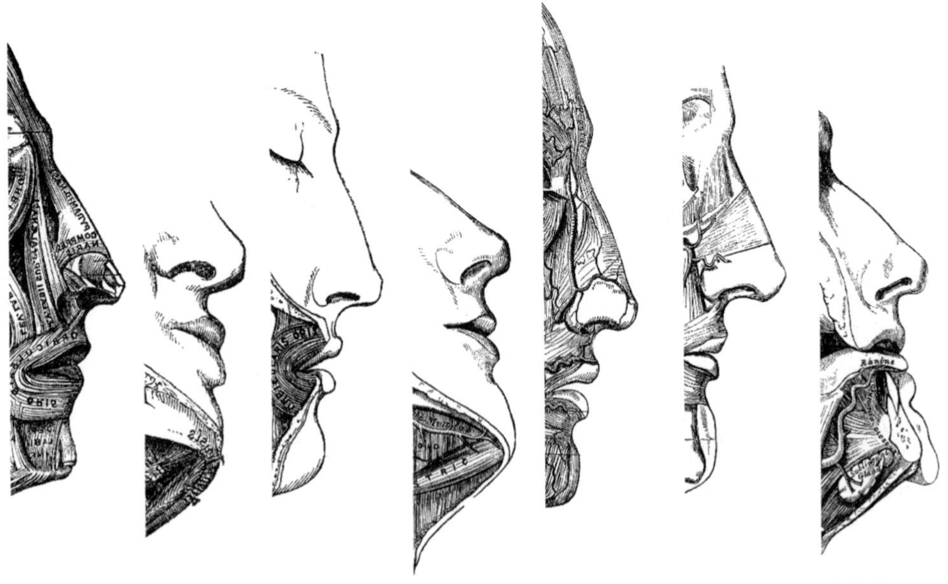

Profiles of human subjects illustrated by Carter. The identities of none of these people can be ascertained with any certainty, as records do not survive. Some of the profiles in *Gray's Anatomy* (especially those from Mascagni illustrating the lymphatic system) may be derivative and/or idealized, but others are original to HV Carter, and may well have been taken from the faces of those dissected at Kinnerton Street in 1856–7. That some of these profiles could derive

schools were closing down, and teaching hospitals like St George's were building their own medical schools, ostensibly playing by the rules.

But the rules were flexible, nevertheless. We know this because, although the Inspector was cautious, and the medical schools careful, just occasionally things erupted into the public press. One such case came to court in December 1857, when most of the work for *Gray's Anatomy* at Kinnerton Street was complete, and the book was being typeset at the printer's. Evidently rumours had been circulating before the matter came to court. A man by the name of Mr Parsley applied at the Lambeth Police Court, seeking the magistrate's advice as how to proceed, as *The Times* put it, under the 'following singular circumstances'. This was a way poor people contrived to obtain legal advice in the days before legal aid:

> The applicant stated that a short time ago his sister... was, by order of the medical gentleman in attendance on her ordered her into Newington Workhouse, she being at the time in an advanced state of consumption. Her admission took place on the 12th of November, and her fatal malady having increased she died on the Thursday the 3rd [December], and, previous to her death, she expressed a hope that her friends would take her body out for decent interment, as at the

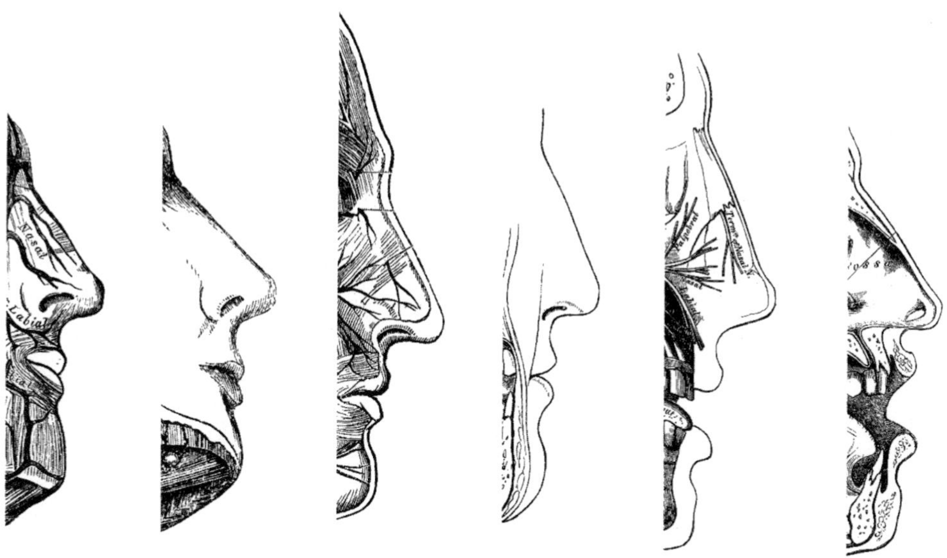

from living subjects cannot be ruled out. Nevertheless, it is worth noticing how respectfully Carter treats the dead (dignified postures, eyes closed, and mouths except when illustrating a particular anatomical structure). The dead in *Gray's Anatomy* exhibit no abject degradation as in John Bell's and other earlier work, there is nothing *winceworthy*: no awkward postures, and no evidence of pain.

> time there was a considerable uproar in the workhouse in consequence of the bodies of several of the deceased paupers having been disposed of for dissection. Her husband and friends promised that her wishes should be strictly complied with, and gave notice to the workhouse authorities of their intention to take away the body. On the following day (Friday) her husband and friends attended at the workhouse for the purpose of taking away the body, but though only 20 hours after death it was nowhere to be found. They were referred to the parish undertaker for information as to what had become of the body, and the only thing they could learn from that person was that it had been buried by mistake for that of a Mr. Bazely. They did not feel at all satisfied with the explanation, particularly from what had taken place in the parish respecting pauper bodies, and used every effort to learn something more on the subject; but they were unable, and had determined to take the opinion of a magistrate as to what they should do.[8]

The parish undertaker and the workhouse master had been operating what appears to have been a complicated scam in conjunction with other parishes, by which younger bodies were substituted for, or were taken to supplement, the older ones which really were unclaimed. Of course there was also money being creamed off:

> many malpractices had been discovered on the part of the late undertaker to the parish, it was found that he had been paid for interment of at least 20 bodies, which he had disposed of to the hospitals for £3.5s or £3.10s each.[9]

The case highlights the manner in which the old system of illegal body procurement continued in other forms after the passage of the Anatomy Act; how undertaker, workhouse master, and anatomist could collude to mutual benefit, whether the corpses involved were claimed or not. It also shows how even the vigilant poor could be duped by institutional functionaries, and their bereavements exacerbated with superadded distress. The body of Mr Parsley's sister was apparently never found.

Many genuinely unclaimed bodies from workhouses were those of elderly people who had outlived their families and friends, or the neglected bodies of vagrants. They did not make brilliant dissection material. Anatomists wanted fresh young bodies on which to teach, especially if they had little fat, and good musculature. In the past, such bodies had been obtained from gallows and graves, but the Anatomy Act had altered the entire market for corpses by providing a supply so cheap as to render both grave-robbing and murder for dissection unprofitable. In a vain effort to de-stigmatize dissection, Parliament had decided to cease the dissection of the hanged—their burial in quicklime within prison grounds dates from soon after the Anatomy Act. Anatomists had been told that the Act would yield plenty of bodies from institutions. In reality they were expected to accept whatever bodies became available, and were always complaining of shortage.[10] It seems clear from the Gillard and the Newington cases that after the Anatomy Act, new kinds of misconduct evolved which served to perpetuate older forms of commerce in the dead. Bodysnatching simply became institutionalized.

The undertaker and the workhouse master at Newington were prosecuted, and were on the receiving end of the public hostility provoked by the case, just as bodysnatchers had been before the Anatomy Act. No real effort, however, was put into detecting which medical schools had profited by their efforts, or which anatomists had bankrolled the entire scam. We do not know if any of the Newington workhouse bodies ended up at Kinnerton Street's back door, or if the medical school at St George's was the beneficiary of other arrangements with other workhouses or other undertakers, which continued undetected.

However, it is known that the anatomists at St George's had been refractory in filling out and returning the requisite paperwork to the Inspector. Certificates of burial for the bodies legally dissected at St George's were frequently late, or missing altogether. The law required dissected bodies to be buried within six weeks of death. Only a short time before Gray had arrived at the medical school, St George's had been at the centre of allegations about delays in the burial of dissected remains long beyond the period demanded by law, and in a notoriously disreputable burial ground, whose certificates of burial few would trust. So outraged were the parishioners of St Marylebone by the medical school's conduct, that the parish refused to send any bodies for dissection at all. Irregularities concerning the burial of bodies from Kinnerton Street had been brought to the Inspector's attention on more than one occasion, so we know that the school's management was no stranger to this variety of illegality. Because little was done about it, we can infer that it was viewed with no more than irritated indulgence from on high.[11]

That the desire to hold on to human remains was not a rare phenomenon among those studying anatomy was expressed quite explicitly in a medical poem of the day. Composed for a sympathetic audience, it was published in the doctors' journal *The Medical Times and Gazette.* The versifier's identity was not provided, but it was entitled

To My Old 'Subject' I am Advised To Bury!

(An imitation of Beranger)

Stay yet awhile, thou dear old friend of mine;
Though mould now gathers on thy shrunken face,
And wither'd are those once strong arms of thine,
And scarce a nerve is left for me to trace:
Alas! 'tis true thou lookest very old,
And much has passed since we as brethren met,
But still thy vessels do injection hold:
So, dear old friend, we must not sunder yet.

Full well I mind the sultry July day
I bought thee, brother (for I wanted one)
When first I saw thee on the table lay,
Not long before your sands had wholly run.
And then I thought—this dead man will to me
Be far more faithful than the worldly set
Who damp my ardour with prognostics drear;
No, dear old friend, we must not sunder yet.

A cut I see thine old subclavian bears,
And the injection oozes from the rent;
So was my spirit cut by many cares,
And all my courage oozed till nearly spent.
But when I saw thy vessels full and red,
I knew some pupils for thy sake I'd get,
Your placid face put brave thoughts in my head,
So, dear old friend, we must not sunder yet.

Together we have spent full many an hour,
Full many an hour of moody silent thought;
But thou, mute friend, did always comfort pour
Into my heart, with dire forebodings fraught.
Thou told'st me that in thee I had a friend,
Who, till I chose, my side would never quit;
Thou'st never leave me for some selfish end,
Oh dear old friend, we must not sunder yet.

I do believe in days when warm blood ran
Where now injection cold and waxy lies,
Thou wast an honest, helpful, kindly man,
With warm, firm grasp and kindly beaming eyes.
Thou had'st a wife, for which that hand did work,
A child whose food it manfully did get,
Which real and honest toil would never shirk:
No, dear old friend, let us not sunder yet.

But still thy work, my brother, is not done;
My wife, my children, look to thee for bread;
Still may injection in thy vessels run,
Still put brave hopeful thoughts into my head.
And when my work comes also to an end,
I trust that heavenward my soul may flit,
Meet thine and thank it cordially, old friend:
But, dear old friend, we must not sunder yet.[12]

The acquisitive intimacy of the relationship between the dissector and the imaginary person whose dead body he works upon is well brought out here. So too is the dissector's recognition of the direction of economic benefit in his use of the working-man's dead body. The poet–dissector seems particularly anxious to stress the companionship and friendship he bears the dead man, masking his guilt towards the exploitation involved, of which he is nevertheless evidently aware. But he seems to be either completely ignorant, or unconcerned, about the legal requirement to bury dissected remains.

The Anatomy Act served to legalize corpse supply to dissecting rooms in teaching hospitals like St George's from two directions: from outside the institution, and from its own dead. The number of anatomy students was assessed annually, and corpses deriving from workhouses, prisons, asylums, or prison hulks were distributed to the schools on a numbers-needing-to-dissect basis: put simply, schools with more students got more bodies. Schools were expected to be honest about the number of students registered. Teaching hospitals had inpatients on their own premises, and were expected to inform the Inspector when an unclaimed body became available from within the institution. But honesty in this respect worked against the school receiving any additional corpses from outside, as every added inpatient's corpse meant the allocation of one fewer workhouse or prison corpse from outside. If patients continued to die, and the school therefore had more than its agreed share of unclaimed bodies (beyond the notional 'fair share' of the Inspector's common distribution system) the bodies of their own patients could be deemed a contribution towards the common pool of corpses, and be allocated to other competitor schools.

There was only a single Inspector for the entire country, so not everywhere could expect to receive an inspection. He was constantly besieged with complaints of shortage. But repeatedly, investigations turned up instances where schools had claimed they had enrolled more students than in reality they had, or, where bodies were found lying in the dissecting room with inaccurate or missing documentation—bodies of which the Inspector had no official knowledge.[13] Such discoveries were invariably explained away as clerical errors, and clerks doubtless knew that their role was to carry the can. But everyone, including the Inspector, knew there was an element of charade in such matters: every school was trying to maximize the number of bodies available for their own dissecting room.

The same was true for the bodies arriving from outside. Since the count of corpses arriving legally via the Inspector was constrained, a medical school wanting to boost its body count had two options. The first, was to do secret deals with other hospitals, workhouses, or undertakers. The hospital in the Gillard case was not a teaching institution, so it had no dissecting room, and sale to a medical school could have brought economic benefits to both staff and institution. The undertaker in the Newington case worked for a workhouse. The advantage of such secret deals was of course that no part of such a transaction need appear in the paperwork, and backhanders probably passed along the way. We do not know if obliging undertakers supplied the dissecting tables at Kinnerton Street, and whether they did or not may never now be ascertained. *The Lancet* was quite clear that the Anatomy Act was in the hands of the undertakers, so it would seem that within the medical profession these arrangements were known to be commonplace.[14]

Alternatively, a school could source more 'subjects' from its own hospital's wards, masking their destination by the management of clerical records. This second course looks likely to have been utilized at St George's. In times of shortage, there would appear to have been a covert arrangement whereby the doctors in charge of clerking-in new cases would knowingly take into the Hospital people who were likely to die swiftly, when it could be ascertained that they had no next of kin.[15] These sick patients were probably people who would not usually be entertained in a charitable voluntary hospital: people so ill that they would normally have been sent off to die in a workhouse. We can infer that this is what was going on when we see numbers of relatively swift deaths peaking early on in the medical school dissecting season, and all of them spared post-mortem examinations. In all likelihood, they were sent intact to Kinnerton Street, though the ledgers do not say so. The arrangement was probably very like the one objected to in Ireland by the distinguished anatomist Benjamin Alcock, Professor of Anatomy at Queen's College Cork, who was told by the Irish Inspector of Anatomy, and the highest Home Office official in Ireland, to procure corpses for the college's dissecting tables by claiming bodies *'in the capacity of a friend of the deceased'*.[16] These unclaimed bodies at St George's would have been fresh, free to the school, and swiftly rendered unrecognizable.

Had the medical staff at St George's been questioned about the moral principles behind such arrangements, they would doubtless have justified them by saying that these patients died in greater comfort inside the wards at St George's than they would have received in a workhouse, or in other places they might have gone to die, such as the Adelphi arches by the River Thames, or elsewhere.[17] One hopes that they really were better cared for than they would otherwise have been. And of course one prays that their deaths were not hastened along by the knowledge that fresh-faced young students were awaiting corpses to purchase/dissect.[18] The inescapable misgiving yet remains that these people were provided with deathbeds at Hyde Park Corner for reasons other than charity.

At this late stage, it is not likely that all the subterfuges employed by Victorian teaching hospitals to obtain corpses for their students can be fathomed. But those I have outlined give an idea of the ingenuity brought to bear upon the problem. The Gillard case probably represents the experiences of many others for which no records survive. For what length of time the parish undertaker had been operating his scam in Newington is unknown: it may well have been proceeding successfully for years. London and its hinterland contained many other parishes, other lying-in hospitals, and many other undertakers, than the single one whose story hit the news.[19] The autumn peak intake at St George's Hospital looks to have been operational while Henry Gray was teaching anatomy there. Although he may not have instigated it, and may not have borne personal responsibility for its efficiency, Gray likely knew all about it. The paramount need for fresh corpses on which to work for the book, means that he was almost certainly both a nominal 'friend' and a beneficiary.[20]

Henry Gray appears to have been among the more honest of the doctors making entries in the post-mortem and case books at St George's Hospital. These were large fat ledgers with metal clasps, which, after a comprehensively tabulated index, opened to a double-page entry on each of the patients' bodies that passed along 'La Via Dolorosa' to the Hospital's dead-house.[21]

On every left-hand page in the ledger, the officiating surgeon–anatomist would enter the patient's details, and a report of the pathological findings made during the post-mortem examination. Opposite, on the right-hand page, the responsible physician would inscribe the patient's case history. When a dead patient was spared a post-mortem on Gray's watch, he invariably gave the reason for the non-performance of this important ritual. His colleagues rarely specified reasons for the lack of a post-mortem report, usually entering a formulaic statement in the ledger, such as: 'This body was not examined' or 'This body was removed unexamined', or simply 'Not Examd.' When a surgeon made an entry of this kind on the left-hand page, the physicians were absolved from having to provide a written case history, and the facing entries in these cases followed a formula, too: 'As the body was not examd. the case is not given', and the physician's initials. The parallel Hospital ledgers relating to the treatment of these patients within the Hospital have disappeared, so in such instances we know little else about them beyond their name and age, the date on which they died, and the surgeon or physician under whom their care had been managed. In the alphabetical index at the front of the surviving ledger, these cases were annotated: 'Not exam^{d}.'

'Not examined' was a conveniently ambiguous formulation of words. It could mean one of two things: that the patient did not undergo a post-mortem examination because relatives objected in time to prevent it, *or,* that the patient's body was unclaimed, and so went straight to Kinnerton Street. We have no way of knowing from the ledgers which of these alternatives applied to the individuals listed; except, curiously, those clerked out by Henry Gray.

In these cases he would sometimes write a lengthier explanation than would the run of his colleagues, in some instances apparently of surprising honesty. On the page for William Stuart, aged 64, who died in the Hospital on 20th July after cutting his own throat, he inscribed: 'This Body by order of the Coroner was not examined.' Gray's report for Selina Owen, aged 30, who died on the 7th November, was annotated simply: 'This Body was sent to Kinnerton St.' and after the death on 9th November of 53-year-old Mary Stichall, he wrote: 'This Body was removed to Kinnerton Street.' These entries were filled out in Gray's characteristically bold hand, and initialled or signed in full by him.[22]

All the patients just named died in 1855, prior to the discussion between Gray and Carter on 25th November that year, regarding the idea for the book. In London, in 1855, the impact of the Crimean War made the price of food so high and work so slack that there were bread riots in London.[23] When Gray specifies, as he does in *Anatomy Descriptive and Surgical*, that subjects to be chosen for study should be 'free from fat', he does not say that the poor of London provided ideal candidates for the position. But the suffering which caused people to riot for food in the midst of mid-Victorian plenty should not be forgotten when we look at *Gray's Anatomy*, for it is the poor who appear there.

During the *next* two years—1856 and 1857, the years in which Gray and Carter were working hard on the creation of the book—curiously, the number of patients' bodies recorded by Gray as having been sent to Kinnerton Street dropped. At the same time, his ledger entries became more like those of his colleagues. Whereas in 1855 four bodies were recorded as having been sent to Kinnerton Street, in 1856 there were only three, and in 1857 only one. Such figures must be regarded as suspect, for a number of reasons.

A mid-Victorian Soup Kitchen.

First, the dissection room at Kinnerton Street cannot have been managed on the small number of bodies admitted to in these ledgers. Mid-Victorian medical students learned their anatomy by hands-on dissection, and the prestige of anatomy schools in part derived from the number of bodes made available to students. Even dismembered, this small number of bodies openly recorded as going to Kinnerton Street would not have been sufficient for the teaching of each class of medical students, especially when we consider that there must have been an additional requirement for bodies in those particular years, while Gray and Carter were at work.

Second, there is a discrepancy between the number of bodies listed in the ledger's synthetic index as having been sent to Kinnerton Street, and those recorded as such in its text. While the rare index annotation 'K.S.' always coincides with mention of Kinnerton Street in an individual patient's double-page ledger entry, on other occasions the annotation 'Not examd.' is used in the front index instead. The discrepancy between the destination data given in the text and in the index suggests that, while 'K.S.' entries are likely to have been accurate, those with the other annotation are not. The listing of dead adult patients as 'not examined' in the post-mortem ledgers for those years was 69, 65, and 40 respectively.[24]

Third, the number of dead recorded as having been sent from the Hospital dead-house to Kinnerton Street not only seems absurdly small, but does not correlate with the figure supplied to the Inspector, who seems to have believed that St George's had contributed substantially more bodies to Kinnerton Street than are admitted to in the ledgers. The Inspectorate's records take the form of statistics, no names at all, just columns of numbers, so we cannot marry them up with any individual entered as 'not examined' or 'sent to K.S.' But the figures alone are sufficient to show that the medical school at St George's must have dissected significantly larger numbers of bodies deriving from the Hospital than are listed in the Hospital's own records as having been sent to Kinnerton Street. The figures simply don't add up. For reasons already

	1854–5	1855–6	1856–7
Deaths	336	298	304
'Not examined' [data from St George's ledger]	69	65	40
Kinnerton Street	4	3	1
Inspectorate's record [data from official records][25]	16	18	17

discussed, the Inspector's figures were more likely to *under*estimate a teaching hospital's body yield, so the discrepancy is curious, to say the least.

Looking at these figures, pondering the discrepancies, and pondering, too, the identities of those on whom the dissections were done, those whose bodies formed the corporeal basis for *Gray's Anatomy*, it is not easy to shut out all suspicion that the book's eponymous hero was telling only half-truth. These bodies may genuinely not have undergone a post-mortem examination. So as far as it went the statement 'This body was not examined' may have been truthful. But what happened to bodies *after* they were not examined?

Cross-checking identities from the post-mortem and case books to other archival sources to validate the assertions made by Gray and his colleagues in the dead-house has so far proved impossible. The St George's dissection-room registers do not seem to have survived from this era, and the Anatomy Inspector's name-registers for the period have also disappeared. The notorious burial ground mentioned above was closed by law in the 1850s leaving no records, and the final burial place of the dissected dead of St George's Hospital at the time Gray was writing his book has so far proved difficult to confirm.[26] We are dealing here not just with the unclaimed, but with the disappeared.

The fluidity of relationships between the school and the hospital at St George's ensured that the same individuals who accessed bodies for their pathology in the hospital dead-house also occupied key teaching positions in the medical school. From 1848 to 1849, Gray, for example, was the Post-Mortem Examiner, the Curator of the Pathological Museum, and the Demonstrator of Anatomy.[27] The system at St George's (as in most Victorian teaching hospitals) ensured that hospital staff successfully rising up the surgical ladder were not merely sympathetic, but alert, to the needs of the medical school, since they had trained at the school and had been socialized into the procurement system as they passed up the rungs of the promotion ladder. In fact, it would be most curious if St George's personnel had not endeavoured to supply their own school to the best of their ability. We should not overlook the fact, either, that more than one Victorian Inspector of Anatomy was a George's man, which suggests good relations with the Home Office, despite the school's substandard paperwork.[28]

Even had all the records miraculously survived, and we could lay them out together to compare names and dates, we could not altogether trust their contents. This is because we also know about the scam at the Newington workhouse, which, while retaining the names and the paperwork associated with individuals legally consigned to dissection, substituted the bodies of other people because they were in better physical condition for the anatomists.

If Gray and his colleagues posed as 'friends' of the dead, there should be no surprise. Posterity may never know whether Gray was being entirely, or just partially, honest in his entries in the ledgers, or whether his statements were a professional charade performed in the interests of his discipline, his students, and his book.

Dissectors are said to have disliked bodies which had undergone a post-mortem, as they were more difficult to inject with preservatives, and the body parts required for teaching could be damaged or incomplete after post-mortem procedures. Yet quite a number of the anatomical dissections whose details are illustrated in *Gray's Anatomy* would not have been prevented even had a post-mortem taken place. Limbs and joints, for example, would not generally have been damaged during such an examination. The procurement of other body parts—especially internal organs—might even have been facilitated by one.[29] Brains, eyes, and other internal organs were more likely to be obtainable in a fresh state from the post-mortem room, than from the dissecting room, where bodies were full of preservative, and needed for teaching. So we ought also to bear in mind that Gray may not have been entirely disinclined to the secondary use of bodies after a post-mortem examination. It is a curious fact, however, that of the bodies listed in the ledgers as having undergone a post-mortem *none* were also listed as having been delivered to Kinnerton Street.

The unlikelihood that every dead body at St George's that underwent a post-mortem was claimed by relatives supports either or both of two inferences. First, that many of the bodies listed as 'not examined' were deliberately sent undamaged to Kinnerton Street, and/or, that an unknown and unrecorded proportion of 'examined' bodies found their way there too, possibly in parts.

There was a culture of body and body-part procurement in medical schools (especially those attached to teaching hospitals) that worked closely in concert with the administration of the Victorian Poor Law to provide bodies and parts for teaching and research.[30] In Cambridge, for example, the Anatomy School received bodies and body parts (such as amputated limbs) from forty different sources between 1855 and 1920, facilitated by covert payments to obliging staff members in supplying institutions.[31]

The Inspector of Anatomy oversaw the licensing system for anatomical premises and dissecting personnel, and endeavoured to appear as if he was credibly in control of the national corpse-distribution system. But the Anatomy Act had no jurisdiction over dismembered body parts: it did not contemplate them, and made no claim over them.[32] On licensed premises, there was nothing to prevent or control this manner of obtaining dissection material by licensed personnel. The law's vagueness in this area was intentional: it had been drafted with the support of Astley Cooper and Benjamin Brodie, who much preferred to use institutional mortuaries than employ bodysnatchers, or import corpses from abroad. Medical museums flourished in the Victorian era. The Inspectorate did not interfere with the commercial sale of bony and other preparations by specialist dealers beyond medical school walls, disallow the sale of body parts to students, or prevent the creation of museum preparations from organs and body parts.[33]

Although it has not proved possible to identify individuals, the kind of people that might have ended up on the dissecting room tables in the teaching hospitals of Victorian London are ascertainable from other sources. The records have not survived for St George's, but materials and records do survive which help provide some indication of the numbers of people, and the types of people, whose bodies might have ended up at Kinnerton Street. The records at St Bartholomew's Hospital, for example, include information such as names, ages, and places of death, from the same mid-Victorian years Gray and Carter were at work. I have sampled these dissecting room records for January to March—the height of the dissecting season—for two years (1856–7). They show that over these two years, the number of bodies (34) requisitioned under the Anatomy Act from workhouses for use in St Bartholomew's medical school dissection room exactly matched the number of bodies taken for dissection from its own wards (34).[34]

The workhouses from which bodies arrived at St Bartholomew's were spread across the capital, from Whitechapel and Mile End in the east of London, to Brentford in the west, but most were fairly local: Holborn, Strand, and St Giles workhouses appear repeatedly. How bad the conditions were in these places is clear from an investigation by *The Lancet* Sanitary Commission, with open toilets on the wards, appalling ventilation, no segregation of terminal, chronic, acute, or infectious cases, and at the Strand

Union Workhouse, for example, 556 people sharing 332 beds. The Bart's records show that the dissection room at Bart's received a preponderance of men's bodies (43 men to 28 women), and that the mean age of those deriving from Poor Law institutions was (as we would expect) generally older than those arriving from the Hospital's own wards (57 from workhouses; 42 at Bart's).[35]

The causes of death listed were quite various: most commonly lung diseases, especially phthisis (TB), chronic bronchitis, emphysema, pneumonia, and quite a number from fever, typhus, and diahorrea. The infection erysipelas claimed several, whose bodies would have been risky dissection material, unless injected well with preservative.[36] Less frequently assigned as causes of death were paralysis, dropsy, debility, decay of nature, and diseased organs. Acute injuries do not seem to have been a major cause of death among 'unclaimed' patients at St Bartholomew's: there was only one case of spinal fracture, and one of concussion. From their names, these people appear to have been mainly of English descent: Smith, Barrett, Masters, Newman, Thompson, Martindale, Watson, and so on, but there are a number which could be Welsh (Williams, Edwards), Scots (Hastie, McDonald), and Irish (Hagerty, Coglin, Conroy, Conoley). Only one obviously foreign name crops up among the sample, that of Guisseppe Franchi, aged only 24, who died of typhus fever at St Bartholomew's Hospital, in February 1857.

We do not have occupational information concerning these dissected patients, but an examination of records which detail the patients taken in at two other London teaching hospitals—patients cared for at Guy's Hospital, Southwark by Richard Bright, and at the Middlesex Hospital by SW Sibley—can provide a good idea of the variety of working lives followed by individuals who sought hospital care.[37] Male occupations were extremely various: stable-boy, night watchman, sailor, gardener, brassfounder, fireman, smith, servant, public house [barman], painter-glazier, shop assistant, bricklayer, porter, sawyer, pewterer, printer, tailor, market dealer, carter, hackney coachman, labourer, greengrocer, shoemaker, collector of skins (currier), ostler, hairdresser, butcher, undertaker, farrier, brewer, plasterer, tin-plate worker, wheelwright and chairmaker. Women's opportunities for paid work were more limited than those of men, and this is revealed in the patient-intake at both hospitals. The women were listed mainly as a wife of someone else, and their occupation that great multi-task, housework. Others worked as servants, servant-maids, needlewomen, washerwomen and laundresses, charwomen, sick-nurses, boot- and shoemakers, and schoolmistresses, but there is also a cook, a milkseller, a shirtmaker, a monthly-nurse, a tassell-maker, a dressmaker, a fruitseller, a governess, a tailoress, a prostitute, and a housekeeper. Women were particularly vulnerable to sickness, malnutrition, and death in poverty, because of their low earning power, and because the care of dependent children made them vulnerable. The pictorial cartoon newspaper *Diogenes* commented on the economic precariousness of mothers in 1854, when it published a visual 'Appeal to the British Nation' of an image of an infantry man in the Crimea at one side, and his poor wife and child pulling the bell at the workhouse door, on the other.[38]

Mid-Victorian hospital patients derived mostly from the working classes, that broad swathe of humanity that stretched from the labour aristocracy of highly skilled craft workers, to the unskilled and casual labouring poor. Many London trades—however skilled—were seasonal, and many people—men and women—were thrown out of work in hard times and during bad weather, especially at the height of the dissecting season, in winter. Ellen Barlee is very clear about seasonality as a cause of poverty and homelessness. George Augustus Sala is just as clear that people were dying from exposure to cold, and from starvation, in mid-Victorian London.[39]

The lives and livelihoods of all working people were crucially dependent on keeping healthy. Their lives were vulnerable to the effects of chance and accident, and their economic position made them more likely to be exposed to working and domestic environments which predisposed them to illness and ill-health. Illness or injury of any kind might precipitate a descent into poverty so disastrous that the workhouse, and the grave, were only a whisper away.[40] Numbers of the poor in hospitals and workhouse infirmaries were probably migrants from the hinterlands of London, without established social networks, and with under-developed immunity to London illnesses or cruel employers, like poor Mary Looch, a young servant-maid aged only sixteen 'lately come from the country', who 'had been put to sleep in a very cold and damp underground apartment' where she acquired the pneumonia which killed her.[41]

The indigenous poor who died in the institutions of the metropolis were liable to undergo post-mortems almost as a matter of course unless relatives could intervene quickly, and they went for dissection if no one arrived to claim the body within forty-eight hours. At St Thomas's Hospital one woman who objected was told 'if persons come to hospitals, they must put up with hospital practice.'[42] The majority of patients in every London workhouse and hospital knew what hospital practice was, and many would have had family, friends, or social networks of neighbours or work-mates outside, who would regard the hospitalization of anyone they knew with concern, and their death with double dread—like the crowds of wailing relatives outside the Middlesex Hospital during the cholera epidemic of 1853–4.[43]

The sick and dying themselves would have known that they were lucky to have people outside who would defend their right to decent burial by claiming on burial insurance policies, go round with a basin to beg from neighbours, or get up a concert in a public house, to raise the money for a funeral.[44] But many, like young Mary Looch, might have been new migrants, socially isolated, ill-cared-for by employers, with no one outside who knew they were ill, and dying. Many would have died knowing their likely fate.

The cultural resonance of the unmarked grave was extremely strong in the Victorian era as an emblem of melancholy contemplation, and fear. Pity for those whose bodies might end up in such a place was expressed in *Household Words*, the magazine conducted by Charles Dickens, in a poem entitled *The Unknown Grave*:

Did this poor wandering heart
In pain depart?
Longing, but all too late,
For the calm home again,
Where patient watchers wait,
And still will wait in vain.[45]

We can see from the dissection-room records at St Bartholomew's that a proportion of the poor population of London, for whatever reasons, fell through social nets, and ended up without a friend in the world to save them from the slab. Such are the folk whom Gray and Carter dissected for their book.

During a period of eighteen months or more, in 1856–7, Gray and Carter dissected enough human body parts for the creation of over 360 illustrations. A single body might have served as the source for a considerable number of dissections, but not that many. A careful examination of the illustrations in *Gray's Anatomy* reveals that, in addition to the bones of the skeleton, and the teeth, the bodies of several adult men and at least one woman, possibly two, and a child were probably used in the dissections to create the illustrations, as well as a variety of body parts from quite a number of other people. It is a sad fact that the identities of the individuals whose bodies and parts appear in *Gray's Anatomy* remain unknown.

There is a silence at the centre of *Gray's*, as indeed there is in all anatomy books, which relates to the unutterable: a gap which no anatomist appears to address other than by turning away. It is the gap between the ostensible subject of the book and of the discipline, and the derivation of the bodies from whom its knowledge is constituted, its illustrations made. In *Gray's*, the legally sanctioned bodies of people utterly alone in the metropolis were the raw material for dissections that served as the basis for illustrations, that were rendered in print as wood engravings. As mass-produced images, they have entered the brains of generations of the living—via the eyes, the minds, and the thoughts of those who have gazed at them.

But nowhere in these books is the human predicament of those whose bodies constituted their basis addressed, or discussed. Nowhere is their native status as the defeated, dismembered, unconsidered, naked poor even mentioned. And in *Gray's Anatomy*, nowhere but in Carter's images, do they receive memorial.

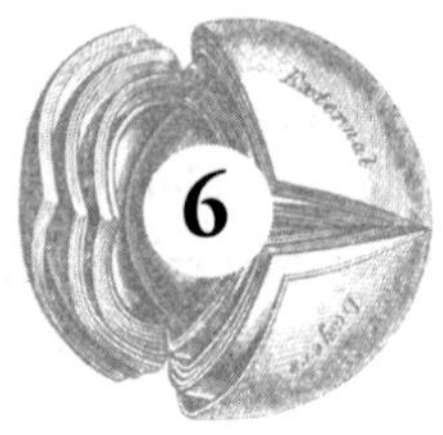

CREATION

1856–1857

Looking closely at Carter's illustrations for *Gray's Anatomy*, it is instructive to notice what a crucial role shadow plays in their effectiveness. Within each specimen, Carter uses shadow to give his two-dimensional line drawings an appearance of three-dimensionality and depth. But unusually for an anatomical artist before 1858, Carter rarely uses shadow *outside* his specimens, to indicate the surface on which they lie. His focus is very close in. And if you look carefully, you can see that in some of his illustrations the light seems to fall upon the specimen from above, while in others the angle is more oblique.

These light effects help us imagine where the work for *Gray's Anatomy* might have been done: whether under the great arched rooflights of the dissecting room at Kinnerton Street, or in some less public room with a window where Carter could sit quietly to observe and sketch, and draft, and erase to correct. For accuracy, he may well have utilized a camera lucida: a small prism by which the detail and the size of the subject could be adjusted, by skilled eyes and skilled hands, to a paper surface.[1] The subsequent work of adding anatomical names to the structures, making a final simplified outline, and transferring the image to the wood may have been done at a later stage. From his diary, it seems that a lot of the work was done at Kinnerton Street. But Carter also speaks of his life as reclusive, so it would appear that for much of the time he was able to retire from the noise and sociability of the dissecting room to somewhere quieter.

The last time we saw Gray and Carter together, they were discussing the possibility of a 'Manual for Students'. Nothing had been decided as yet. The conversation, you will remember, had taken place at the end of November 1855. Gray was brimming with enthusiasm, and Carter was doubtful as to whether he really wanted to become involved. Carter's concern at the time was that the project would be 'too exacting', too demanding, 'for would not be simple artist'.[2]

Carter seems to have had a sense that Gray was trying to entice him into something he would not have chosen to do. He had a ruminative habit of taking time to think things over, which meant that sometimes he would berate himself for lacking initiative. In this case, two months would pass before he made a written commitment to Gray's new project.

There were good reasons for Carter's apprehensions. The most pressing was his own need for clear space in which to concentrate on finally qualifying as an MD, a doctor of medicine, from London University. Carter had to be ready for the tough week-long examination scheduled for the following November, 1856. Carter was a qualified apothecary and surgeon, but a doctorate in medicine would make him a fully fledged physician. He could diagnose and prescribe as well as undertake surgical operations, and dispense medicines only if he so chose.

This was not what is laughingly called 'diplomatiasis', or over-qualification. There may have been an element in Carter's retiring personality that partook of life-avoidance by study, but to aim for this qualification in medicine was also a sensible life-choice. Carter did not want to settle for an assistantship to an apothecary–surgeon, which was all he was trained for at the time. To qualify in all three disciplines was an unusual route to medical professionalism, and is partly due to the historical era in which Carter found himself. English doctors of his generation were caught between the old monopolies of the ancient guilds: of physic (intimately associated with the established church and the old universities), surgery, and the apothecaries (and their traditional training of long apprenticeships), as against the newer professionalism and more academic focus of the sciences. The modern and more forward-looking London University was breaking the medical cartel operated by the old English universities and the established church, which had operated a policy of social and religious exclusiveness for centuries.[3] London graduates had established legal parity professionally with those of Oxbridge only since 1854, so Carter's resolve was fresh.[4] When the General Medical Council was established in 1858, it recognized both varieties of medical qualification.

Two of Carter's meticulous illustrations of the human spleen, created originally for Gray's entry for the Astley Cooper Prize Essay competition. They show that Carter's illustrative style was well developed before his work on *Anatomy, Descriptive and Surgical.* Carter's spleen images reappeared in the book version of the *Essay* published in 1854 by JW Parker and Son. In both cases Gray omitted to attribute the illustrations to him. The best images, including these, were only credited to Carter when they appeared a third time in *Gray's Anatomy.*

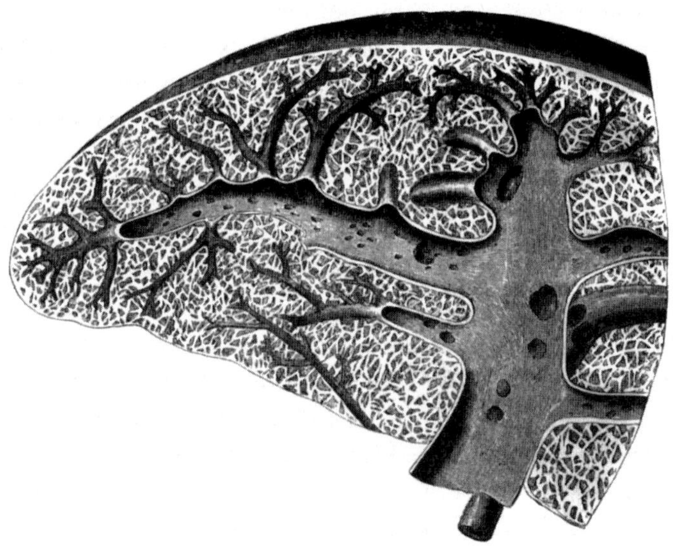

To some extent Gray and Carter personify old and new approaches to medical success. Under the old system success was backward-looking, largely associated with insider status, patronage, and funds, whereas the newer way was less costly, less bound by social class and social convention, and had a greater emphasis on intellectual ability, testing by examination, and the merit and effort of the individual. There were numbers of young doctors like Carter pushing at the door, but each had to fight their own battle, find their own path. Entry to the Civil Service, and imminently, to the Indian Medical Service of the colonial government in India, would be by examination, not appointment. In the 1850s, tradition and the patronage system in medicine were visibly on the wane. The battle against them had been long and arduous, and—especially in established hospital hierarchies—still had some way to go, but the spirit of the age associated meritocracy with modernity, and Carter had bought in to the idea in a serious and life-changing way. Being qualified under both old and new systems would mean that for the rest of his life, Carter could avoid subservience in his profession. He was suffering financial difficulty now for this prospect. With an MD, he would be *Doctor* Carter, not the traditional *Mister* of the surgeon or apothecary.

At the time Gray first approached him about the book Carter was pondering all this. He had already obtained his bachelor of medicine (MB) with honours, which he needed to enter for a doctorate. Carter knew he had only a year's further study before the doctoral examination was due. We know how devastated he had been at his earlier failure, so it is not difficult to sense his anxious determination not to fail now.

While he had been working at the College of Surgeons Carter had matured quite a bit, and understandably enough, was now seeking to focus on his own career. Things had starting moving slightly in his favour in the last few months: he was now a member of the Pathological Society of London, and he'd done work for other colleagues at St George's that would result in publications in the following year.[5] Yet despite his collaborations and diplomas Carter still felt an outsider. But he had worked hard to get where he was, and now was looking for some recognition himself.

There was another reason, too, for Carter's reluctance to become involved with Henry Gray's anatomy manual, one that had been simmering below the surface for over a year. Carter's meticulous illustrations for Gray's prize essay and learned papers had helped Gray's career prospects. Gray was the golden boy of the school: Fellow of the Royal Society, Hospital Governor, seemingly on a preordained route for institutional power and professional success. But

Gray had been quite ungenerous to Carter: he had never credited him with an illustration, and had even failed to attribute to him the beautiful illustrations in his recent book on the spleen. There may have been unrecognized intellectual debts, too, in which Carter did not wish to involve himself again. This was perhaps what Carter had meant when he noted 'for would not be simple artist'. He was genuinely dubious as to what was being asked of him.

It was perhaps understandable that Gray had not named him as the main illustrator for his Astley Cooper Prize essay, since there had been great emphasis on candidate anonymity in the competition rules. But when it was published, *The Spleen*'s preface had expressed Gray's fulsome thanks to several other George's men—to his friend Mr Timothy Holmes for 'much valuable assistance' on the historical portion of the work, to Henry Minchin Noad and to Henry Pollock for all the chemical analyses, to his 'esteemed friend', Mr Athol Johnson, who Gray said had 'with much trouble revised the work in its passage through the press', and even to the Royal College of Surgeons, for the use of the dissected specimens which Carter had spent hours drawing for him.[6] But there was no mention at all of Carter himself.

Henry Gray's silence felt very public. The named George's men—all of whom surely knew the best of the illustrations were Carter's—would have been aware of it. Even *The Lancet* review of Gray's *Spleen* had noticed individually those named in his acknowledgements.[7] Gray had either entirely forgotten, or had knowingly omitted mentioning Carter. It could have been an oversight, straightforward and simple forgetfulness; but it could also be taken as a social, professional, or a personal slight, or a denial of Carter's contribution from some other motive. And so far, there had been no apology.[8]

Carter's diary records his incomprehension, and his pain, at the time: 'see Gray's Book on Spleen takes no notice of my assistance tho' had voluntarily promised … rather feel it.' Being unable to account for Gray's silence, Carter attributed to it 'jealousy p'raps', and added his own verdict on Gray: 'not candid quite'. Later, Carter added in his forbearing way: 'Gray's book very creditable—make[s] <u>me</u> emulous.'[9] But Carter had been hurt.

The subject was not a happy one, and Carter had no desire to confront Gray. He had not mentioned the matter at the time, and had avoided it ever since. But now, his assistance was being sought on another of Gray's career-enhancing projects.[10]

There is a bit more to this part of the story than Carter's sense of unease. Behind the scenes there was a rumbling controversy concerning Henry Gray's receipt of the Astley Cooper Prize. Another competitor, Dr Edwards Crisp, was engaged in a campaign to clean up the Prize administration.

Carter and Crisp certainly knew one another: it was Dr Crisp who had suggested to Carter that he advertise his artistic abilities in *The Lancet*, back in 1853.[11] But whether they had ever discussed this matter is not known. Carter's diary is silent on the matter, so it seems unlikely. The young man's personal reticence and apparent closeness to Gray may have precluded it. Whether Gray had any knowledge of Crisp's allegations at this stage is also unclear, but that there was gossip elsewhere concerning the matter seems probable.

Crisp looks to have entered for the competition in all innocence, and worked hard on the spleen over the same period as had Gray. At some time before the submission of his own essay, Crisp somehow gained inside knowledge of what was going on at St George's in advance of the prize award in 1853. He was rewarded by seeing the prize go to a person whom he suspected was presenting the work of others as if it were his own. Crisp had requested to see the winning essay, and had been denied. He was aware that three of the Prize judges were Fellows of the Royal Society, and that the same body had awarded Gray a research grant of £100 of public money for research on the spleen. Scenting unfair advantage, favouritism, or old corruption, and perhaps hoping that the Cooper Prize judges might want to assert their own honour and that of the award, he had written to them making his concerns explicit. The judges ignored him. When *The Spleen* came out as a book, Gray's *Preface* gave Crisp some real evidence to go on, but the judges still did not wish to address the matter. 'My complaint', wrote Crisp,

> is, that the essay was concocted by four persons, and that Mr Gray was furnished with a grant of the public money for presenting the investigation—that contrary to law, he introduced a paper previously published in the Transactions of the Society which supplied the grant, and that the adjudicators, and other authorities of Guy's Hospital, kept this essay in the dark for nearly twelve months (although I had made repeated applications to see it) so as to prevent the early protests that would have been made against the legality of the award.[12]

It appears from surviving correspondence that one of the younger Prize judges, John Hilton, was sympathetic to an investigation, but that his senior, Benjamin Babington, quashed his doubts:

> It is no business of ours how many persons concocted Mr Gray's Essay or what public money was furnished for prosecuting the investigations. An essay

> was sent in with a motto, thus satisfying the requirements. It proved in our opinion the best—the motto was owned by Mr Gray—As a matter of course he obtained the prize.

Some published material, and the advantage of a grant for research costs, in Babington's view, did not render Gray's essay legally ineligible according to Sir Astley Cooper's will. Concerning keeping the essay from Crisp, Babington argued:

> he cannot know how many essays competed for the prize, or what was the length of each ... the essay, preparations, and drawings, are, according to the will, the property of Guy's Hospital, and surely the authorities may do what they like with their own.[13]

Crisp comes over as a man tenaciously nursing a fierce sense of injustice, and he was probably regarded by the judges at Guy's Hospital Medical School as a crank. The Museum curator at Guy's, Samuel Wilks, who had received Gray's bottled spleen preparations (which had been preserved only in saline, and had to be re-bottled in alcohol) almost cursed the annoyances of the Prize, 'both in costs and Crisps!'[14] Dr Crisp, however, was no fool: he subsequently won the Astley Cooper Prize himself, *twice*: in 1859 for an essay on the thyroid, and again in 1862, on the pancreas.[15]

Crisp had probably guessed that the finest illustrations in Gray's prize essay, and in Gray's book, were Carter's work. He would have viewed with irony Gray's assertion that his interest was in 'the minute structure, and chemical composition, of the most essential elements of this Gland', since Crisp had surmized that the minute structure had been seen and drawn by Carter, and the analyses of chemical composition had been done by Noad and Pollock.[16]

If Crisp had ever managed to get a sight of Gray's prize essay, all his doubts would have been reinforced, because although the entire manuscript is written in longhand, the handwriting is not Henry Gray's. He had employed a professional copyist.

These unwelcome thoughts were out of character for Carter. He did not wish to think badly of Gray, and in many ways still admired him. When the idea for the new book was raised, Carter was working on a series of forty-one enlarged display drawings of microscopical subjects Gray had commissioned him to create for the medical school, as well as some other diagrams, for none of which he had yet been paid. Other work was irregular and scanty. While he was drawing, Carter probably turned Gray's proposition over in his mind. The idea may have improved with contemplation: here might be a way to survive financially until he had finally qualified.

By the 9th of December, the new project was well-formed enough for Gray to be able to promise a regular income of £10 sterling a month for a period of fifteen months, which would at least cover Carter's rent and food while he worked towards the MD.[17] Gray may have hoped that the prospect of a regular income might be persuasive after the irregular and thin pickings to which Carter had become accustomed.

The winter had really set in. Writing to his sister Lily, Carter attempted a light-hearted tone, bemoaning the cold and the smoky fire in his dingy lodgings, 'pity the poor, say I', but the usual good-natured fun in his tone is flat, too near the truth of his own situation.[18] By late December, the big series of drawings for the School was complete, and although he felt 'precarious' in his mind, Carter looks to have expressed an interest in becoming involved in the book. His diary records a renewed conversation with Gray concerning the 'proposed Manual of Anatomy wh. I am to illustrate', indicating that he had come round to the idea, and explaining why: 'may end in something: Gray shrewd but considerate: the proposal seems Providential.'[19]

But Carter still had signed nothing when he went back to Scarborough for Christmas 1855. The money for the micrograph series had still not materialized: his parents had sent him the railway fare.

Carter needed the break: 'constant work in the dissecting-room and mental exertion have somewhat wearied me', he noted.[20]

Negotiations were renewed after the Christmas break, and a verbal agreement reached. But Carter's diary entries for early January show him completely unable to get down to work, and Gray apparently uncomprehending. On 8th January Carter put it like this:

> long chat w. Gray who cannot understand that anyone should really wish to work, & yet not be able to begin. He is altogether practical—do it—his aim 'money' chiefly: as for self need energy & right counsel. Mind certainly not healthy or balanced.[21]

The following day, Carter (who had probably had a restless night) confided in Prescott Hewett, an established senior surgeon at St George's who seems to have been an occasional mentor figure for him: 'Call on Hewett in his room & gave way to feelings ... very low spirited & yet long for sign of God.' Hewett warned him to be careful what occupation he got into. There followed further discussions with Gray.

Carter felt at a disadvantage in negotiating with Gray. He refers to him as 'very shrewd' which I take to mean astute in business terms, but the word did not have an altogether positive connotation, implying cunning, ruthlessness, or manipulation. Aware of Carter's economic circumstances, Gray seems to have conceived the idea that the medical school might appoint Carter to a nominal position while the work was being done. But the expenditure required the agreement of the other teachers, which was not forthcoming until some tough debates were had. On the 14th, Carter notes: 'meeting of Med Off to consider Gray's offer wh. much concerns me—no decision—money wanted. Gray like Owen, wants to use the facts, of course, for self.' On the 21st: 'nothing settled at St G's about my affair—they demur at giving the money yet Mr Hewett assures me from no personal considerations. Gray pushes hard and is like Owen at the College: he feels his power.'

Things remained uncertain until the very end of January, and there was still no money forthcoming for Carter, even for the work he had completed before Christmas. His writing hereabouts is noticeably diminutive.

After another ten days in this limbo, and without any further explanation in the diary, on the evening of the 31st January, Carter called on Gray at his home and selected the first illustrations he would create for the new book.[22] He found Gray already 'at work & employing others'. The following day, Friday 1st February, Carter noted: 'made 1st Drawings for the work.'[23] It looked as though Carter had yielded to Gray's terms. But events conspired to precipitate some kind of a scene between the two men.

Both Gray and Athol Johnson gave scientific presentations that week featuring material created by Carter. Gray's came first. At the Royal Medical and Chirurgical Society, on Friday 1st February, Gray made use of ten of Carter's micrographs for a scientific paper on points of discrimination between cancerous and non-cancerous tumours of bone.[24] Carter was not a member of the Society, and, since his diary is silent, it appears he was not invited to attend. On Saturday, Carter noted in his diary: 'fear the 40 mic drawings will not be paid for by the school. Gray has annexed his own hard proposition to my rights & both are rejected.' On Sunday, Carter took communion, but described himself as 'lifeless'. On Monday, he wrote making an agreement with 'Parker the Publisher'. 'Now fixed for one year at least', he wrote.[25]

Athol Johnson's paper was delivered the following day (Tuesday 5th February) to the Pathological Society. It concerned an unusual case of cystic disease of the testis in a child. Johnson supplemented his own paper with a Carter drawing, a careful pathological analysis by Carter reporting his own microscopic examination of the specimen, and a literature search by him which had revealed a single precedent for Johnson's discovery in Wedl's *Pathological Histology*.[26] Gray may have been present, Carter certainly was. The entry in his diary recording this important professional moment is typically brief:

> Tu. to Pathology Soc—Johnson read my report on his case—wh. was carefully done—will be printed—no debate.

On Friday (8th February) the new drawings Carter had created for the school arose in conversation. 'Talk w Gray about the mic. drawings', Carter wrote.

> He will not pay the full sum. I act foolishly and … rather speak too much.[27]

Carter's telegraphic prose implies he had said things he might later regret. Feelings, pent up for so long, may have found voice badly. But the following day, he noted simply: 'Splendid day', as though he hadn't felt better in a long time.[28]

Of course we have no access to what actually passed between them, but this event feels like a tipping point in their relationship. As far as we know, before this, the two young men had never had a cross word; after it they were never as close. Carter had waited patiently for over a month for this news, and was stony broke. Cordial relations had been based on Carter's silent forbearance; now the truth had come tumbling out. At last, it seems, Carter told Gray he was fed up with not having *'the full sum'*. The money was only a part of it—it looks as though he also told Gray that he was not in the business of drawing micrographs and other anatomical images to promote Gray's career and feed on air: Carter needed recognition, too. Gray at last witnessed the damage he had inflicted on his quiet young associate. Despite his doubts, Carter was prepared to remain involved in the book, but it could not be on the old footing. The force of Carter's words may have taken Gray aback. He appears to have accepted that if he wanted Carter to continue on the book, appropriate funds and recognition would have to be agreed for the full extent of his contribution.[29]

Lightness of heart was only ever a temporary phenomenon for Carter. After the relief came the aftershock. Within another day he was recording his own sense of unworthiness again. 'I am falling daily in own esteem and respect & that of others ... it seems sometimes like breaking up of whole self' ... 'Drawing is a subsidiary employment'.[30]

Whatever the difficulties of their relationship may have been, Gray and Carter saw eye to eye when it came to the sort of book they wanted to create. Both were dissatisfied with existing books for students.

If the medical school Librarian at St George's had been diligent in collecting the best ancient and modern anatomical works, the Library's anatomy section would have carried quite a number of fine texts beyond those in daily use in the teachers' office at Kinnerton Street Dissecting Room. Anything more recent, or more esoteric, might be obtained from the medical circulating library of HK Lewis, with whom St George's held a subscription. When the project to create a new anatomy textbook was first raised, both young men might have been found frequenting either library at odd hours, carefully examining texts: Gray for words, Carter for pictures.

They knew that a student book must be affordable, and it must be well organized. It would need a simple structure, be easily navigable, and must be well-indexed. Clear verbal descriptions were important, and it couldn't do without illustrations, which must be clear and large, with no proxy labelling or picky footnotes: straight off the blackboard, as it were.[31] Students needed help as to where to cut, and to be told what was what. The new book would tell them.

Both men had worked as Demonstrators of Anatomy at St George's, teaching the basics to younger students. Each would have had a good awareness of the books available to students and teachers, and their relative strengths and weaknesses. Having taught anatomy courses for several years, and discussed things with Carter and Parker, Gray would have had a clear idea what sort of a book he wanted to make. He had his own lecture notes to utilize, and his own ideas about how to encourage students to appreciate the significance of the anatomy he was teaching. Gray had understood for himself the profound importance of anatomical knowledge in his various assistantships at private surgical operations, so the real-life surgical application of dissecting-room anatomy was quite fresh for him. The writing itself would have presented no difficulty to the winner of the Cooper Prize, and author of scientific papers.

The illustrative styles of most existing texts would have been familiar to Carter. His own strength, he knew, lay in drawing, and in the sharpened powers of observation it brought. Having graduated via dissecting, his Membership examinations for the Royal College of Surgeons, as well as the demonstrating and teaching at Kinnerton Street, Carter knew what needed to be taught, which books were good for which bits of the body, which illustrative styles were helpful, and which were not. Even before his work at the Royal College of Surgeons gave him the wide perspective and discipline of comparative anatomy, Carter had been a self-confident creator of original anatomical images. By 1856 his own style was already well developed. He had worked to others' commission, and he had his own sketchbooks and portfolios of lecture diagrams to help him. For Carter, how a better book might be made was rooted in the images it provided, and he knew he could provide better ones than had yet appeared in any work for students.

Had they colonized a table and examined everything the school had to offer, Gray and Carter might have passed an interesting time, weighing up the idea, much as Gray had done with Parker. This was an opportunity for making new. They would have discussed what an ideal student anatomy book might look like, what they would like to emulate, and what to avoid. For both of them, the student experience was not so long past that they had forgotten their own difficulties and incomprehensions. Doubtless some of the volumes they referred to went back as trophies on loan to Wilton Street.

Carter's diary is pretty well devoid of any direct information as to how the work for *Gray's Anatomy* was actually done. Some sort of a regular work process developed over the time the two young men were working together, but we cannot be sure of the details. Carter was evidently working extremely hard for sustained periods of time. His diary entries are no longer daily, but written retrospectively at weekly—and often at longer—intervals: he had little time off to reflect on things. He seems only ever to have written anything substantial when he managed to get a break from the unremitting workload, when he fled London. At such times he hardly mentioned the work except to put it behind him, finding a refuge in landscape, and open water, or the sea.

References to Gray himself are scarce in Carter's diary during this time, even though they were probably collaborating on a daily basis. Gray's workload for the book was doubtless onerous, and he was also on duty in the Post-Mortem Room, and the Anatomical Museum. Both of them were also teaching, and in the period of the book's creation, remarkably, Carter also passed his M.D.

When *Anatomy Descriptive and Surgical* was finally published, the book had 720 pages of text, plus contents pages, and featured 363 illustrations. To create a book of such size in twenty months—between December 1855 and July 1857—was a prodigious achievement for the two of them.

Being the older figure, the more forceful, and the intended author, Gray probably took the driving seat in the shaping of the book from the outset. He had probably been working on it for a while before Carter began drawing, so it is likely that the structure of the book was largely planned by the time Carter started work.

The most likely manner of proceeding—especially with that 'subsidiary employment' comment in mind—could have been something like this: following their initial discussions about the character of the book, Gray would have sketched out the general schema, estimating how many illustrations Carter would need to create, what space they should occupy, and where these images would fit in the greater plan. Gray would have had a good idea which of Carter's existing anatomical diagrams he would like to see adapted for the book and had begun working on the text by the time Carter visited him to select the first images to work on. Drafted passages were probably already in hand ready to be illustrated when Carter committed himself to the project, so he would have had a good idea what anatomical landmarks Gray particularly wanted the images to feature, together with page references to Quain (a copy of which Carter owned) and perhaps with other books ready bookmarked, to which Gray expected Carter might want to refer.

Gray's estimate in December 1855 had been that the book would take fifteen months—so the intended completion date at that stage was planned to coincide with the end of the dissecting season in March 1857. A detailed timetable of work was probably already planned out, with deadlines for each of them to meet.

When Carter had visited 8 Wilton Street at the end of January 1856 to select these first illustrations, he had found Gray 'at work & employing others'. At the time, Carter had thought Gray's method of work was something to learn from, so exactly what these others might have been doing is interesting to ponder. Although Gray's handwriting and choice of words in the surviving post-mortem ledgers is fluent, I imagine him dictating to an amanuensis. His decisive script somehow makes me think he'd have been happy walking about declaiming. Were the 'others' perhaps assisting Gray by transcribing or reading aloud passages concerning the same structure from a number of sources, and/or helping Gray distil out their essence by taking down his dictation? The sense of a busy cottage industry in Carter's description of Gray's method of work puts me in mind of Dr Johnson's hive of activity at Gough Square, when he was making the great Dictionary.[32] Whatever writing technique Gray had devised, our witness suggests it was not sitting down alone before a sheet of paper, and waiting for inspiration.

For the passages on surgical anatomy intended for dispersal throughout the book, alongside the appropriate gross anatomy, Gray may also have received the help of a highly competent older colleague, Timothy Holmes, the man who had done the work for the historical introduction to Gray's *Spleen* essay. Holmes was later to become a vitally important figure in the story of *Gray's Anatomy*. Like Carter, Holmes was from a humbler background. His father was a warehouseman, and he lacked eminent patronage, but he was exceedingly bright, and a hard worker. Holmes had become a surgeon via an exhibition to Cambridge, and was already a Fellow of the Royal College of Surgeons, without having first taken the examination for Membership. Holmes may have been brought in as a third pair of eyes for checking the accuracy of Gray's prose, and Carter's illustrations.[33]

In this scenario, Carter would use Gray's text to plan and sketch out each illustration. He might conceive images directly from a fresh dissection or an existing preparation, from his own existing diagrams or his mind's eye, from a previous authority or, most often perhaps, from a process of mental interweaving of anatomical insights from a number of sources. Carter's sketch-plans did not concern simply topography—the form of anatomical structures—but also their naming. His accuracy was therefore doubly important.

The first illustrations Carter worked upon do not appear to have been done from dissections. We don't know what the subjects were, but he decided which they should be at Wilton Street on the evening of 31st January 1856, and finished them the following day. They may have been from bones or preparations in his own or Gray's collections, or from his own drawings. The likelihood is that they were distillations, just as Gray's prose must often have been—of his own and others' anatomical insights.

When the book was published, Carter's list of illustrations gave credit for specific illustrations to quite a number of earlier anatomists. While the great majority were his own original conceptions, 22 per cent of his illustrations gave credit to the work of others. Carter's openness in indicating the derivative roots of these illustrations accomplished several things. Above all, he was listing debts, debts of information or of inspiration. He was also making clear the extent to which he wanted those who used his illustrations to recognize that his original work was rooted in the efforts and perceptions of anatomists and artists who had visited these structures before: how much all human anatomy is a palimpsest of knowledge. Neither Gray nor Carter made any claim that they had made any new discoveries. None of the structures shown and named in *Gray's Anatomy* are original in themselves: as far I am aware, all had been described and illustrated before. In his scholarly way, Carter was also being courteous, and scrupulous, in knowing contrast to his experience of Gray, perhaps.[34] It is from this list that the real authorship of the beautiful drawings from Gray's *Spleen* is manifest, since they are listed exactly as all the others 'to be considered original' by Carter himself. Although throughout the book the ratio of debts to original illustrations is generally low, in the lymphatics five out of six illustrations were after Paolo Mascagni, while in the arteries and circulation, of thirty illustrations, no debts were listed at all. Perhaps it was images like these that Carter had visualized in Paris.[35]

A good proportion of the sketches for the illustrations were probably done from materials and teaching aids already existing in the Dissecting Room or the Anatomy Museum at Kinnerton Street. All dissecting rooms had boxes of bones, teeth, and a dissecting room skeleton, and Carter may have been able to access preparations of body parts in the Anatomy Museum, some created in earlier dissecting seasons by Gray or Carter themselves. The engraving style in the book looks cruder at the outset than it does later on, but the order of engraving may not have been the order of drawing. Carter would have wanted to utilize the season—the work began in winter after all—to undertake as many dissections as possible while classes were still running, and to use bones and preparations for slacker times, when fresh bodies were in short supply, or the weather too warm to dissect.

One body, that of a Mr Henry Perry, was signed out of the Dead-House at St George's by Gray during January 1856, in an unusual way. Gray wrote in the Post-Mortem book: *'This body was used for operations'*, an uncommon entry in these volumes. Mr Perry was a young man of 28, for whom no cause of death was given. The ledger's index records his body as 'Not examined'.[36] My suspicion is that Gray removed this young man's body to Kinnerton Street to inject and preserve for use by himself and Carter for the book, and perhaps incidentally for teaching. If Gray was truthful about the use to which the body was put, Carter's surgical anatomy illustrations in the book may have been among the earliest he made, but this seems doubtful, as the surgical images are extremely sophisticated and complex, and look to have been done at a later stage. If Mr Perry's body was injected, his body would have lasted a great deal longer than had it been left fresh—even in cold weather—and he may have served as the model for a number of illustrations, particularly the muscles and fasciae, the arteries, the brain, and/or the joints or articulations, and even ultimately, some of the bones. Injection took time, as the substitution of body fluids was accomplished by gravity, from bladders hung up above the body. So Carter might have started working on fresher body parts until the preservation process on Mr Perry was complete. Formaldehyde was not in use until much later in the century, so for such an important and useful body Gray may have used alcohol, which although expensive, is also effective, and is used even today in special cases as it allows a body to remain supple.[37] Gray was in charge of supplies, so did not have to ask anyone's permission.

During the dissecting season, class-work on other bodies could be utilized to verify many aspects of the work, for example, the routes taken by specific nerves, blood vessels, or lymphatics, the precise layering of muscles and ligaments, and the accuracy of muscle attachments shown in the book marked upon the bones (utilizing 'the plan recently adopted by Mr Holden') and which Gray said in his *Preface* had been verified on 'recent dissections'.

Separable and portable bodily material—such as amputated limbs, eyes, and organs/viscera—could well have been procured from the Hospital operating theatre, post-mortem room or the dead-house, especially when whole bodies were in short supply, and when human material in the dissecting room was needed for teaching. Amputated limbs would have been useful for muscles, vessels, hands and feet, articulations, and eventually, bones. The post-mortem room may have been the source of the illustration of the child's neck area used in the illustration of the 'Surgical Anatomy of the Laryngo-Tracheal region'[38] which shows an excavation through a portal in the neck, carefully cut as if it were to be replaced, the rest apparently untouched.

If either Gray or Carter questioned or needed to confirm an authority, were unsure themselves about a particular feature, differed between themselves, or wanted to improve upon traditional presentations by creating a new view of an anatomical structure, they could verify anything by dissecting. Both of them were highly skilled—and by now, swift—dissectors, and two are quicker than one. One body could be made to serve for quite a number of dissections, descriptions, and illustrations. They might go ahead with describing and illustrating the simpler or more easily accessible structures first, and wait for a number of problematic topics to accumulate for a day's intensive work, whenever a fresh a body became available, or Mr Perry's body could be excavated at leisure, to reveal and verify repeatedly on its way to oblivion.

Teaching time could have been utilized when opportunities arose, creating dissections for later use, helping Gray clarify the detail of his verbal descriptions, and Carter to see exactly what to draw, or exploring problematic anatomy as a classroom task. Both men were working daily, six days a week in the dissecting room at Kinnerton Street, so at various times they may have dissected turn and turn about, or together. Gray was a busy man with many responsibilities, and it was in his nature to delegate, so the lion's share of the dissecting for the book probably fell to Carter, but the title page of *Gray's* credits both of them with the dissections, and uses the term *jointly*, which implies at least a parity of effort, and at best much work done together and perhaps too, a good working relationship over the bodies of the dead.

They probably depended upon each other for checking the accuracy of each other's work. The presumption must be that since exactness in an anatomy book is crucial, Gray would not have made a finished statement about an anatomical structure unless he was fairly sure it was correct. If Carter, however, discovered an error while he was drawing, he would surely have mentioned it; there might also have been times when a question, a doubt, a wrong name, a misspelling, or a visual infelicity found its way into one of Carter's anatomical drawings, which Gray would have noticed. Before finally being drawn in detail on the block, Carter's sketch-plans might have needed acceptance or tweaking from Gray, a process perhaps as reassuring for Carter as it was for Gray. Both young men were probably agreed on the importance of such a mutual correction process so as to ensure consonance between text and image, which in *Gray's* feels to have been prioritized from the outset.

How the process of checking was actually organized is unknown, but since both men were working a good deal at Kinnerton Street, they may have instituted a dedicated portfolio system in the Dissecting Room office to prevent the loss of important original scripts and images, whereby checked images were ticked off by Gray, or put from one folder into another when he had seen them, together with his newly written texts for the next batch. At times of urgency, one of his 'others' might walk down out of Belgravia to shabby Ebury Street with a correction in the evening, so as to keep Carter supplied with work. Whether this process, or something like it, worked on a weekly, daily or an ad hoc basis, we do not know. At one point Carter charted his weekly timetable, which shows how packed his own working weeks were, with teaching, dissecting, drawing, and studying. A period on Saturday afternoons is specified when he was to visit Gray at home, so perhaps this scheduled weekly meeting is how the whole project was coordinated.[39]

Dissecting was a daytime activity, because of the need for good light. Carter's diary makes clear that much of his time was spent in dissecting and drawing, mostly at Kinnerton Street, probably under the high curved glass roof of the dissecting room. But the atmosphere in there was not always conducive to the kind of close work Carter was required to do, day in, day out. The sociability of dissecting rooms in the Victorian era was related to the male conviviality of London medical students, whose general reputation was associated with drinking and carousing, good humour, mischief, and perhaps anatomical mayhem. There isn't much evidence to associate such misbehaviours with St George's, but a dissecting room song has been preserved by a contemporary of Gray's, Frank Buckland, with the comment: 'These are exceeding clever verses we used to sing at St George's Hospital & in Kinnerton Street dissecting room':

The Medical Student's Alphabet.

A was an Artery filled with injection;
B was a Brick never caught at dissection.
C were some Chemicals—lithium & borax;
D is the Diaphragm, flooring the Thorax.
Lol de rol…
E was an Embryo in a glass case;
F a Foramen that pierced the skull's base.
G was a Grinder who sharpened the fools, and
H means the Half-and-half drank at the Schools.
Lol de rol…
I was some Iodine made from sea-weed;
J a Jolly cock, not used to read.
K was some Kreosote much over-rated;
And L were the Lies that about were stated.
Lol de rol…
M was a Muscle—cold, flabby and red
And N was a Nerve, like a bit of white thread.
O was some Opium, a fool chose to take;
P were the Pins used to keep him awake.
Lol de rol…

Q were the Quacks, who cure stammer and squint;
R was a Raw from a burn, wrapped in lint.
S was a Scalpel, used to cut bread-and-cheese,
And T was a Tourniquet vessels to squeeze.
Lol de rol…
U was the Unciform bone of the wrist;
V a vein which a blunt lancet missed.
W was wax, from a syringe that flowed;
X, the 'Xaminers' who may be blowed!
Lol de rol…
Y stands for You all, with best wishes sincere;
Z for the Zanies who never touch beer.
Now I've finished my song without missing a letter
And those who don't like it may grind up a better.[40]

One can imagine the assorted male voice choir boisterously rising and falling under the high barrel-vault roof to get a sense of the noise that occasionally erupted in the dissecting room. The echoing glass above would not have made it any quieter. There may have been more peaceful places in which to work, like the adjoining Anatomy Museum during daylight hours, but even there, mischief might follow. There is the funniest little card glued into Frank Buckland's scrapbook, which made me laugh when I came upon it in the College of Surgeons' Library:

A singularly large Hydatid,
extracted from a woman in St George's Hospital by Mr F Buckland.
The rest was cooked and eaten by the Operator.

Below it in Buckland's hand, a note explains that it was 'one of the labels Lloyd used to stick upon my things at the Deanery'. Buckland (whose father was the Dean of Westminster) was a great collector and bottler of things botanical, piscitorial, and biological; Charles Lloyd was a friend from St George's. The sense of good humour and fun surrounding Buckland gives an idea of the undergraduate prankishness which Carter might have wanted to avoid.[41]

The regular school dissecting season ran through the coldest months, when days were short, which implies early hours to catch what sunlight there was, and lengthening days with the approach of spring. Carter certainly—and perhaps Gray, too—dissected far beyond the season, working later into the warmer weather of the summer months, when the students were gone. Both men also worked at home in the evenings—we know this of Gray because Carter found him there hard at work, and of Carter because his brother Joseph tells us so. Joe was following in his father's footsteps to become an artist. Between July 1855 and June 1857, the two brothers were sharing lodgings at number 33 Ebury Street, Pimlico, while Joe was studying at the Royal Academy art school in Somerset House. Joe's letters home are not profoundly informative, but nevertheless provide a glimpse of Carter's life we would not otherwise possess.

> The room is a very pleasant one, high up, two windows and two doors, and two cupboards: contains a sofa and an easy chair, the last has attracted the favourable notice of Harry, and the first is dedicated to drawings, folios and other untidy objects; we always dine at home, now, which is much more to Henry's taste, and generally more convenient … [Henry is] not very sociable … As Henry requires our apartment frequently for his visitors, I have got a bona fide attic upstairs for my studio. Henry & I have a very (too) quiet life at present, he is at home drawing … etc a good deal, and I am out studying pictures a good deal, we generally are both home in the evening, reading, smoking (not I, yet) drawing, retiring (and reappearing next morning).[42]

The 'visitors' may have been Carter's own students (he was doing some private tutoring to keep himself financially afloat, and not entirely dependent on Gray) as well as Gray himself, or others in Gray's employ, such as Holmes. Perhaps too, Carter was visited by staff from Butterworth and Heath, the engravers the Parkers had selected for *Gray's Anatomy*, bringing the proofs of illustrations they had already engraved for correction, and collecting the blocks on which Carter had drawn new images. Engravers regularly employed runners to collect artwork and deliver proofs to artists, so this is probably how their association with Carter was largely conducted.

Carter recorded nothing about his relationship with Butterworth and Heath, which is a pity. Their premises were quite close to the Parkers, on the topmost floor of 356 Strand. There is no sign of them in the London street directories until 1856, the year we are looking at now in the creation of the book, which suggests that their partnership was relatively recent, and that *Gray's Anatomy* was among their first major commissions.[43]

While the partnership may have been relatively recent, their skills were not. Heath was probably the senior partner, one of a clever family of copper and steel engravers involved in the precision engraving of bank-notes, whose versatility included engraving on wood.[44] Parker senior may have known him since his own copperplate-printing days. Charles Butterworth was probably a clever pupil who had served his time, and had joined his master in the new business.[45]

Part of the excellence of *Gray's Anatomy* resides in the quality of the engraving. The firm may have been newish, but their skills were outstanding. The clarity of line and lettering, the careful delineation of shadow, and the adept use of contrast show that whoever the engraver was—and it looks to have been mostly done by the same hand—was a master or mistress of the craft. There can be no higher accolade for the quality of Butterworth and Heath's wood engraving than to notice that soon after finishing the work on *Gray's*, they were employed by the *Art Journal*, an institution with a finely tuned sense of excellence in engraving skills, and which chose its engravers from the pick of the London workshops. There were several Butterworth and Heath engravings in the 1859 volume of the *Art Journal*, but a single commission suggests how highly they were thought of: they were asked to engrave from an original by the Victorian artist Birket Foster, whose exquisite foliage and scenes were demanding of great delicacy and judgement.[46] This commission alone is sufficient to reveal Parker's great sagacity in employing Butterworth and Heath on *Gray's Anatomy*.

There is a lot of snobbery in the art world about Victorian wood engraving, because it was a semi-industrialized process, with less 'arty' and cheaper results than the older techniques of engraving on metal. It derived only in part from the crudity of the mediaeval woodcut, which was a flank surface carving method. Wood engraving worked across the rings of the wood, utilizing the end-grain, producing a harder and more durable image. It was especially useful for the long print runs generated by steam printing presses.[47] Prior to wood engraving, illustrated books had separate plates, usually engraved on copper or steel, or lithographed, all of which had to be printed separately, and glued-in individually by hand ('tipped-in') at places convenient to the bookbinder. Wood engraving developed out of metal and glass engraving in the late eighteenth and early nineteenth century, and worked best on hard slow-growing woods, such as box, the end-grain of which holds detail really well. It allowed great delicacy, and required close work and considerable skill.[48] Its great advantage in printing terms was that the engraved woodblocks could be set in the same page set-up as type (letterpress), so instead of having to be printed separately, text and illustrations could be printed together in one operation.[49]

Early examples, like Thomas Bewick's beautiful engravings, were generally small, with curved outlines to fit within the shape of the boxwood trunks and branches from which the blocks derived. Many of Parker's early books show this kind of illustration. It was soon realized that blocks could be squared off and fixed together in multiples, to make much larger images. Parker senior was using multiple blocks in the 1830s, to publish large wood engravings in *The Saturday Magazine.*[50] Illustrations in other early nineteenth-century magazines—like the *Penny Magazine*, *Punch*, *The Builder*, and *The Illustrated London News*—were carved by engravers working on boxwood. So too were all the illustrations in *Gray's Anatomy.*

The disadvantage of Victorian wood-engraving was that two skilled processes were interposed between artist and printed page. The original image had to be transferred *in reverse* to the boxwood surface, so that when printed, it appeared the right way round. Only after this transfer process was the image engraved, by which all the areas which were intended to appear white in the finished image were carved away by hand, leaving only the lines standing proud to take the ink, as it was rolled over the surface of type and block. These two processes—reversed image transfer and carving—were usually executed by different people, with differing skills. Engravers' draughtsmen were expert at reverse drawing for engravers, using a hard-edged simplicity suited to the black/white contrasts of the printing process. Wood-engravers were generally another breed, adept at the hand-carving technique required for translating—with the greatest skill and nicety—two-dimensional surface drawings into sculpted printing blocks.[51]

As a boy in the 1850s, the artist Walter Crane worked for a Victorian artist-engraver, WJ Linton, whose office occupied the garrets of an old house on Essex Street, Strand, which might well give a flavour of Butterworth and Heath's workroom, too:

> [It] was a typical wood-engraver's office of that time, a row of engravers at work at a fixed bench covered with green baize running the whole length of the room under the windows, with eye-glass stands and rows of gravers [cutting tools]. And for night work, a round table with a gas lamp in the centre, surrounded with a circle of large clear glass globes filled with water to magnify the light and concentrate it on the block upon which the engravers (or 'peckers' or 'woodpeckers', as they were commonly called) worked, resting them upon small circular leather bags or cushions filled with sand, upon which they could easily be held and turned about by the left hand while being worked upon with the tool in the right. There were, I think, three or four windows, and I suppose room for about half a dozen engravers; the experienced hands, of course, in the best light and the prentice hands between them.[52]

Crane's experience is invaluable, since he also explains how drawing on wood was done. Linton respected the boy's artistic abilities, and set him to make a pen and ink sketch:

> on a small block of boxwood, showing me the way to prepare it with a little zinc-white powder (oxide of bismuth was generally used) mixed with water and rubbed backwards and forwards on the smooth surface of the boxwood until dry. On this the design was traced in outline, and then drawn with a hard pencil to get the lines as clear and sharp as possible for the engravers. I did not find the 4H pencil put into my hands a very sympathetic implement.[53]

Henry Vandyke Carter's early finished artwork was executed on paper, but he quickly made the decision to bring his originals closer to the reader by becoming his own draughtsman. He taught himself to work in reverse, drawing directly onto the surface of the woodblock. "Work as usual" he wrote in the early June of 1856:

> pretty assiduous at Kinnerton Street drawing from Nature & on wood: result of latter 50:50 require practice: improving.[54]

Carter's willingness to do his own reverse drafting may have been influenced by the likelihood that as a child Henry witnessed his father preparing the illustrations for the delightful *Guide to Scarborough*, for local publisher Solomon Theakston, in 1841. Henry Barlow Carter's originals, showing the sights of the fine seaside town in which the family lived, were smallish oval wood engravings.[55] Carter would have been under ten years old at that time, so might have been familiar with—and perhaps undaunted by—the technique when he later came to take on the work for Gray. His father may well have given him helpful advice, especially on how best to master reverse drafting.[56]

Most likely Carter's images went off to be engraved in batches as they were done, to allow the engravers to keep abreast of the work. Interestingly, Butterworth and Heath were advertising in *The Times* in April 1856 for an assistant wood engraver 'accustomed to first rate book work', possibly to help cope with their usual workload while one of the partners was occupied on the new textbook for Parker.[57] What prompted Carter to make the transition from artist to artist–draughtsman a few months into the project is unknown. He may simply have disliked how his drawings were emerging as engravings, perhaps the firm itself found the double-drafting too slow and difficult.[58] He may have been invited to visit, or was drawn to Butterworth and Heath's workrooms simply from curiosity, wanting to see how the transformation of his images was being accomplished, and was perhaps encouraged while there to try working directly on the wood.

With his first month's earnings, Carter was at last able to put down a deposit on something for which he had yearned for several years: a good microscope. The lenses and other attachments he acquired over time.[59] Gray and he worked assiduously all the spring and through the summer of 1856. In mid-June, Carter wrote an entry which offers the best glimpse we have of his thoughts during their working relationship on the book:

> see most of Gray who may do one much good, and indeed, considering how highly have regarded him in by-gone times, his energy and perseverance, and almost envied him, it does seem the act of a kind Providence to have brought me so much into contact with such a character—fruits yet to come.[60]

This is a positive verdict on the two men's relationship once the work was really underway. Whatever the earlier tensions may have been, it shows Carter weighing up Gray's personality and concluding it curate's eggish: good in parts. Carter's high regard of Gray is spoken of as a thing of the past, however, so it is clear that he was aware that his own immature hero-worship had subsided, and was reassessing what he still found admirable in Gray.

At the end of the month, Carter mentioned being lonely at work, so it would seem that Gray was not always about. Carter had to attend the funeral of his maternal grandfather in Hull in early July, and grasped the opportunity to spend a few days in Scarborough with his mother. He found her altered, and looking quite unwell, but she seems to have improved under his care. While he was at home, he did no drawing at all; but he had a letter from Gray: Carter had been appointed Demonstrator in Histology for the coming year. This position had been newly created for him, and was comparatively unusual in London at the time. The news must have felt like a small victory for science within the medical school. The job was unpaid, but it was 'a step', he noted.[61]

The work he and Gray had embarked upon resumed at redoubled pace: the break may have delayed his schedule, and Carter's working days during the rest of the long summer hours of daylight were eight to nine hours at a stretch at Kinnerton Street.[62] Gray was granted a month's leave of absence for September 1856, and Carter took an autumn break too, walking in the Lake

District. After so much close work in smoggy London, the wide landscapes and clear air of Wordsworth country were a delight for him.

Both of them were back at full spate again at the opening of the winter session in October.[63] Carter was working to the punishing schedule of the original deadline, having hardly any social life outside the rigorous weekly routine of dissecting, drawing, and studying for his MD. Doubtless Gray was pushing himself, too. Carter's one reliable break was Sunday, which he continued to observe, as he had promised his mother, but it isn't clear that he kept the whole day for rest. This unrelenting regime continued through the rest of the autumn of that year and into 1857, with a swift Scarborough visit for a few days at Christmas. In early December Carter had at last obtained his Doctorate in medicine, and both he and his family were over the moon: Dr Carter at last!

Back at Kinnerton Street, the schedule continued in the Dissecting Room through the dark days of the coldest months. In early March, Carter was sounded out as to whether he might be interested in applying for the Curatorship of Pathology, which would be coming vacant later that year. This was an important post, with a small salary, and certainly a step up on the Hospital ladder. As usual, Carter took his time to decide. To try for it would be to throw in his lot at the Hospital, to serve as the body procurement officer for Kinnerton Street, to spend hours doing post-mortems in the dark basement next to the dead-house, and to express an intention to remain at St George's for the rest of his professional life. It would also be to make a serious bid to become a top figure at the Hospital. Other colleagues—such as the supportive Hewett—thought he could do it, but Carter had his own doubts. At the end of the March dissecting season, Carter noted in his diary that he was 'Fagged with work ... thoroughly wearied'.[64] A few days afterwards, Carter received an urgent summons to return home.

> Dispatch which contained the words 'Your Mother died this morning. Come.'[65]

His 'dearest Ma' had died the previous night in her sleep, lying beside his father. Her death was not entirely unexpected, but it was not the less dreaded for that. It was a painful blow for Carter in his exhausted state. Everything came to a halt. He rushed to get a ticket for the long railway journey home to Scarborough, and arrived in time to kiss her farewell, before the funeral.

The return to dissecting and drawing the dead would have been tough for Carter, and if in his grief he became low, it would be unsurprising. He finished what was left of the academic year. The original deadline at the end of March passed, and he was still drawing. We do not know whether Gray had met his own deadline. The MD award ceremony in May left Carter unmoved. If anything, he was curiously detached: 'M.D. conferred at Lond Univ. took no part: am daily becoming anonymous', he wrote.[66] By the month's end, Dr Carter noted his own morbid despondency, ascribing it to 'a single, constant, solitary, occupation, this drawing on paper and wood'. The work had served its purpose and was becoming a chore.[67]

Mid-May 1857 was significant for Gray, too, for he received a letter from a Mr Loch of Albemarle Street, agent for the Duke of Sutherland:

12 Albemarle St
15 May 1857.

Dear Sir
I have a note from my friend Dr Kingsley encouraging me to believe that you would not object to call here that we may have a talk of matters to which you and he have already given some consideration, in connection with the arrangement of which I am glad to hear is in contemplation between you and the Duke of Sutherland.

If therefore you can call here tomorrow Saturday morning as soon after 10am as convenient, I shall have great pleasure in seeing you.

I am, Dear Sir,
Yours faithfully,
G Loch.[68]

Dr George Kingsley and Gray had remained friends since their student days at St George's. We have met him before, when pondering the possibility that his novelist brother Charles facilitated Gray's introduction to the Parkers. Dr Kingsley was now medical adviser to an aristocratic family, the Ellesmeres, cousins to the Sutherlands.

There can be no doubt that the receipt of this letter represented a moment of unalloyed pleasure for Gray. This kind of service was exactly what he had hoped for—an association with a family of the highest social status, and of wealth reputedly greater even than Queen Victoria.[69] An appointment like this put the seal on his upward professional career: six months' service for such a high aristocrat would mean all Belgravia and St James's—the richest districts in London—would be at his feet. The social cachet for Gray was enormous within the Hospital. Outside, news of the appointment would serve as a magnet for paying patients.

Gray was swift to apply for leave of absence from his duties. He was to attend the Duke of Sutherland on his private yacht around the coasts of England and Scotland and at his estate at Dunrobin Castle, for the next six months, June to November 1857. Suitably impressed, the Hospital Governors—chaired by Brodie—granted extraordinary leave, and Gray was off.[70]

Carter was left to toil on alone, without Gray's bracing demands or encouragement to keep him at his task; except by proxy through Holmes, who now shouldered responsibility for seeing the book through the press while Gray was away. The mail service for the Duke's entourage was doubtless efficient, but the leisurely coastal progress of the noble yacht may well have affected the frequency of letters plying between the author and his minions.

Work on the book continued, just as it had the previous year, through the long mid-summer days until, in mid-July 1857, it was at last complete. Carter finished his last drawing, and the job was over.

In Gray's absence, he had reached a critical decision, more important to him than the completion of the book:

> Have now taken some decided steps which have severed my connection with the Hospital—no longer Demonstrator—did not apply for the Curatorship ... the thing is done and I am not certain how wisely, monotony & uselessness of my present life, as I have myself made it, was the final inducement... one occupation gone, and none other selected.
>
> The Book is finished: health pretty good. I trust, I hope.[71]

What had brought him to this point is not explained in his diary, but it may be that the submission of the final illustration was the cue for payment of the balance of his fee by the Parkers, and the sense of freedom it brought helped precipitate his escape. Carter had been teaching pupils privately, and creating occasional artwork to commission, so he had been earning money without having an opportunity to spend much, and he may have been in the unfamiliar position of having accrued some savings. The Parker ledgers record a total payment of £286.0.0 to the Artist.[72] Over half would have been deducted had Carter been paid £10 a month out of this amount, even if only to the original deadline of the first fifteen months. Even so, he might have had a lump sum of over £100 to count on, plus his savings.[73]

A diary entry some time later conveys a clearer notion of events. Carter had felt stuck in a 'torpid chronic studentship' at George's, always in a subordinate role, and desperately wanted to strike out on his own. The feeling of *not* wanting to work at St George's for the rest of his life was suddenly intense.

Carter investigated a number of job opportunities that autumn, mostly outside London, but he finally settled for one last examination. In the autumn Carter was back at his old lodgings, attending Bethlem Hospital for a period of medical practice in lunacy, which he needed so as to qualify for the Indian Medical Service. 'Its practical life one needs', he wrote. 'I really believe India is the right place, should one be so fortunate as to get attached to one of the medical schools.'[74]

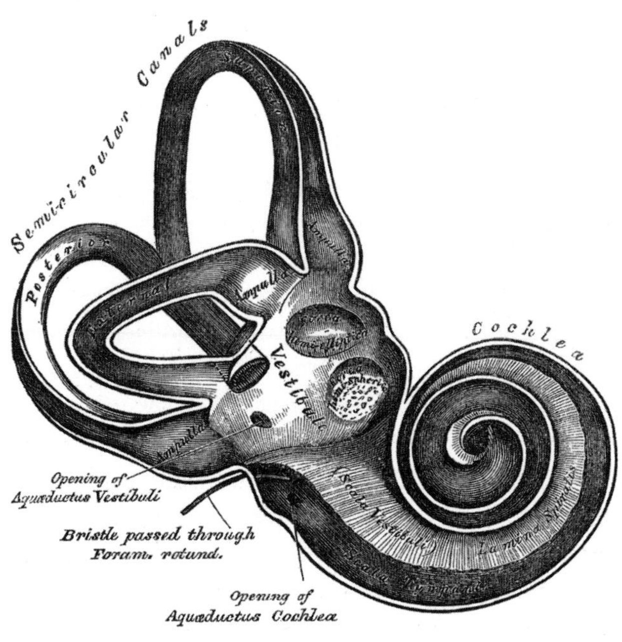

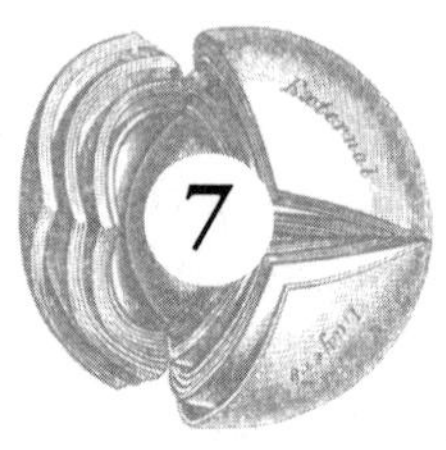

PRODUCTION

1857–1858

The manuscript had been delivered, and everything was going according to plan when, in early November 1857, the Parkers received a serious shock. It was suddenly discovered at this late stage in the project that the pictures were too large for the space allotted to them on the page. The boxwood blocks on which the illustrations had been engraved so beautifully by Butterworth and Heath were the *wrong size.* They were much too large to lie inside the agreed page margins; their dimensions were wide of the mark for royal octavo paper. The illustrations would not fit in the book!

Nothing at all is known about how this discovery was made. It could simply have been that Parker senior noticed the problem when he set eyes on the blocks during a courtesy visit to Butterworth and Heath to arrange for the first tranche of images to go to the printers. Or, Butterworth and Heath might have delivered the first box of blocks direct to John Wertheimer, the printer of *Gray's Anatomy*, who would probably have sent a messenger to the Strand bookshop with a handwritten note explaining the difficulty as soon as it was noticed. Young Parker usually kept the shop, so if this is what happened, he would have known about the problem ahead of his father.

In whatever manner the discovery was first made, young Parker would have been thunderstruck by the news. Someone had made a serious miscalculation.[1] Parker knew enough about printing to realize what it might mean: the whole printing process would have to be suspended, and the difficulty must somehow be sorted out. The upheaval would be expensive, and inevitably, would delay publication. The printer would have other jobs to do, and the halt and subsequent delay would lose the Parkers their place in the printing programme. The whole project would have to be rescheduled. The paper had been purchased; everything had been ready to roll. This was a catastrophe. *Gray's Anatomy* was the Parkers' flagship book for the following year, on which they had staked so much. The responsibility was his own: but *how could such a thing have happened*?

An entry in Henry Carter's diary at the time tells us enough to infer some of what had gone wrong. These brief lines provide a key to an understanding of why the first edition of *Gray's Anatomy* looks the way it does, and why its famous illustrations were created as they were.

These are the words he used, for his own perusal, to deliver his own verdict:

> Parker's business really serious—cuts too large for book—but his neglect is as great as my ignorance at least—Gray will clear himself. The error might have been remedied had I acted up to conviction at the beginning. [2]

Carter says the matter was serious for Parker, that the cuts were too large, that Parker had been neglectful, that he himself had been ignorant, and that Gray would escape blame. It looks as if, at this stage, blame for the problem was being directed towards Carter himself, though he doesn't exactly say so, or from where it derived. We don't know if Gray had returned yet from his six months' jaunt with the Sutherlands, or if Carter had heard the news from Holmes, or indeed from Parker himself. Clearly, Carter had inadvertently made the illustrations too large, but why? Carter usually loaded himself with self-blame on the smallest pretext, and he was probably doing so here. Yet briefly, too, he endeavours to spread the responsibility in a gesture of self-defence—balancing his own ignorance with Parker's neglect, which suggests perhaps that he had been informed of the problem in such a way as to feel—or be made to feel—responsible. His response is significant, especially his allegations. He seems to have felt that his own expectations of a publisher had not been met.

The first point to make is that he refers to a singular Parker, which I take to be young Parker, who was commissioning medical books for the firm, and, being the key figure in the relationship with Gray, was probably overseeing the entire job. Carter's diaries record very little personal contact with the publishers. At the beginning, he had committed himself to the project by letter, and the Parkers appear to have respected his employment at St George's under a direct arrangement with Gray, which kept artist and publisher at arm's length. Carter's diary records a single meeting months later, at the end of June 1856: 'to Parker's conversation tol[erably]. pleasnt', and the receipt of a letter from them while he was in Scarborough in early July for his grandfather's funeral. But in neither case do we have further details.[3] Carter's disenchantment is nevertheless palpable in his charge of neglect.

We must recollect, in examining his expectations, that as a child Carter may have witnessed a quite different relationship between artist and publisher. His disappointment derives from a sense of neglect: neither Parker, father nor son, had sought to confer with him in the same sort of way as Solomon Theakston (also a printer-publisher) had done with his own father, over Henry Barlow Carter's beautiful vignettes for *Theakston's Guide to Scarborough*.[4]

The claim of neglect conceivably also reflects Carter's belief that oversight of the engravers had been wanting. Being an old printer, if Parker senior had even glimpsed these blocks at an earlier stage, he would probably have recognized the problem immediately, before things had got too far ahead. The engravers' workshop was only a minute or two from the Parkers' shop. Parker senior had delegated responsibility for this book to his son, and had apparently resolved not to meddle. The problem revealed that young Parker still had much to learn.[5]

The coda to Carter's note—that he should have acted up to conviction at the outset, and didn't—suggests that he'd had an early intimation that something was wrong. Something in the way Carter puts this, 'The error might have been remedied had I acted up to conviction at the beginning' along with 'Gray will clear himself', gives the impression that Carter had questioned the dimensions he was working to, but had accepted Gray's assurances. The likelihood of some kind of interaction of this sort is of a piece with what we already know of the personalities of the two men, and the past history of their relationship. Even though Carter had sensed something was not quite right, he had deferred to Gray's apparently superior knowledge. The older man, upbeat and impatient, not one to entertain doubt, had no idea what a grave problem was storing up for the Parkers. Now, though, Gray was seemingly excluding himself from the equation. There seems to have been a flicker of the old resentment in that 'Gray will clear himself.'

The fact that the error was discovered at such a late stage reveals that despite Carter's doubts, Gray had not actually conferred with the Parkers. This, in turn, suggests that Gray harboured some disdain for the practicalities of the printing trade, which also looks to be of a piece with what we know of Gray, and his habit of delegating the bother of the printing process to his hospital subordinates. Gray liked the kudos of publications, but was too busy with his career to have much interest in the practical side of publishing them: just as he had delegated *The Spleen* to Athol Johnson, Gray had briskly delegated to Holmes the business of seeing *Anatomy Descriptive & Surgical* through the press.

At the publishers, young Parker appears to have been working on the assumption that Gray was familiar with publication procedure, and didn't require reminding. He would doubtless have informed Gray of the expected page size in royal octavo, but the difficulty probably arose because he failed to specify the need for alertness concerning the printer's needs: the space required for gutters, margins, headers and footers, running heads, etcetera. Young Parker probably assumed that Gray knew that the page space available had to include room for captions and framing: after all, Gray was already a published author. Parker junior was neither as blunt nor as practical as his own father, and was either over-trusting of Gray's knowledge, or wary of stating the obvious to a valued author. The young surgeon's self-confidence and executive flair conveyed an air of competence; but it belied the fact that although Gray should have known, he clearly didn't. Parker himself was not experienced enough to foresee the problems that might arise if the printer's needs were not made explicit. Between them, bluffing and deference had bred misunderstanding.

Only the artist—conscious of the differences between the work for *The Spleen* and for *Anatomy*—seems to have been alert to the problem. *The Spleen* had been a completed manuscript when Gray won the Astley Cooper Prize. There had been no difficulty in the engraving and printing of its illustrations, because when the Parkers received the manuscript the illustrations were already in existence, their dimensions already fixed. The book's letterpress had been designed around them.

The process for *Anatomy* was not at all the same. This was a commissioned book. The illustrations did not exist until they were created, to commission. The work of drawing had dovetailed chronologically with the writing, over the joint dissections. The text was being assembled and written, and the illustrations were probably being engraved, in parallel with the drawing process. The engravers may well have been forewarned there were to be some large-scale images, which would probably have been a relief to them by comparison with the usual kind of finicky little illustrations textbook publishers demanded, especially in anatomy. These ones had called for careful and detailed carving, especially in the lettering, but they were mostly of a goodly size for an engraver, featuring a great deal of white space, which engravers appreciated. As Carter's ready-drawn blocks began to arrive, Butterworth and Heath got on with the work, not questioning the block sizes they had been given to engrave.

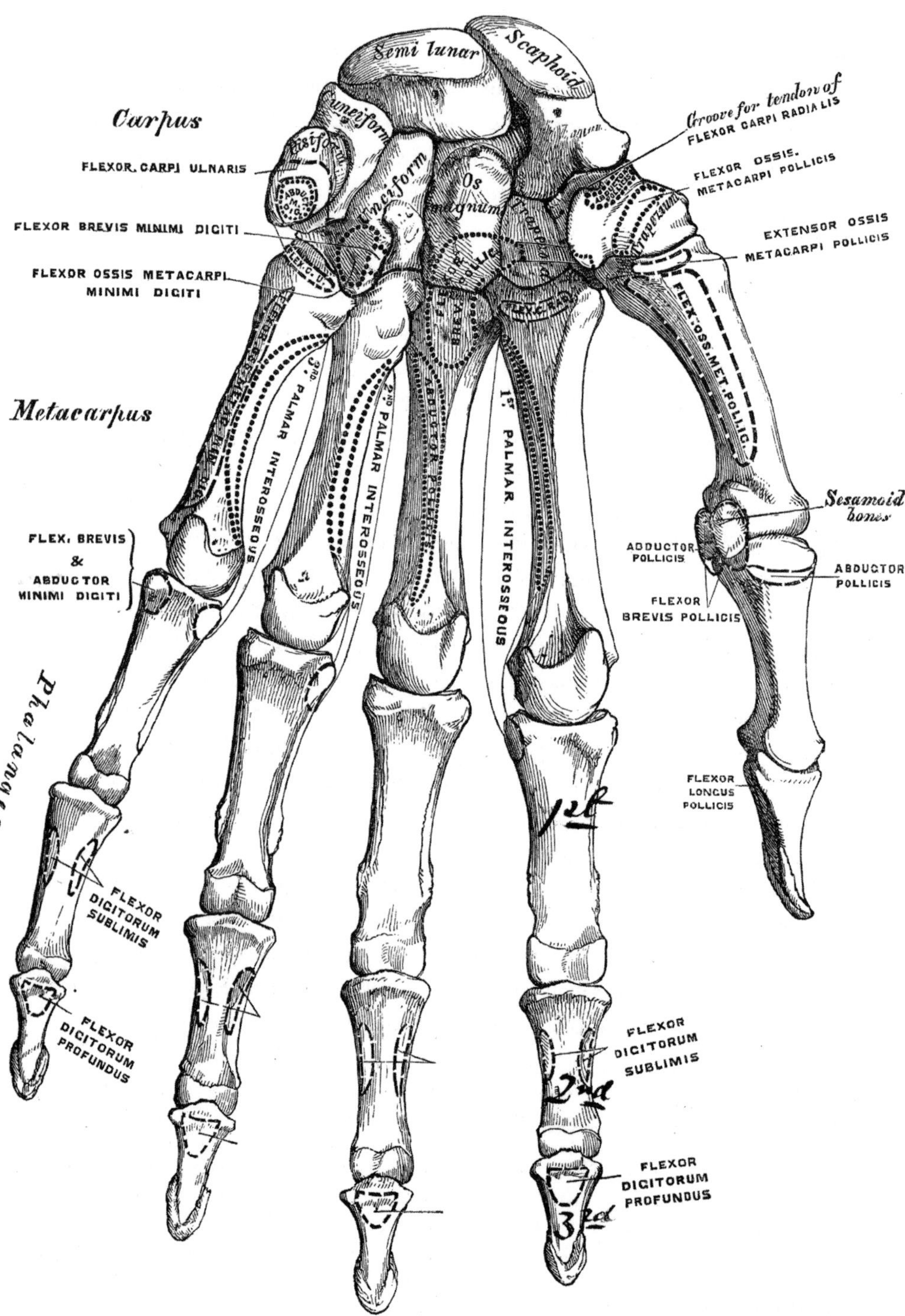
Semi lunar
Scaphoid
Carpus
Cuneiform
Pisiform
Unciform
Os magnum
Trapezoid
Trapezium
Groove for tendon of
FLEXOR CARPI RADIALIS
FLEXOR. CARPI ULNARIS
FLEXOR OSSIS.
METACARPI POLLICIS
FLEXOR BREVIS MINIMI DIGITI
EXTENSOR OSSIS
METACARPI POLLICIS
FLEXOR OSSIS METACARPI
MINIMI DIGITI
Metacarpus
3RD PALMAR INTEROSSEOUS
2ND PALMAR INTEROSSEOUS
1ST PALMAR INTEROSSEOUS
ABDUCTOR POLLICIS
FLEX. OSS. MET. POLLIC.
Sesamoid bones
FLEX. BREVIS
&
ABDUCTOR
MINIMI DIGITI
ADDUCTOR
POLLICIS
ABDUCTOR
POLLICIS
FLEXOR
BREVIS POLLICIS
Phalanges
FLEXOR
LONGUS
POLLICIS
1st
FLEXOR
DIGITORUM
SUBLIMIS
FLEXOR
DIGITORUM
PROFUNDUS
FLEXOR
DIGITORUM
SUBLIMIS
2nd
FLEXOR
DIGITORUM
PROFUNDUS
3rd

So it was probably not until the majority of the blocks were finished, and perhaps not until they had arrived at the printer's—by which time the great reams of royal octavo paper had already been purchased, and the compositors had already begun to set up the text—that anyone noticed something was wrong, and the mismatch was perceived to be serious. An order must have been given swiftly to halt typesetting until the difficulty could be sorted out.

Looking closely at the detailed layout of the first edition, it is evident that there *is* something curious about the size of the illustrations. Carter's note about them being too big for the book is evidently an accurate statement of fact. Reviewers congratulated the publishers on the large illustrations, never realizing that their dimensions had been arrived at by mistake.

It is not just that the images in *Gray's* are larger than the run of anatomical illustrations of the day, which they were: some are between three and five times the size of the generality of *Quain's.* But once the reader has recognized the veracity of Carter's note, and begins to look carefully, it becomes apparent that many of the images in the first edition are indeed arrestingly large: so large that they extend beyond the regular dimensions of the type in the book. One has only to lift a leaf of the book slightly to be able to perceive the proper dimensions of the letterpress page: it generally shows through from the page beneath. Once these key dimensions are registered in one's mind, it is not difficult to see that Carter's illustrations jut out into the page-margins much more often than they should.

It is only after many careful measurements and comparisons, that the extent of the accommodation subsequently managed at Wertheimer's printing office becomes apparent. On the vertical axis, it seems that the main difficulty related to the figure numbers and captions originally hand-engraved on the illustrated blocks: they took up too much vertical space.

The bones of the human hand from *Anatomy Descriptive and Surgical.* The magnificent size and informative quality of Carter's illustrations is evident here. Carter's bone illustrations were inspired by Holden's, but Carter provides the student with much more information. A comparison with Erasmus Wilson's version of the same subject (shown on page 105) reveals the open generosity of Carter's work, and helps explain the visual appeal of *Gray's Anatomy* to prospective buyers. The length of this image was not a problem for Wertheimer, but the width certainly was: it invades both right and left margins, leaving only a quarter of an inch clear at the inner gutter/hinge edge. The manuscript is typical of Victorian owners' emendations. Actual size. Author's collection.

This difficulty is interesting not only because of the trouble the captions caused, but because of the insistence on having captions at all. Other anatomy books, like Quain or Wilson, had none: they used a figure number centred over each illustration followed by an asterisk, referring the reader to a footnote where the illustration was explained. Part of the innovation in *Gray's Anatomy* had been to dispose of footnotes, but this in turn elevated the importance of captions. One has only to think of the enormous number of illustrations in the book, all needing adjustment of their captions, for a sense of the Parkers' anxiety and perhaps the printer's consternation to become palpable.

As if this wasn't enough of a headache in itself, they were confronted with another problem that compounded everything: many images were also too wide. The size of the illustration of the hand-bones on page 104 of *Gray's Anatomy*, for example, suggests that Gray and Carter had been working in the belief that the *letterpress*—that is, the printed area on the page—was six inches wide, whereas in fact, that is the entire *page* width after binding. This is a considerable error, and one for which it is difficult to account.

Looking at it in the context of the actual page sizes of *Gray's Anatomy*, and putting oneself in the position of the printer's compositors, one can perhaps appreciate the seriousness of the difficulty. The printers were being asked to accommodate illustrations that overran the textual area of the pages they were printing by almost an inch. The illustrations were one sixth wider again than the textual area of these royal octavo pages. What Carter had thought of as an illustration of a half page width was nearer two-thirds, encroaching on the amount of letterpress that could be accommodated on an illustrated page.

Seeing this together with the caption problem, we must suppose that there had been a serious disjunction of understanding between publisher and authors concerning the page area available for printing in a book of royal octavo size. Put simply, the page dimensions to which Gray and Carter

---

The muscles of the hip and thigh seen from behind, from *Anatomy Descriptive and Surgical*. This large figure has been nicely accommodated horizontally by means of a narrow column of type. Vertically, the crest of the ilium (the hip-bone) at the upper edge of the illustration is rather high relative to the caption. Happily, the caption's second line only extends half way, and has been ranged leftwards to accommodate the curve of the hip-bone on its rise. Without a caption the illustration would have fitted neatly within the expected head and foot margins of the letterpress area. It certainly seems fortuitous that the caption typeset by Wertheimer allowed for a short second line. Even so, the lower edge of the wood-engraved image extends below the letterpress into the margin at the foot of the page by the equivalent of one full line of type (3 x-heights).

fascia covering its outer surface. The fibres gradually converge to a strong flattened tendon, which is inserted into the oblique line which traverses the outer surface of the great trochanter. A synovial bursa separates the tendon of this muscle from the surface of the trochanter in front of its insertion.

Relations. By its *superficial surface*, with the Gluteus maximus, Tensor vaginæ femoris, and deep fascia. By its *deep surface*, with the Gluteus minimus and the gluteal vessels and nerve. Its *anterior border* is blended with the Gluteus minimus and Tensor vaginæ femoris. Its *posterior border* lies parallel with the Pyriformis.

This muscle should now be divided near its insertion and turned upwards, when the Gluteus minimus will be exposed.

The *Gluteus Minimus*, the smallest of the three glutei, is placed immediately beneath the preceding. It is a fan-shaped muscle, arising from the external surface of the ilium, between the middle and inferior curved lines, and behind, from the margin of the great sacro-sciatic notch; the fibres converge to the deep surface of a radiated aponeurosis, which, terminating in a tendon, is inserted into an impression on the anterior border of the great trochanter. A synovial bursa is interposed between the anterior part of the tendon and the great trochanter.

Relations. By its *superficial surface*, with the Gluteus medius, and the gluteal vessels and nerves. By its *deep surface*, with the ilium, the reflected tendon of the Rectus femoris, and capsular ligament of the hip-joint. Its *anterior margin* is blended

172.—Muscles of the Gluteal and Posterior Femoral Regions.

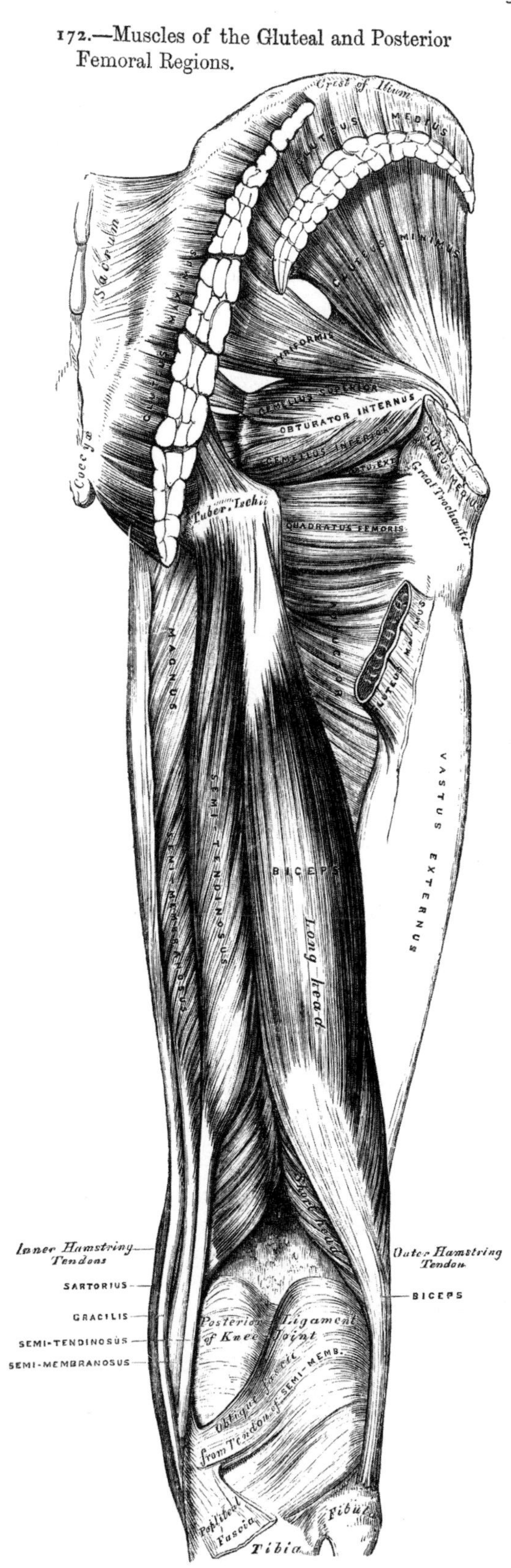

(and presumably also Butterworth and Heath) had thought they were working, were quite wrong. How this could have happened remains a mystery.

The predicament the Parkers found themselves in, essentially, is this: mid-Victorian wood-engraved images could not simply be enlarged or reduced in size like a photograph, with the rotation of a lens, or made to fit to a specific size by the touch of a couple of keyboard buttons or the click of a computer mouse. If an illustration was found to be too large in 1857, it had to be scaled down by hand, and hand-drawn again on a new and smaller woodblock, and then re-engraved by hand. This would take the skills of both draughtsman and engraver, all over again; and of course, each craftsman would take time on each block.

With a book containing so many illustrations, to have chosen to go for re-engraving upon such a discovery would have been not only prohibitively costly, but would have delayed publication by many months, possibly years. Carter had begun producing his extraordinary anatomical drawings at the beginning of 1856, but it was not until the November of the following year, 1857, that the size problem was noticed. By then, the majority of the blocks had probably already been engraved. There were no drawings to go back to: Carter's meticulous finished drawings had been done directly on the blocks, so, but for those which yet remained uncut, they had been obliterated by the engravers' carving. The book had to be ready for launch in the late summer of 1858—in time for its promotion before the start of the medical school year in the autumn—or an entire year's sales would be lost.

Looking at the surviving financial accounts for the first edition of *Gray's Anatomy*, it's clear that the engraving was the greatest single item of expenditure, amounting to more than double the cost of the printing.[6] The blocks were always the biggest investment in a book like this: they were planned to be good for years of future use. The thought of breaking the news to Butterworth and Heath, after they had worked so hard to produce such fine engravings, must have seemed daunting.

Parker senior and his son would have had some very hard thinking to do. Re-engraving the illustrations was in reality unthinkable on grounds of cost, and the other alternatives open to them were severely restricted. Compressing the words on the page, the letterpress, by using a smaller typeface than had been planned for, would certainly have created more space for these large illustrations, but such an expedient would also have put all the pagination out of kilter. The book had been carefully designed so that each illustration lay alongside the appropriate text: a smaller text would have involved a near-complete redesign. Not only that, but smaller type would have defeated one of the key design features of the book as originally conceived: it would compromise the book's legibility. A smaller typeface would also mean that in place of typographic harmony, there would be an unhappy disjunction between text and image—small type, large illustrations—attracting attention to the mismatch.

To decide to accommodate the larger illustrations by using a larger format—to make the book itself bigger—would have involved purchasing a fresh consignment of larger paper, and using the paper already in hand for some other volume. Such a choice would have entailed a large and unplanned outlay of extra capital the Parkers probably didn't possess, or could not spare. The book had been planned as a royal octavo volume, a tall and stately format. Any further enlargement would have made it unwieldy, and would also have defeated the original intention of making *Gray's Anatomy* a large but practical book for the dissecting room. Giving the volume a larger format would have changed the book's entire concept. It would also generate extra costs—with larger paper, more ink, higher printing costs, extra sewing and binding costs, as well as extra weight in postage, carriage, and the need for added space for storage—all of which would certainly push up the cover price.

There was hardly any room for manoeuvre, not even in truth the option of raising the price. Publishers recognize that student sales are particularly price-sensitive: high prices drive down sales. Most Victorian medical students forked out for an anatomy textbook at the outset of their medical school training, and the Parkers wanted a good prospect that if a student's pocket stretched to a new textbook, *Gray's Anatomy* would be the one they would choose. The entire project had been planned on that basis. Twenty-eight shillings was a standard price for good books, reasonably expensive, but nevertheless not an unreasonable outlay for a basic text required by a first-year medical student. Quain's latest edition of 1856 was just out at a new lower price of 31 shillings, so it was crucial to set the price of Gray's single volume below that.[7]

The Parkers' customary bound-in leaf of advertisements tucked in at the back of *Gray's Anatomy* when it was finally published showed a selected list of twenty-two books they hoped might be of interest to the book's likely readers, only two of which cost over 28 shillings. Both were specialist medical textbooks, whose buyers were likely to be more established in age or income than the new medical student for whom *Anatomy* itself was intended. The average price of the rest was under 10 shillings.[8]

Young Parker probably had sharp recollections of needing money for books when he was studying at King's College, so father and son look to have been of one mind when it came to wanting to offer good value. Both had a keen sense of affordability and price-sensitivity. They were both painfully aware that few students and not even many doctors from either Charing Cross or King's College hospitals, both on the Strand, had been into the shop to buy Henry Gray's book on the spleen, which had cost only 15 shillings. Raising the price on *Anatomy* was not an option: it was imperative that this one *sold.*

Alongside all these considerations, the financial situation was grim in 1857. Business was slack because of an economic downturn caused by the Crimean War. Booksellers, hit by the slump, were selling up at an increased rate—entire stocks were being auctioned off—and several printers had already appeared in the bankruptcy courts.

When the Parkers had done their sums after the initial meetings with Gray, they had decided on a print run of 2000 books: an ambitious but not over-optimistic number of textbooks to shift. This was a niche market, and the existing books were stale. With good advertising, this one could make a splash and sell out quickly, if it was as good as they hoped it would be. The fees for author and artist had been agreed on that basis, the paper had been purchased on that intention, and all the other costs had been planned likewise. The Parkers' profit margin on the project was already wafer thin, if it existed at all. This book was a speculation, as all books are. But because a new anatomy textbook was so enormously costly to produce, the opportunity to break even on the first edition was already slight. The size problem now threatened to capsize all their calculations. Carter was not over-dramatizing when he said the matter was serious. We know this, because before the woodblock difficulty, the sums would have looked like this:

		£	s	d
Author	*[Gray]*	300	–	–
Drawings	*[Carter]*	286	–	–
Woodcuts	*[Butterworth & Heath]*	752	–	–
Printing	*[Wertheimer] [included typesetting]*	352	16	–
Index	*[? Holmes or ? Grub St indexer]*	8	–	–
Paper	*[? Spalding & Hodge / ? John Dickinson]*	218	11	–
Binding	*[? subcontractor for Wertheimer]*	110	8	4
Advertising	*[medical press]*	111	12	6
Total costs:		£2139	7s	10d

Projected earnings for a print run of 2000 books at 28 shillings each was £2,800, so after deducting production costs, takings would have been about £660.0.0. But this amount was not clear profit, for three reasons. Prior to publication, all these figures represented investment in advance: the above sums were predicated on outlay and deficit, not cash in hand, until the entire first edition had sold out. Second, a number of copies had to be sent out free of charge to reviewers. Last, the 'takings' figure included the share of the cover price expected by every bookseller.[9]

What sort of deal the Parkers offered their booksellers is not known, but in the mid-nineteenth century the average bookseller's cut was 25 to 30 per cent of the book's cover price—at least 7 shillings or more per volume of *Gray's*, or about £700 in all. Counting this into the calculation, we can see that even before the size of the woodblocks became a problem, the Parkers were expecting to carry a deficit of about £40 on the first edition, less whatever proportion of the bookseller's cut they were able to make up themselves by retailing their own books from their West Strand bookshop. They would have had to sell over 100 copies from the shop to break even. What they lost on this edition, they would of course hope to make up handsomely on the next, but nothing was guaranteed, and (especially if we bear in mind the slow sale of Gray's *Spleen*) their commitment to bring *Anatomy Descriptive and Surgical* to publication represents an act of faith of no small proportion.

We can see from these figures how very swiftly the Parkers' profits could have been engulfed had this project run in such a direction as to require either re-engraving the illustrations, or switching to a larger format. As a modern publisher has memorably commented: 'publishing is more dangerous than wrestling with lions.'[10]

The printer, John Wertheimer, would have been concerned by the discovery of the size problem, because typesetting and printing time had no doubt already been scheduled for this print run. The disruption of pre-laid plans would have thrown his print-shop into temporary crisis. He would also have been worried about the amount of type already set up in the typeset formes (the metal frames which hold the set-up pages of type and illustrations in the printing press), since it meant that during the suspension of printing, those types couldn't be used for anything else. Normally proofs would be run off, corrections added, a revised proof finalized and the 2000 run of both sides of that particular sheet section (signature or 'gathering') would be printed off, allowing the types to be used again for another set of pages, or for some other job. The enforced delay might well have knock-on effects preventing the completion of other jobs in addition to this one.

Before we go any further, we must look at what this might have meant for Wertheimer. Evidence survives which allows us to know, that for him—even more than for the Parkers—any delay was not just serious, but extremely grave.

John Wertheimer's predicament was this: like many other businessmen of his day, he was experiencing financial difficulties in 1857–8 so severe that his entire business was under threat. He and his staff had weathered the periodic ups and downs of trade cycles for years, but a conjunction of adverse circumstances in 1857 meant that he was faced with business collapse. In November 1857, when the discovery was made concerning the over-size illustrations for *Gray's Anatomy*, Wertheimer's business was on a knife-edge, and bankruptcy was imminent. We have no direct evidence to suggest that the Parkers had any inkling that this was the situation.

The full story behind Wertheimer's financial near miss is unknown. Who foreclosed upon him, and set the entire proceeding going, or even how much he owed them, probably cannot now be ascertained. Bankruptcies, especially in the printing industry, often resulted when others who were going under called in their debts, putting stress on regular relationships which had served well in better times, and bringing panic—sometimes disaster—to those whose liquidity was similarly tenuous. Fortunately for Wertheimer's business, and for *Gray's Anatomy*, the majority of those men to whom Wertheimer and his business partner George Littlewood owed money arranged to meet. Rather than throw Wertheimer and Littlewood into a debtors' prison, and all their fine typesetters and pressmen on the parish, these men decided by a majority that an enforced sale would mean not just idleness for the printers: it would yield nothing but the sale price of Wertheimer's printing equipment, which would not have been high at such a time. Instead, they agreed a financial arrangement whereby all those involved in the partnership's financial problems would be better served. The presses at Circus Place, Finsbury, would yield real money if they were to continue in use; none if they were silent. Earnings would cover at least a proportion of the debt.

The great parchment document recording this collective arrangement survives. It names those appointed to oversee the procedure, and those who accepted the payment of seven shillings in lieu of every pound sterling they were owed (35%). The document offers a glimpse behind the scenes of the Victorian printing industry, exposing a cluster of relationships which in less difficult circumstances might never have come to light, either because they derived from word-of-mouth deals, or because the documents recording them have long since disappeared. The total amount to which Wertheimer and Littlewood were jointly indebted was in the region of £6,500. This sum was owed to a group of twenty-three companies and individuals, but probably touched the lives of scores of others.[11]

In mid-Victorian bankruptcies, two principal creditors—generally those to whom most money was owed—were appointed 'Inspectors'. Being major creditors, they were also interested parties. Their official task was to ensure that the correct amount of money materialized within one calendar year of the agreement, for a fair distribution to all creditors. In Wertheimer's case these key figures were William Caslon, of the famous family of typefounders, and Frederick Edwards, a manufacturer of printers' ink. For some unknown reason, Edwards's name is entered only in pencil on the great parchment schedule that records the details of these matters, and the amount he was owed was never finally filled in.

Several of Wertheimer's other creditors remain unidentified, but those that I have been able to track down (or take an informed guess at) derive, as one would expect, largely from the printing industry. The table shown in Appendix 2 gives their names, the amounts they were owed, and their likely identities, derived from searches in a variety of mid-Victorian London trade directories. They were wholesale stationers, manufacturers of paper, ink, and of printing press equipment, typefounders, bookbinders, lithographers, an engraver, and the well-known publisher, Charles Knight. Two Finsbury tradesmen—a grocer and a wine merchant—perhaps reflect more personal and local debts of a domestic nature, and show how the tragedy of business failure could implicate every aspect of a printer's life.

All these events being so long ago, it is impossible now to interrogate anyone as to what Wertheimer and Littlewood went through at this time. The closest I have managed to arrive at appears in an account by the well-known newspaper proprietor and philanthropist John Passmore Edwards (1823–1911), who went through a similar crisis in the early 1850s, in which he faced the prospect of imprisonment for debt in the Queen's Bench prison. He was saved from such a fate by the decent men with whom he had done business. His creditors met, and the majority, he later wrote,

> took a generous view of my position, and thought that I had suffered a reverse of fortune from no moral fault of my own, and that I was rather entitled to sympathy than condemnation.[12]

They voted to accept five shillings in the pound (25%). Passmore Edwards was not at all happy to pay back only a quarter of what he had owed. He spent the next thirteen years paying back the full amount to all those to whom he had been indebted. In 1866, to celebrate the final repayment of all his past debts, his ex-creditors threw a grand testimonial dinner in his honour at the Albion Tavern in Aldersgate Street. Mr Hodge, of the Drury Lane paper manufacturer and supplier Spalding & Hodge, took the chair at this celebratory occasion, and admirable sentiments were expressed on all sides.

John Wertheimer, too, somehow avoided being bankrupted. We know about the private arrangement with his creditors only because of the fortuitous survival of the great old parchment document, in its unfinished state, with the name of the second Inspector drafted only in pencil. The whole document looks convincingly official, but no reference to the matter seems to have appeared in the official pages of the *London Gazette*, so it looks as if the public shame of near-bankruptcy and Inspectorship was somehow avoided. The parchment detailing Wertheimer's debts bears a number of dates, inserted at various stages during its drafting. The earliest is 16th December 1857, the date of the original petition for bankruptcy; and the last, when it was signed and sealed by John Wertheimer, is 4th June 1858. The final payment to creditors was scheduled for 6th March 1859.

Henry Vandyke Carter's words in his diary note about the outsize illustrations convey the impression that the process of printing may only just have started at that time: the discovery sounds fresh. It was written on the 6th November 1857. We know that the book's page proofs were still being worked on through to the end of June or early July the following year, and that *Anatomy Descriptive and Surgical* was finally published at the very end of August or possibly the first week of September, 1858. So it seems that during almost the entire period in which the typesetting and printing of this huge and complex textbook was proceeding, the printing works at Circus Place Finsbury was the scene of a slowly unfolding financial crisis, and possibly also perhaps, its positive resolution.

It may be that we now know more about Wertheimer's knife-edge financial situation than the Parkers did, living at the time. It may be, that to retain his customers' confidence, Wertheimer had breathed no word about the sword of Damocles hanging over him and his printing works. But equally, the possibility exists that Wertheimer contained his angst up to a critical point, when he spilled the beans to his fellow printer, Parker. If such was the case, the discovery of the woodblock problem might have been the occasion

Wertheimer chose to take Parker into his confidence. Parker may already have received intimations about Wertheimer's difficulties from mutual friends: a number of people on the creditors list probably knew both men. Charles Knight of 90 Fleet Street, for example, was a very well-known figure in the London educational publishing and printing world, publisher of the *Penny Magazine*, and connected by marriage to Parker's old boss William Clowes. Parker's old printing partner, Harrison in St Martin's Lane, was another possibility. Alternatively, he may have heard through the paper manufacturer Mr Hodge, of Spalding & Hodge of Drury Lane, who regularly supplied Parker and Son with printing papers for their books, and may indeed have supplied the machine-made paper used for the first edition of *Gray's*.[13] If Parker did know of Wertheimer's predicament, it may be that *Gray's Anatomy* was printed at Finsbury Circus because Parker had been prompted by sympathy for him. All sorts of scenarios are possible. We simply do not know.

It was probably Parker senior who came up with an old printer's solution to the outsize problem. Somehow, a way could be found to accommodate these excessively large blocks in the planned royal octavo format.

Having worked long in the trade, the elder Parker knew all about old printers' expedients: *trimming* (shaving areas from blocks or illustrations) and *forcing* (encroaching upon margins to accommodate unconventionally large material). No self-respecting printer enjoyed doing such work: it smacked of makeshift. It really went against the grain for a printer–publisher like Parker senior, who loved a beautiful book. Trimming and forcing *Gray's Anatomy* would seriously affect the appearance of the book, and would diminish the pleasure an experienced printer would find in handling the final result.

But we can safely surmise that while the presses were suspended, at some stage soon after the discovery of the problem, publisher and printer reluctantly agreed that in the circumstances in which they found themselves, there was no practical alternative. Trimming the existing blocks, and/or somehow accommodating their odd dimensions by adjustments of the letterpress, was really the only choice.

Trimming and forcing would involve Parker in further costs: that he knew. Printers' conventions demanded additional pay for such fiddly extra labour. Wood engravers would have to be paid for trimming, and for recutting if that was deemed necessary in particular cases. This last could only be a final resort. Only when an estimate of what was involved was made could it be seen how costly even forcing might be.[14] At some stage, most likely Parker senior, because of his printing knowledge, sat down with John Wertheimer, to see in detail what might be done. A methodical process of going through the page plans for the whole of *Anatomy* followed: measuring the blocks and the spaces available for them, looking out for tabulated data that had to be kept together, examining the problem in close detail to agree how to proceed. During this process, choices were made that influenced the appearance of the book for ever.

The necessity of an engraver's skilled knowledge in this process suggests that George Littlewood, Wertheimer's partner in the bankruptcy document, or someone senior from Butterworth and Heath, may have been party to this important discussion, to ensure that the printers' decisions could realistically be put into execution, and also to assess likely costs in time and money. The decisions taken demanded considerable skill and nicety on the part of both engraver and printer—each would be pushing skilled capability in his field. Young Parker may have joined the discussion so as to witness and learn from his elders. Carter would surely have mentioned the event had he been there, but it looks as if he was not called upon. Where additional editing was required, Holmes (or Gray, if he had returned from the Sutherlands' service) could be asked to advise on the feasibility of cutting any superfluity of information from images, captions, or text. But this was basically a printing matter.

Three key policy decisions look to have been agreed pretty early on. First, not to re-engrave, but to keep the illustrations as they were, and to accommodate them by forcing margins, footers, headers, and letterpress where necessary. Second, to replace all the main illustration captions already created by Butterworth and Heath with new ones in cast type smaller than the main body of the text. This would involve trimming a large number of blocks. Third, to ask Butterworth and Heath to use a smaller format for engraving the remaining illustrations they still had in hand.

Several landmark engravings were scattered through the book, which occupy an entire page each. These were probably recognized as fixtures, between which the intervening text and illustrations had to be adjusted to fit. Careful attention was paid to precisely where crucial fractions of an inch had to be saved to accommodate a line of type, which woodblocks must be trimmed down, or where images could be rearranged, split or piggy-backed to save space. Discussions centred upon where a hand-engraved caption on a block could safely be cut away and substituted by a smaller one set in type, or where even a new typeset caption might be truncated. We do not know if the text itself was edited down in places, or if the smaller compressed typesetting of the sections on surgical anatomy was an original design feature, or a function of the pressure on space. On many pages, to save a single line of type was an achievement.

Although this procedure was probably tedious, in a way it may also have been reassuring, because everybody came to perceive the problem was resolvable. And, after the initial shock and anxiety, it became clear that the costs—though substantial—were not impossible to meet. Furthermore, principles could be laid down for adoption consistently throughout the book. As matters proceeded, there may even have been a curious feeling that the process might eventually produce a better book than the one originally planned.

Most of the rectification work was to be done by multiple small adjustments of typography and spacing, and by the trimming of blocks. Trimming was a difficult and painstaking job, one that no engraver particularly enjoyed, but because of the danger of damage to the engraved image, it demanded considerable care, involving skill and judgement. It called for the use of a special kind of wood-plane to shave away the edges of the woodblock, permitting the overall dimensions of engraved blocks to be reduced down to new proportions, while allowing the engraved edge detail of the image to remain, clean-cut and strong enough for printing.[15]

Some of the engravings in the first edition of *Gray's Anatomy* show signs of block trimming, without damage to the image. In one instance, for example, one of Carter's characteristic boundary lines was removed for the printing of *Gray's*. His image of the branching Malpighian Corpuscles (see page 142) had originally appeared in Henry Gray's book *The Spleen*, inside a circle drawn and engraved in black line, signifying both the frame of the illustration and the circular boundary of what was visible down the lens of a microscope. When the block was reused in *Gray's Anatomy*, the original circular frame to the illustration had disappeared. The excision had the effect of leaving the anatomical structure intact as the unbounded object of attention, but permitted the block to occupy a much smaller space. The amount of space saved by the excision was a significant six lines of type.

Parker and Wertheimer had probably completed the replanning of the book before Carter (feeling a 'fraud', a 'humbug', and an 'impostor', he said) took and passed the exams for the Indian Medical Service in January 1858. Before he left the country, Carter spent some of his savings on books, tropical kit, and traded in his old microscope for a better one. By the time he had boarded the ship *Sultan* at Southampton in February for the journey to Bombay, the block trimming and typesetting of new captions was already in full swing.[16]

Carter's optimism about a future in India had somewhat faded. He recorded 'sharp pangs of regret' on leaving London, and confided to his diary: 'were God in his Providence to open up a way for me in this country I should indeed prefer it'.[17] His arrival in Bombay on 23rd March probably coincided with the production of early page-proofs in London. The final correction process was well on its way towards completion when in May, Carter was officially gazetted as Professor of Anatomy & Physiology at Grant Medical College, Bombay.

There is little news of Gray for these early months of 1858, except for a bereavement. Towards the end of May 1858, Gray's younger brother Robert, died at sea on board a merchantman, *The Indomitable*. The news probably took several weeks to arrive in London. Gray may have remained at home in Wilton Street with his grieving mother, or may perhaps have gone away again. Wherever he was, in June the final proof sheets from Wertheimer's found their way to him.[18]

On a summer's day in the early twentieth century, soon after the end of the First World War, the descendants of Charles Butterworth, of Butterworth and Heath, did a wonderful thing. They donated the early proof sheets for the engravings of Carter's illustrations for *Gray's Anatomy* to the Royal College of Surgeons. The firm had looked after these proofs carefully for sixty-five years, and during several moves of premises. Ever since the day of their gift, the priceless portfolio has remained safely in store in the archives at Lincoln's Inn Fields. Thankfully, they were not destroyed by the direct hit suffered by the Museum next door in the London Blitz. So it is by good fortune, and the care of a chain of people who recognized the importance of these beautiful monoprints, that most of them still survive.

Opening up the original cardboard portfolio reveals the inscription on the inner paper wrapper:

First Proofs of the Engravings on Wood
by Butterworth and Heath
for the First Edition of
Gray's Anatomy.
Published by J.W. Parker and Son, West Strand,
1858.
Presented to the Royal College of Surgeons, London, by F.R. and A.W. Butterworth,
August 1923.

Unwrapping this, one can see the sheaf of cartridge sheets on which the prints were mounted, hand sewn down one side to form a tall home-made album. Cut-outs of the 'first proofs' have been carefully mounted on these sheets.[19] The images were printed directly from the newly engraved blocks, on a very thin almost tissue-like paper known as 'India' paper, which was a favoured surface for Victorian engravers, because of its ability to pick up the finest of detail. These are pristine working prints, which have suffered over time, but their blacks are clear, their edges sharp and clean.

A view of the interior of a Victorian wood engraver's studio I have seen shows an engraver at work at his bench, before a window.[20] Behind him, by the wall, stands a small cast-iron hand-press with a solid base and arch, and a great turning handle at the top, whose fat screw mechanism stands poised to lower a heavy surface downwards onto the base. It is very like presses still in use nowadays by traditional hand-bookbinders. The delicate prints in the album were probably made on just such a hand-press at Butterworth and Heath's premises at 356 Strand, one by one, or in small groups, soon after they were engraved. Many of the prints have been carefully dissected, so that the areas of purest white (which had undergone the process known as 'lowering') have been excised from the proofed images, and the remaining fragile skeletons of the images have been carefully mounted on the cartridge pages.[21] Sadly, the album's home-made binding has deteriorated. But, although it is no longer complete, or in perfect condition, the majority of the pages survive: quite enough to assess the nature of the changes Parker and Wertheimer had to make when the size problem was discovered.

A close examination of Carter's illustrations as they were engraved by Butterworth and Heath reveals that the boxwood blocks were *not* pierced for metal type, as in some botanical and biological illustrations of a similar date. All the words featured on the anatomical structures in *Gray's Anatomy*, and the captions above them were originally engraved by hand. Occasional uncut shavings, and the shading lines within the engraving itself which pass behind the letters on numerous occasions, further attest to this. In a few instances, an engraving was proofed before the lettering was quite finished, and the method by which the letters were formed is revealed: straight-line outlines of the letters were cut first, then the non-printing areas carefully excised. The delicacy of the lettering on the illustrations in *Gray's Anatomy*, as well as the morphology of the anatomical structures themselves, makes manifest the care and skill with which the engraving was done.

These proofs confirm that the most frequently verifiable difference between the blocks when they left Butterworth and Heath's premises in the Strand, and their publication in the book, is the replacement of the hand-engraved main illustration captions with new ones in small close-set cast type. But for the excision of the hand-cut captions, all the images in the Butterworth and Heath proofs are the same size as they appear in *Gray's*. The block-trimming accomplished at Wertheimer's was done very skillfully—not a single image was damaged or curtailed, although as we shall see, several blocks originally featuring two adjacent images had to be cut at the midline.

Comparing the proofs against the printed first edition, it is possible to see that the decision to replace the main captions was highly significant, and that it accomplished a number of beneficial things at once. Because the type captions took up less height, the change saved significant vertical space. Crucial lines were saved for letterpress text. The substitution was alone almost sufficient in many cases to prevent noticeable forcing of illustrations into the margins at the foot of the page, or at least to minimize whatever forcing was needed. Had the original captions been left as they were, numerous illustrations would have been visibly forced into the footer margins: now, only a few cases remain. Again, however hard an engraver tried for uniformity, hand-worked lettering cannot fail to exhibit minor evidences of craft, revealing the variety of idiosyncrasy; so the new captioning made the illustrations appear more 'professional' because printer's cast type is both more uniform and more formal in appearance. The substitution of hand-cut captions by type was therefore as much a stylistic change as a space-saver, and in addition, the use of type blurred the sharp demarcation between Gray's words and Carter's images: subtly blending text and illustration, and allowing the illustrations to look more at home in and alongside the letterpress.[22]

A process of careful comparison reveals that the saving of lines for the letterpress seems to have become less urgent as the book progressed. In the earlier part of the book, subject sections had run on without much of a break, without starting at the top of a new page. By page 132, almost half a page was allowed to remain blank at the conclusion of a section, and the next section—'The Articulations'—began at the top of a new right-hand facing page, suggesting that the problem had partly been resolved by then. Particular pages survive even later in the book to show evidence of the compositors' difficulties: page 353, for example, is a pig's ear of a page, with an illustration of the surgical anatomy of the upper arm which aligns with the tabulated plan of the brachial artery at the bottom of the page, but not with the adjacent short (and thin) column of letterpress (see overleaf).

One of the most exciting finds this study of the Butterworth and Heath proofs has brought to light is an illustration for *Gray's Anatomy* which did not appear in the book. It is an anterior (front/interior) view of the human spine, absolutely straight and rather stately (see page 192). It would have formed an accompanying illustration to the lateral (side) view of the spine in the book's opening osteology section. It was certainly intended for *Gray's Anatomy*, drawn and engraved ready to go in. It looks to have been discarded at an early stage in the adjustment process, to create room for other oversized illustrations. The jettisoning of this illustration would have put all the subsequent figure numbers out of kilter, so may have been what initially precipitated the idea of changing all captions and figure numbers to metal type.[23]

The proofs contain a number of twinned images which could not be shown together in the book, because the original conjoint block was too large for the page.[24] In one case, two small illustrations of nasal bones (figures 35–36) appear in Butterworth and Heath's album on a single block, one above the other, with individual hand-engraved captions, while in the first edition of *Gray's* they have been separated—their joint block split and captions trimmed away. But placed beside each other, one ranged right and the other left, they were printed with a narrow vertical column of type between them. The new caption line for each is exactly parallel, so both captions occupy only a single line, and in small type. Only a modest saving of space was made, but the illustrations were situated together on the same page with the text that relates to them. In several other instances of twinned images, the compositors who set up the pages were unable to employ this expedient because the illustrations were too large to be placed beside one another, and the second of the pair has been moved to the following page. This latter expedient is particularly evident in the osteology section at the beginning of the book.[25]

Certainly the most inventive and dramatic methods of saving space and of its course, this vessel lies internal to the humerus; but below, it is in front of that bone.

Relations. This artery is superficial throughout its entire extent, being covered, *in front*, by the integument, the superficial and deep fasciæ; the bicipital fascia separates it opposite the elbow from the median basilic vein, the median nerve crosses it at its centre, and the basilic vein lies in the line of the artery for the lower half of its course. *Behind*, it is separated from the inner side of the humerus above, by the long and inner heads of the Triceps, the musculo-spiral nerve and superior profunda artery intervening; and from the front of the bone below, by the insertion of the Coraco-brachialis and the Brachialis anticus muscles. By its *outer side*, it is in relation with the commencement of the median nerve, and the Coraco-brachialis and Biceps muscles, which slightly overlap the artery. By its *inner side*, with the internal cutaneous and ulnar nerves, its upper half; the median nerve, its lower half. It is accompanied by two veins, the venæ comites; they lie in close contact with the artery, being connected together at intervals by short transverse communicating branches.

200.—The Surgical Anatomy of the Brachial Artery.

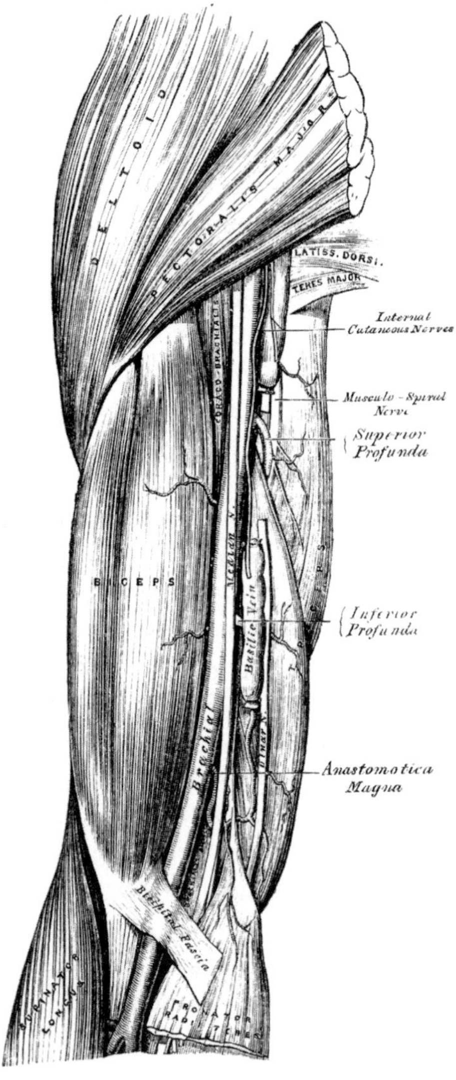

PLAN OF THE RELATIONS OF THE BRACHIAL ARTERY.

In front.
Integument and fasciæ.
Bicipital fascia, median basilic vein.
Median nerve.

Outer side.
Median nerve.
Coraco-brachialis.
Biceps.

Brachial Artery.

Inner side.
Internal cutaneous.
Ulnar and median nerves.

Behind.
Triceps.
Musculo-spiral nerve.
Superior profunda artery.
Coraco brachialis.
Brachialis anticus.

accommodating the larger illustrations appear early on in the book. The typesetting of captions benefited only the vertical dimension, so where images were too wide there was more of a problem, and greater ingenuity was required. In figures 88 and 115, for example, hand-engraved labels extrinsic to the actual image have been excised by significant block-trimming, and replaced by more compact typeset versions, releasing space for almost half a column of letterpress. Elsewhere, illustrations which *should* have been half a page wide, but which occupy nearly two-thirds of the page-width, are nevertheless permitted to fit within the letterpress margin by the typesetters' device of narrower than usual columns of type, apparently normalizing the excessive size of the illustration (figures 47, 59, 92, 93).

'Forcing' is more evident, too, in the first half of the book, and where nothing could be done to mitigate the width of wide blocks. Illustrations are shunted downwards into the margin at the foot of the page (figures 80, 87, 88) into the gutter margin at the hinge, (figures 56, 63, 85, 103, 109) into the margin on the loose edge of the page, known as the foredge (figures 74, 120) or sometimes in two directions, as in figure 84 (foredge and gutter) or figure 87 (foredge and foot). In their other books, the Parkers usually tended to favour wide margins, so it is very likely that the page margins originally planned for *Gray's* were intended to be broader, and that they were narrowed to allow more space for letterpress text. It may be, too, that the Parkers agreed an extra line or two of letterpress per page, to accommodate text displaced by the large illustrations. We don't know what the intended dimensions of the printed area originally were, so these adjustments cannot be ruled out.

Probably the most important typesetting manoeuvre employed at Finsbury Circus was to preserve the running head and the header margin at the top of the page pretty well uninvaded by these large images. This visual expedient allowed the entire book to retain a feeling of unity. The consistent banner to every page gave the book a look of manifest intentionality, which allowed the oversize illustrations bursting the bounds right, left, and centre to radiate an aura of normality, and even uniformity, in their generosity. The book gave the impression that it was *intended* to be the way it was, to look the way it did.

With the extent of the size problem in mind, it is now possible to look at the book as a whole, and appreciate the intrepidity of the block trimmers and compositors at Wertheimer's in tackling such a complex and awkward job. Apart from Wertheimer and Littlewood, the names of these men are

---

One of the less successful page layouts in the first edition of *Anatomy Descriptive and Surgical.*

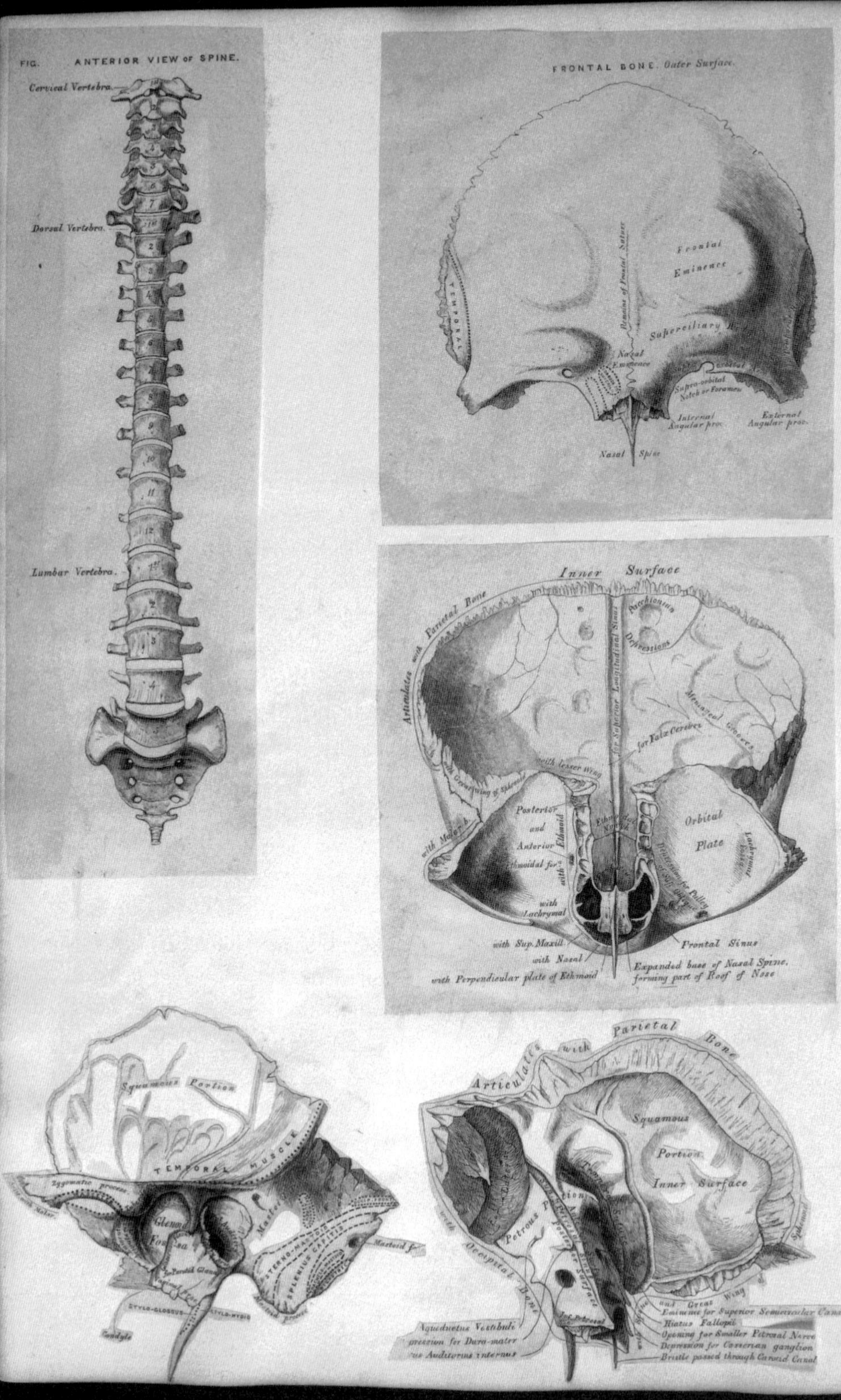

FIG. ANTERIOR VIEW OF SPINE.
Cervical Vertebra.
Dorsal Vertebra.
Lumbar Vertebra.
FRONTAL BONE. Outer Surface.
Frontal Eminence
Superciliary R.
Nasal Eminence
Supra-orbital Notch or Foramen
Internal Angular proc.
External Angular proc.
Nasal Spine
Inner Surface
Articulates with Parietal Bone
Pacchionian Depressions
for Superior Longitudinal Sinus
Meningeal Grooves
for Falx Cerebri
with lesser Wing
Posterior and Anterior Ethmoidal for.
with Ethmoid
Orbital Plate
with Lachrymal
with Sup. Maxill.
with Nasal
with Perpendicular plate of Ethmoid
Frontal Sinus
Expanded base of Nasal Spine, forming part of Roof of Nose
Squamous Portion
TEMPORAL MUSCLE
Zygomatic process
Glenoid Fossa
Mastoid
STERNO-MASTOID
SPLENIUS CAPITIS
Mastoid for.
STYLO-GLOSSUS
STYLO-HYOID
Condyle
Articulates with Parietal Bone
Squamous Portion
Inner Surface
Petrous Portion
with Occipital Bone
Aqueductus Vestibuli
Great Wing
Eminence for Superior Semicircular Canal
Hiatus Fallopii
Opening for Smaller Petrosal Nerve
Depression for Casserian ganglion
Bristle passed through Carotid Canal

not known, but we ought nevertheless to pause for a moment to admire their skilful efforts in trimming and accommodating the text and images for *Gray's Anatomy*, in such a way as to have allowed the difficulty to have gone unnoticed for so long.

In the Parker accounts, two amounts—for printer's alterations, and for re-engraving woodblocks—are listed separately:

Alterations &c	£ 69	10s	0d
Cutting blocks &c	£ 4	18s	0d

These sums show that the necessary alterations resulting from the oversize illustrations cost the Parkers a total of £74. 8s. 0d, an amount which works out at nearly 10 per cent of the total originally charged by Butterworth & Heath, or 20 per cent of Wertheimer's original printing bill.[26] The extra costs of the alterations deepened the expected shortfall the publishers had intended to bear, and threw the entire *Gray's Anatomy* project into deficit. But, through negotiation, and the deployment of the printing skills of Wertheimer's staff, such costs were bearable: a long way from having to re-engrave or redesign the format from scratch.

The two amounts, listed as they are, give the appearance of having been billed separately. By far the greater sum covers the printers' alterations, and suggests how much of the work was done at Wertheimer's. The extra typographic costs would have been charges to cover the costs of the extra fiddly labour by Wertheimer's compositors, who, instead of producing a regular book by regular methods, had to make up many pages in a bespoke individualized manner, adjusting margins to accommodate outsize blocks. The printers' men had been asked to do unconventional and time-consuming work, and for this they had to be paid. A scale of fees operated across the metropolis between employers and men, which Wertheimer would have ignored at his peril. Over the book's 750 pages, the average added cost of the alterations worked out at one shilling and tenpence farthing extra per page.

A page from the remarkable Butterworth & Heath portfolio containing the first proofs of their wood engravings for *Anatomy Descriptive and Surgical.* The image of the spine here was never used. Reduced.

The sum designated for 'cutting blocks' probably refers to re-engraving, so what I have called *trimming* was probably accomplished in-house at Wertheimer's under Littlewood's supervision, and included in the larger sum for printers' alterations. We can assess how little re-engraving proved necessary by looking at the cost of wood engravings from Butterworth and Heath before the size problem became apparent. The average cost of a single engraved illustrated block in *Gray's Anatomy* was £2. 0s. 9d. So the total sum for 'cutting blocks'—£4. 18s. 0d—reveals how very little new or altered engraving was required: the amount might have covered a couple of blocks damaged in the trimming process. The joint conference had minimized the need for much re-engraving at all.

After page 400 or so, the number of forced margins and trimmed blocks drops noticeably, so it would seem that by the time the size problem was discovered, Butterworth and Heath had still to engrave many of the illustrations which appear towards the end of the book, and were able to transfer Carter's larger drawings to smaller blocks. Carter makes no mention of doing this work himself before he left for his new life in Bombay, but it is possible that he and/or a good draughtsman at Butterworth and Heath's accomplished the transfer. But for a few exceptions, which were probably drawn and engraved at an earlier stage, or were small enough not to require replacing, the illustrations after page 400 look generally more diminutive (smaller by about a third) and have a slightly more wooden, stiffer, and less fluent feel about them.

The Butterworth and Heath proofs are remarkable and very special documents in the history of Victorian printing. But probably the most extraordinary and illuminating find made during the research for this book has been a part-set of the original first edition page proofs of the text of *Gray's Anatomy*, held in the archives at the Royal College of Surgeons of Edinburgh.[27]

These are the final proofs of pages already typeset by the compositors at Wertheimer's. They were printed up for last revisions and error-checking, just before final printing, and several are marked 'Press' agreeing their onward progress. They are on thin paper of a slightly heavier gauge than that used by Butterworth and Heath for the proofs of the engravings. The Wertheimer page proofs were printed in sets of eight on each side of the large sheets of royal-sized, and folded down to book size, to produce sixteen pages at a time, just like real pages. They are rather rough and ready, with hand-cut edges, revealing ink and pressure show-through from the letterpress, and they have additional hand emendations. There are also some inky fingerprints, which make studying these pages curiously intimate, though it is not clear if they are in printing or writing ink, or whose fingers might have left them there.

The pages they represent are mainly from the last quarter of *Gray's Anatomy*, but there are one or two signatures (gatherings of 16 pages) from earlier on, and the very last pages and index are not present. The page proofs for the rest of the book appear to have been lost. How these ones survived separately is not known, but there can be no doubt whatever that they are genuine.[28]

Someone alive at the time recognized their importance, and kept them from being destroyed. My deduction is that they were retained by somebody at Finsbury Circus, as they must have gone there for the next stage of work to be done. But after that? Possibly they remained at Wertheimer's, and were rediscovered by someone when his business archive was cleared out; or, they may have found their way to 445 Strand. Why should they have been kept?

Three of the signatures carry handwritten dates: 'QQ', (pages 593–608) is dated 1st June 1858; then 'SS' (pages 625–40) is dated 12th June 1858, and 'TT' (pages 641–56) is dated 14th June 1858.[29] These dates are rather a special find, because the intervals reveal the pace of work (on average 3.5 days to a signature) which allows us to project a schedule for completion of printing of early July. This feels very late on in the publishing sequence for a book due out in August. It would seem that when this particular batch was made, the entire final proofing process was quite close to completion, since the title page and preface are usually among the last parts of a book to be finalized, and this surviving batch of proofs contains both. Wertheimer had probably delivered or sent the original first proof correction sheets direct to Holmes at a slightly earlier stage. These seem to contain Gray's last word on the book—his final revision, before printing went forward. If Gray was away somewhere as he had been the previous year, Holmes or Wertheimer had found him.

While these proofs are in some respects illuminating, they also add somewhat to the mystery surrounding the matter of who did what on *Gray's*. Directly after thanking Carter in his *Preface*, Gray thanked Holmes specifically for his 'able assistance' in 'correcting the proof-sheets in their passage through the press'. There is evidence of two hands writing on these proofs, Gray's is easily recognizable in black, the other is in brownish ink, but it does *not* belong to Holmes, and there are no initials to go by. Writing in this hand is generally confined to the top margin of a few pages, and since it is limited largely to dates, looks to be that of someone at Finsbury Circus coordinating the printing of Parker's work, possibly Wertheimer himself. Let's say it is his. This handwriting has light pressure, and a rather shapely thin-nibbed style, slanting forward; Gray's also slants forward, but he used a wider nib, and a lot more pressure is visible in his downstrokes, showing a firm hand with a hasty but elegant and decisive quality.[30]

All the corrections on these pages seem to be in Gray's writing. Most of them concern details: spellings, capitalizing initial letters or italicizing words, small deletions or insertions, and the highlighting of possible printer's errors. Across the top edge of page 305, which shows fractures, Gray wrote firmly:

> The cuts of 'The Arteries' I wish to be as good as is possible. H.G.

Page 305 began a new gathering, labelled X, and the section on the arteries began overleaf, filling the rest of the gathering, so Gray was asking for particular attention to be paid to the careful printing of the illustrations for that section of the book. Carter's artery drawings are among the most original and spectacular material in the entire volume, and Gray would have known it. If they were based on the drawings Carter had visualized in Paris, they may even have been what had prompted the entire idea for the book. Wertheimer's men would have known that by 'good' Gray meant printed well, with good inking and adequate pressure on the blocks to give the images their due: what were known as *good impressions*. In the *Preface*, the word 'woodcuts' was to be replaced by 'wood engravings' throughout—although, interestingly, Gray still used 'cuts' in his own script to describe the illustrations of arteries. He also asked for the word 'numerous' to be changed to the more generous 'abundant'.

The most significant corrections appeared on the title page, which the compositors had typeset according to a schema laid down by JW Parker and Son. It resembles many title pages in Parker books. A handwritten note—in the *other* script that I have attributed to Wertheimer—runs along the top edge of the proof, stating: 'This form of title appears to be approved by Mr. P.'

In this original title page proof as set up at Finsbury Circus, the book's name appeared in a comely serif type in large capital letters, with **ANATOMY** centred at the very top, and **DESCRIPTIVE AND SURGICAL** centred at an appropriate distance below, rounded off with an authoritative full stop. There follows below, after a decent space, BY in much smaller capitals, and underneath: **HENRY GRAY F.R.S.** in bold capitals modestly smaller than the title itself. Gray's occupational status—LECTURER ON ANATOMY AT SAINT GEORGE'S HOSPITAL—follows below in smaller capitals, slightly less emboldened. After another well-judged gap, THE DRAWINGS BY appeared below in the same size type as the BY with Gray's name, and then: **H. V. CARTER M.D.** in the same size type as Gray. Carter's new post PROFESSOR OF ANATOMY AT GRANT COLLEGE, BOMBAY; was typeset below his name, followed on a new line by his old job-title and affiliation, LATE DEMONSTRATOR OF ANATOMY AT ST GEORGE'S HOSPITAL. After another gap, a line of larger small capitals ran right across: THE DISSECTIONS JOINTLY BY THE AUTHOR AND DR. CARTER.

Those without an interest in typography might view these details as unimportant, but their significance ought not to be overlooked. From this title page we may perceive the views of both Parker and Wertheimer concerning Carter's role in the book. Moreover, these proof sheets allow us also to understand something significant about Henry Gray's character.

Carter's name appeared a goodly distance beneath Gray's, in the same size type. He had been assigned a secondary position with regard to the book's authorship, but given equal standing with Gray in its creation: the positioning and the parity of type-size says as much. This is what we would expect from the Parkers, whose estimation of Carter's importance to the book was rooted in an experienced publisher's understanding of the labours involved in the creation of those extraordinary illustrations, and a publisher's recognition of Carter's contribution to the originality of the volume about to be issued under their imprint. He was not a simple artist.

JW Parker and Son were aware, too, that Carter had received only a one-off fee for his work, and that he had been excluded at the outset from the royalty agreement from which Gray was to receive £150 for every 1000 books sold.[31] They perhaps may have wanted to express their own sense of generosity towards Carter, by honouring his importance to the book on its title page.

Gray, however, had quite different ideas. He took up his pen, dipped it in black ink, and deliberately drew a slashing line right through Carter's name and doctorate. Then he did it *again*. Beside Carter's doubly struck-through name, Gray drew a large bracket with the direction 'Type size of the name below', and arrowed downwards to the lower line of type about the dissections, just over half the size.

He drew another line, even thicker, through Carter's new job-title, obliterating it almost entirely. Another thick line went through Wertheimer's handwritten note. So as to avoid any misunderstanding, Gray wrote next to

Send a revise before finally printing it off

ANATOMY

DESCRIPTIVE AND SURGICAL.

BY

HENRY GRAY, F.R.S.

LECTURER ON ANATOMY AT SAINT GEORGE'S HOSPITAL.

THE DRAWINGS BY

Type size of the names below

~~H. V. CARTER, M.D.~~

~~[illegible] COLLEGE, BOMBAY,~~

LATE DEMONSTRATOR OF ANATOMY AT ST. GEORGE'S HOSPITAL.

To be omitted H. J.

THE DISSECTIONS JOINTLY BY THE AUTHOR AND Dr. CARTER.

LONDON:

JOHN W. PARKER AND SON, WEST STRAND.

1858.

the expunged professorship: 'To be omitted' and he initialled the direction firmly with his own imprimatur, 'H.G.' The excision of Carter's appointment in Bombay meant that all that was left remaining below Carter's name was the position from which he had walked away: *'LATE DEMONSTRATOR IN ANATOMY AT ST GEORGE'S HOSPITAL'*. This residual title announced his status in aspic as a subordinate of Henry Gray, the 'Late' making it seem almost as if he were dead.

If one wished to think well of Gray, one might try to explain away the last modification here by suggesting perhaps that he wanted the book to appear as a proud product of St George's, or that he hoped to avoid jumping the gun about a new appointment that was not yet public knowledge. But there was no secret about Carter's new job. It had been gazetted in London in May 1858, and these alterations were made in June. Carter's good fortune would have buzzed around St George's as soon as the announcement was made. That a George's man should have landed such a posting was a great fillip for the school, and a source of buoyant hope, especially for students and staff outside the magic enclosure of high patronage. The likelihood is that Holmes himself had informed the Parkers or Wertheimer of Carter's prestigious new job.

The second alteration—calling for Carter's name and doctorate to appear in smaller type than that used for Gray's name and his fellowship of the Royal Society—denies us the opportunity for a charitable view to be taken of either of these proof 'corrections'. Smaller type for Carter can have been demanded for no other reason than that Gray wanted to gainsay the equality the Parkers believed to have been there in reality.

Diminishing a type-size down was a small gesture, and might be regarded by many as of little consequence, if it had not accomplished a number of things. It introduced an exaggerated and unwarranted sense of distinction between word and image, and between intellectual and manual skills, casting Gray as the authoritative figure, while at the same time also presenting Carter at the outset of the book as a mere illustrator, promulgating the notion that all illustrators are social and intellectual subordinates.[32]

It was clearly insufficient that Gray's name and FRS, and his superior position at St George's, should appear in prime position: all indicators of Carter's hard-earned status must be diminished or expurgated. The smaller type-size downgraded Carter's role, his importance to the book, his identity,

---

The original proof Title page of *Anatomy Descriptive and Surgical* showing Henry Gray's corrections made in June 1858. Shown at 79 %.

his doctorate, and he was stripped of his professorship into the bargain. Professor Carter could not upstage Lecturer Gray.

Gray seems to have felt that by downsizing Carter he was enhancing his own reputation; but the contrary was in fact the case amongst those who knew him. Wertheimer certainly, Parker surely, and Timothy Holmes probably, would have seen these corrections, and would have noted the character they reveal. There may have been a suspicion that Gray might intervene to minimize Carter's importance to the book, hence the need for the clear statement of Parker's preference at the top of the title page proof. Had Parker or Mr Bourn perhaps attempted to forestall it, seeking to strengthen the likelihood that the original might be left unchanged? All these men had sharp eyes, and would not have missed the meaning of what Gray had done.

Whether Holmes knew of the change, and whether he remonstrated with Gray or kept his own counsel we do not know. Being so closely involved in the proofing, he would have been aware that Gray's *Preface* referred to the absent Carter, and indeed to Holmes himself, as 'friends', and would perceive such an action as a questionable mark of friendship.[33]

These title page alterations offer a motive behind the survival of these proofs. Perhaps Wertheimer retained them so as to be able to show Parker why his original title page schema had not been adhered to; or someone else unknown held on to them in a quiet form of protest, as documentary evidence of the publishers' and printers' original intentions, and who was responsible for countermanding them. If this was why they were kept, whoever was responsible perceived what an injustice may be perpetrated by a simple type size.

Doubtless, the Parkers would have noticed and pondered the meaning of Gray's alterations. So too would Wertheimer's men, who had to do by hand the finicky work of making the changes Gray had demanded: breaking up the well-set original, extracting the offending type, inserting the author's preferred version, accommodating the missing line of print and the smaller typeface, and printing off the second revise Gray had asked to see before the final printing.

8

PUBLICATION

1858 and on

The last proof corrections for *Anatomy Descriptive and Surgical* were probably completed by late June or early July 1858, and the final printing was begun. The index had been created in parallel with the paginated proofs, and was probably among the last sheets to be typeset and printed, while the collators and folders and the stitchers were warned to be in readiness for the arrival of the printed sheets. Wertheimer may have been a large enough printer to employ his own expert folder-collators at Finsbury Circus, and almost certainly had his own heavy standing press for the final flattening of the folded quires, but stitching and the rest of the work of book-assembly were tasks done at the bindery. Folding, collating, and stitching were women's work, so just for a short period, during the assembling of the artefact of the book itself, the work fell into female hands.[1]

The manner of binding for *Gray's Anatomy* was typical of its day. The folded gatherings were placed in collated order in a sewing press, and three woven tapes spaced at intervals across the spine. These were oversewn in position by the stitcher, using strong thread and a special binders' stitch, which went through the back of the folded paper gatherings and over the tapes, through the gatherings and back again, zigzag style, to pull and hold both strongly together.[2] Once oversewn, the tapes were trimmed to leave about half an inch loose overhang, for reinforcing the connection of the book with the front and back boards. The stitched spine was then painted with natural glue (probably heated) and swiftly overlain with a linen or cotton gauze bandage—'mull'—the length of the spine, again with a half inch overhang at each side, and a paper or card fillet was laid in place smartly along the spine. This, adhering to the glue oozing through the gauze mull, was designed to form the inner curve of the hollowback spine, and helped firm the bond between the gatherings and around the threads, holding all firmly in place as it cooled. The gauze mull overhang on each side was trimmed at each corner at an angle, ready for glueing into the cover casing. The casing itself was made separately, from milled boards covered in bookcloth, embossed, and title-gilded on the spine. That too was given an internal fillet along the spine (to form the other curve of the hollowback) ready for assembly by means of the mull and tapes, glued in position across the hinges between book and boards. Finally, the fixed endpapers were neatly glued in place.

The Parkers would have been invited—probably by Wertheimer—to express a preference from a selection of publishers' binding cloths, so they would have had their say concerning the volume's outer appearance. The endpapers would have been their choice, too. The binding of the entire first edition of 2000 copies of *Gray's Anatomy* cost £110. 8s 4d., one shilling and one penny-farthing per book, which gives a good illustration of the low-wage economy of Victorian London, especially for women, and even for skilled work.

Where *Gray's Anatomy* received its binding is not known. There were a number of large edition binders in mid-Victorian London to whom Wertheimer could have subcontracted the work, whose production lines of manual workers were capable of turning round an entire publisher's edition in a matter of days. One, Westleys, was employing two hundred women in 1843, for folding and sewing.[3] Another of these large firms was Burn of Hatton Garden (roughly halfway between Wertheimer's and the Parkers) whom we have mentioned before, and who were responsible for binding two books for Wertheimer which survive in a special collection of rare bindings now in the British Library, one of which—like *Gray's*—has brown endpapers.[4] Numerous smaller binding workshops were tucked away in London's back streets, which could have been using a similar range of binding cloth and endpaper styles. Either of the two bookbinders on Wertheimer's list of creditors—George Astle of Coleman Street, or William McMurray of Lillypot Lane—might have been responsible. Both were within easy walking distance of Wertheimer's premises at Circus Place, Finsbury. *Gray's Anatomy* was probably bound by someone with whom Wertheimer had good relations, because when the Parkers asked him to print another scientific book in 1860, the binding was almost identical.[5,6]

Because *Gray's Anatomy* has always been a popular book, and much used, most surviving copies from the two thousand printed by Wertheimer at the time of its first appearance have been rebound by libraries, or by appreciative owners. So few copies of the first 1858 edition of *Gray's Anatomy* have survived in their original binding, that for a long time I feared I might never set eyes on one, to discover what the original binding actually looked like. But a fine copy is securely housed in the Library of the Royal College of Surgeons, in Lincoln's Inn Fields. Its first two owners kept it in good condition, and it was then donated to the College. It has not needed repairs, and it now has its own secure box. A tiny sticker by the front hinge shows that at some time in its biography, the volume was sold by the London medical bookseller, Kimpton's. Its back free endpaper is hand-inscribed discreetly at the top corner in pencil with the book's original selling price, twenty-eight shillings.

Gray's Anatomy: the original embossed bookcloth binding of the first edition.

JW Parker and Son's final tally for monies spent on publishing *Gray's Anatomy* included £111 for advertising, a few shillings more than the cost of binding, and a considerable sum for a publisher already facing deficit. If the first edition sold well, this money would be well spent.

It is difficult to track exactly where and when the Parkers' advertisements were placed with periodical publishers. There was probably a separate office ledger for such details which has not survived. Most paper copies of old medical journals which remain in libraries have been separated from their advertising pages during binding. However, two advertisements for *Gray's* have been found, both in the *Medical Times*, and interesting they are.

The first appeared just in advance of publication on the very front of the journal, on 28th August 1858.[7] It announced that *Anatomy Descriptive and Surgical* by Henry Gray FRS would be appearing in a few days, and listed Carter just as Gray would have had him appear, shorn of his current post. But it nevertheless managed to give Carter's work considerable prominence by stating: 'This work is illustrated by 363 woodcuts, from original Drawings, chiefly from Nature, by H.V.Carter, M.D., late Demonstrator of Anatomy at . . . .' The second advertisement was placed in a prime position, on the inner side of the journal's title page, in the following week's issue, on 4th September. It occupied an entire half page, and featured Carter's Professorship at Grant Medical College, Bombay, as in the original proofs.

When *The Lancet* reviewed the book on 11th September 1858, it gave full details of Carter's new posting. The lead time for placing advertisements in medical journals, especially for the busy time just prior to the start of a new medical school year, may have been a contributory factor in how Parker advertisements were worded, but there had been time since June (when Gray had altered the title page) to conform to the author's directions, had the Parkers wished to do so in their advertising. The survival of Carter's job-description in the larger half-page advertisement and in *The Lancet* review suggests either that these were sent out before the smaller advertisement in the *Medical Times*, which seems unlikely, or that the Parkers were averse to the emendation.[8]

Supporting the latter possibility is the fact that Gray's requirements concerning the emendation of the book's title page were never put into effect. The title page he wanted is not what emerged from the press room at John Wertheimer's. What did emerge as the final printed title page of the first edition of *Gray's Anatomy* does not conform to Gray's directions. Someone in a position to intervene saw what had been ordered, and countermanded it. The design layout of the page was altered slightly, and Carter's new job-title was deleted as had been demanded. But Carter's name and MD was set in a larger typesize than Gray had indicated, smaller than Gray's own, but not the insignificant size he had specified. What was ultimately printed was a compromise. Carter's apparent role with regard to the creation of the book was still diminished, and his current job-title erased, but the new version suggested a much closer approach to parity than Gray had wished for.

Royal octavo makes a tall book. The volume you are reading is in modern royal octavo, which is shorter than the Victorian variety by about an inch, paper sizes having changed in the intervening period. The advance copy of his own book that Mr Gray would have held in his own hands fresh from the publisher at the end of August 1858, was ten inches by six, it had 750 pages including the index, plus 32 pages at the front for the dedication, preface, contents, and illustrations. It weighed 3lbs 3oz. (1.45 kilos).

Had Henry Gray experienced a wave of satisfaction when holding his own book, and gazing on its gold-embossed title for the first time, he would have been more than justified. It was a fine volume. Flipping through the pages, he would have been able to confirm the correction of things he'd asked for, and appreciate how it looked, how it read, how *new* it felt, and smelt. The result of all that work could now be seen: rather a splendid thing, substantial and comely. It had a fine modern feel about it—and made *such* a bold contrast to fusty old Quain. The short title was gold-embossed, about a third of the way down the spine:

GRAY'S
ANATOMY
DESCRIPTIVE
AND
SURGICAL

with the publisher's imprint, 'JOHN W PARKER & SON' discreetly in gold, at the very foot. Its boards were clothed in a fine mid-brown embossed cloth, with a regular all-over pattern of small raised bumps, blind-embossed with a picture-frame motif to both front and back boards. The casing gave it a good feeling in the hand, solid, and not too superficially shiny. The pattern of embossed bumps (called 'bead grain', resembling an embossed version of the dotted 'half-tone' screens used for later newsprint images) was neither discreet nor low profile, but chunky and well spaced, giving the book at once a plain and practical but stylish feel.[9] The endpapers were in a Parker favourite light chocolate brown, harmonizing with the cover cloth.

It had taken some nerve to put Gray's name assertively like that on the spine. It invited comparison with Quain's triple decker (which had *'QUAIN'S ANATOMY'* inscribed on each of its volumes) and those of other anatomists whose names were integral to their spine titles.[10] This title looked like a bid for immediate fame, or at least a bit of clever marketing to promote the new book in contradistinction to existing standard works. Was it also perhaps a modern Victorian play on words, a deliberate literary echo of *Gray's Elegy*, in the mischievous way publishers and authors have when they endeavour to make book titles memorable? Resonances of the famous Georgian poem about a peaceful country churchyard may have lent the new book a sympathetic boost in some obscure way, its gentle irony possibly softening fears of death, lessening repugnance towards dissection, emphasizing the inclusive humanity of the book's subject, making the new title easy to recall.

The title page looked *almost* as Gray had wished, as did the dedication to Brodie, with a page to itself right at the start. Gray's *Preface* was quite clear, stating at the outset that the book was designed primarily for students, and that its emphasis was the practical application of anatomical knowledge. The first two sentences convey Gray's explicit intentions, and his businesslike brevity:

> This work is intended to furnish the Student and Practitioner with an accurate view of the anatomy of the Human Body, and more especially the application of this science to Practical Surgery.
>
> One of the chief objects of the Author has been, to induce the Student to apply his anatomical knowledge to the more practical points in Surgery, by introducing, in small type, under each subdivision of the work, such observations as shew the necessity of an accurate knowledge of the part under examination.

The redesigned titlepage at publication, in the first week of September 1858. By moving Gray's name upwards, and making it more prominent in appearance, it had proved possible to give Carter a decent sized typeface. Emphasizing THE DRAWINGS, and THE DISSECTIONS, also served to emphasize the parity of his contribution to the book, and his occupational by-line was now in the same-size type as Gray's, making it longer. Notice, too, that the small 'r' in Dr. on the first proof had been upgraded to a full capital. The entire shift upwards gave added space for the Parkers' insignia. The titlepage now registered the labours of four people, and Wertheimer's colophon on the reverse (the show-through lies between Carter's two roles) added the entire printshop. Gray had asked for Carter's name to appear in lettering the same size as the line of type above the insignia. Shown at 79%.

ANATOMY

DESCRIPTIVE AND SURGICAL.

BY

HENRY GRAY, F.R.S.

LECTURER ON ANATOMY AT SAINT GEORGE'S HOSPITAL.

THE DRAWINGS

By H. V. CARTER, M.D.

LATE DEMONSTRATOR OF ANATOMY AT ST. GEORGE'S HOSPITAL.

THE DISSECTIONS

JOINTLY BY THE AUTHOR AND DR. CARTER.

LONDON:

JOHN W. PARKER AND SON, WEST STRAND.

1858.

The *Preface* itself was short: only a page and a half. It gave a brief guided tour of the book, highlighting what Gray wanted the reader to be aware of, even if only from a cursory examination. All the major sections of the book were covered, from the inner bony structure of the human body to its covering and contents. The regions of the body, important to surgery, Gray explained, were interspersed with the bodily systems, and surgical anatomy also had its own section. He stressed that the book contained some microscopical anatomy, which would have made it feel very up-to-date.

The care involved in the book's descriptions was emphasized, as was the concise detail provided, and the surgical relevance of the anatomy. Gray followed up his descriptions of each section with mention of the illustrations, highlighting their abundance, their accuracy, their accurate lettering, and their size. He referred, repeatedly, to the accuracy, verity, and freshness of the book's information, and to its direct verification from ocular examination: 'taken from, or corrected by, recent dissections' is a key phrase. He pointed out what was new, such as the diagrams showing the lines of incision for muscle dissections, and what—hitherto complex—had been made more simple, such as bone development and the articulation of joints.

Some sections of the book, Gray said, presented material 'as in ordinary anatomical works', which placed his book within an existing tradition, but also prompted the idea that this book was extraordinary. The *Preface* was open about Gray's use of other authors' visual work, and used the same term 'copied from' to explain illustrations whose content was verified from the dissected body or were borrowed from or inspired by other printed sources. Reference was made to the use of material from three named authors: Luther Holden for his manner of showing muscle-attachments on bone, Gilbert Breschet for the veins of the spine, and Paolo Mascagni, for his elaborate work on the lymphatics. These credits placed Gray's book in the same league as the finest modern British and European works.

The impression given by Gray's choice of language is one of generosity (it is intended to 'furnish the Student' ... 'abundantly illustrated by engravings') a notion reinforced by particularity (each bone, each joint) and topic listings (in the sections devoted to neurology and regions) designed to convey the book's comprehensive scope. Emphasis on the practical value of anatomical knowledge is conspicuous. Gray concluded his *Preface* with a short passage of acknowledgements, which is best quoted in full:

> The Author gratefully acknowledges the great services he has derived, in the execution of this work, from the assistance of his friend, Dr. H.V. Carter, late Demonstrator of Anatomy at St. George's Hospital. All the drawings from which the engravings were made, were executed by him. In the majority of cases, they have been copied from, or corrected by, recent dissections made jointly by the Author and Dr. Carter.
>
> The Author has also to thank his friend, Mr. T. Holmes, for the able assistance afforded him in correcting the proof-sheets in their passage through the press.
>
> The engravings have been executed by Messrs. Butterworth and Heath; and the Author cannot omit thanking these gentlemen for the great care and fidelity displayed in their execution.
>
> Wilton-Street, Belgrave Square,
> August, 1858.

The *Contents* pages, which follow, present a clear and logical inventory of the human body, laid out in all its parts for the reader to discover the entire landscape of the body as presented in the book, in a listing of sections and subsections over the next sixteen pages, the main divisions being:

> Osteology (bones)
> Articulations (joints)
> Muscles and Fasciae (their coverings)
> Arteries
> Veins
> The Lymphatics
> The Nervous system and Organs of Sense
> The Viscera—Digestion, Circulation, Voice and Respiration, Urinary organs, Organs of generation
> Regional / Surgical Anatomy.

The subsequent eight-page *List of Illustrations* follows the same structure, presented as a simple numbered list of figures and their titles, giving credit if derived from another source, and pagination.

Anyone flipping through the book would have noticed not just the number of illustrations—almost every other page had a figure of some sort—but also their sheer size and variety. To an unknowing eye, the forcing of the images accomplished at Wertheimer's would not have been noticeable: what stood out was simply the number and size of the illustrations, and their bold originality.[11] Compared to any other book for the price, the illustrations in *Gray's Anatomy* were enormous, sometimes near life-size, and they explained themselves: they had caption headings, and carried their own legends. No proxy labels, no asterisks, no footnotes: this was extraordinary by comparison with other student works. We shall be considering Carter's contribution to the book shortly; here, what is needed is to imagine the feeling of a first impression, perhaps that of someone picking up the freshly printed book in a bookshop, and simply flicking through its pages, wondering whether or not to buy.

In a bookshop, a reader would have been able to compare the new *Gray's* physically against new copies of existing manuals of anatomy: against the new three-volume edition of Quain, whose pages looked just the same as the old edition, or single-volume works like Ellis's, Holden's, or Steggall's dissection manuals (all text only, lacking illustrations) against Wilson and Knox, both good, but with illustrations as small as Quain. Next to *Gray's*, they all looked disappointing. *Gray's* also stood up beside specialized books like Holden's *Osteology*, since it utilized the blackboard diagram technique he had used for bones, but for the whole of human anatomy.

Why a buyer would choose to purchase it is not difficult to understand. There really was no competition. The first edition of *Gray's Anatomy* knocked its competitors into a cocked hat. The daunting pages of type in all the other books were enough to put anyone off: you would only buy a book like that out of necessity. Gray's was quite different: larger, more readable, no footnotes, and with such good illustrations: illustrations which showed you where to cut, what to see. It lay well in the hand, it felt substantial, it contained everything required. And to top it all, it was cheaper than Quain!

The generosity of the book's conception conveys itself swiftly to those who look inside. The book's simple organization, and the simplicity of its page design, were matched by the simplicity of the pictures and of the prose. Carter's illustrations, by unifying name and structure, enable the eye to assimilate both at a glance. The distinct self identity of bones and arteries, muscles and nerves could be appreciated even by non-medical readers—anyone could lay their hand over the life-size diagram of the hand-bones and sense its rightness. Parents could more easily be persuaded to pay for such a text when they could see how good it was, new students could sense immediately that they could learn from it. Older students raised on Quain, and now coming up to finals, saw how confidently they might revise from *Gray's*; and seasoned practitioners fearing memory-loss clouding their own hard-won knowledge, could refresh their memory in an instant. Anatomy had never been so *legible*.

The sense of excitement for the subject such a fine new work generated cannot easily be conveyed. There was a feeling in the air in 1858 that the new era under the General Medical Council—just established that year, under the presidency of Sir Benjamin Brodie—was unifying the profession, while

Gray's was unifying a discipline; summing up what needed to be known, between two boards, concise and generous, professionally accurate, up-to-date, accessible and beautiful in its simplicity.

The reviewers were of one mind. Within a week of its appearance, *Gray's Anatomy* was acclaimed in *The Lancet* in such a way as to gratify all those whose efforts had contributed to its production. The book was facilitative,

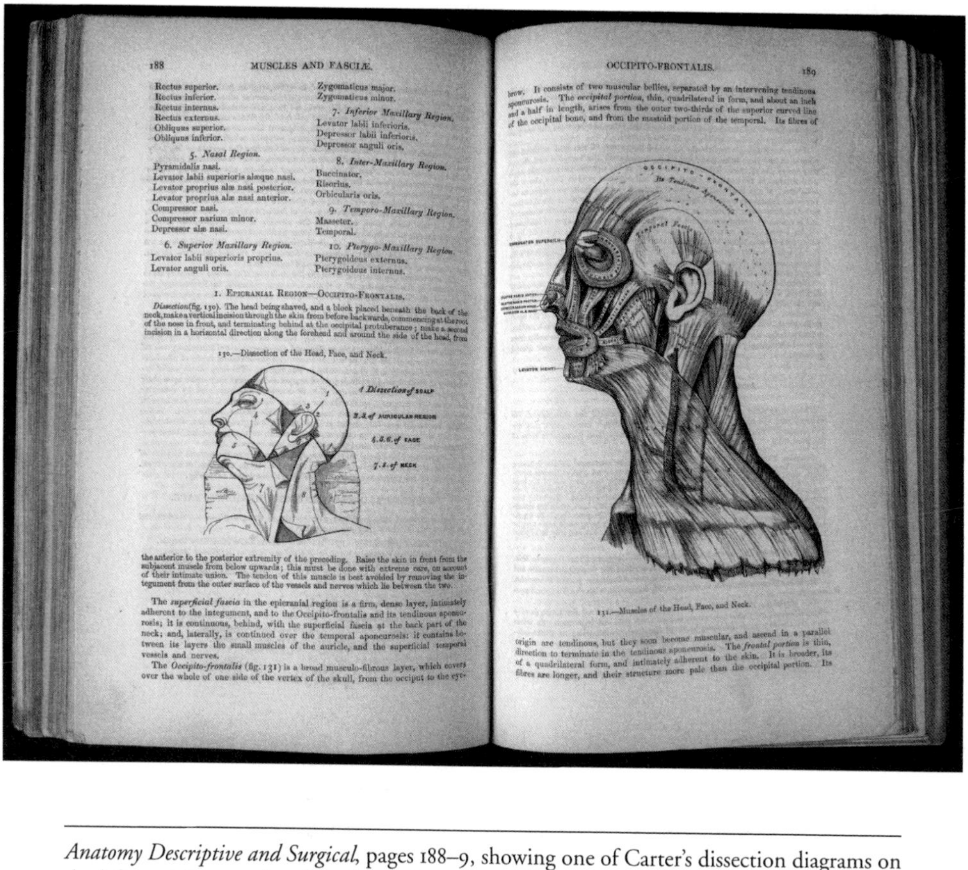

Anatomy Descriptive and Surgical, pages 188–9, showing one of Carter's dissection diagrams on the left, and the muscles of the head, face and neck on the right. The helpfulness of the one and the beauty, informativeness, and nobility of the other typify Carter's contribution to *Gray's Anatomy*. Note the rare use of proxy number labels on the dissection diagram, and the equally rare use of extraneous props in the wooden block under the head, which somehow manages not to look uncomfortable. Note, too, the Butterworth and Heath hand-engraved lettering of the key. No other dissection manual featured incision diagrams.

scientific, and practical, it reflected the 'highest credit upon the author and the draughtsman'.

> It is a work of no ordinary labour, and demanded the highest accomplishments, both as anatomist and surgeon, for its successful completion. We may say with truth, that there is not a treatise in any language, in which the relations of anatomy and surgery are so clearly and fully shown.[12]

The unnamed reviewer said the descriptions were 'admirably clear', and the illustrations were 'perfect'. The regional anatomy was praised as 'concise in expression, and full in detail', and the description of bone development hailed as 'most original and simple'. The reviewer emphasized the work's verification on fresh dissections, and stressed that only by actual dissection in the dissecting room, and by study in the hospital ward, could anatomy and surgery be 'practically learnt', hastening to warn the student who trusted only to books—'however ably written and accurately illustrated' that he 'will find himself, in the hour of trial, theoretically learned but practically inefficient.' Gray's book was designed to 'cherish and foster a diligent study of the subject'. The reviewer concluded:

> As a full, systematic, and advanced treatise on anatomy, combining the various merits of the volumes of many countries, scientifically excellent, and adapted to all the wants of the student, we are not acquainted with any work in any language which can take equal rank with the one before us.

The *British Medical Journal*, which was slower to arrive, concurred entirely, characterizing *Gray's* as 'far superior to all other treatises on anatomy', adding:

> the woodcuts, from the drawings of Dr Carter, a former distinguished pupil of St George's, are excellent—so clear and large that there is never any doubt as to what is intended to be represented. The dissections have been made jointly by Mr Gray and Dr Carter; and those who know these gentlemen will not need a better guarantee for their accuracy. In conclusion, we cannot avoid congratulating St George's Hospital on the production of such a book by two of its old pupils—a book which must take its place as *the* manual of Anatomy Descriptive & Surgical.[13]

Both Gray and the Parkers must have been well pleased to receive such reviews, and awaited with anticipation what the *Medical Times* had to say. In the meantime, sales had already started off briskly, and Gray began to think about the next edition.

Gray's words are the text of the book, the letterpress, the book's voice. His method of conveying information, and his choice of words favours brevity and clarity. The concision of his *Preface* is characteristic of the book's entire text. He makes no gesture of welcome, declares no manifesto beyond the commitment to surgical understanding in anatomical teaching. He simply dives in. At the end there is neither conclusion nor farewell. Were one to run his prose through a computer, as some literary scholars have done with Shakespeare and other great writers, I have the impression that—excluding the terminology relating to the scientific naming of anatomical structures—the vocabulary in the book would reveal itself as quite limited, mainly focused on directional terms—*behind, above, over, into, below, inner, superior, adjacent, upper, anterior, lateral, outer*, etcetera; morphological terms—*triangular, bulky, solid, oval, elastic, concave, smooth*; and passive verbs—*opens into, serves to strengthen, passes directly, passes through, arises from, transmits, lies along, lies between*, and so on.

There is no attempt at charm or narrative, some at variety, but the subject itself seems to defeat the attempt. Anatomy in this kind of detail is deadly unless you study it, need to know it, get to understand it intimately structure by structure, and find the interest in it. While doing it, the descriptions assume a value not accessible to the ordinary reader, and the meticulous directional and morphological terminology and rote-like phrasing of the prose can then assume a kind of poetry in its particularity, assisting memory, and prompting visualization. After a while, no doubt, every student might be able to describe a structure as Gray does, if they were to think of it in the same way, in the same order, and in the same terms. Gray had been studying and teaching anatomy for twelve or more years, so he may have had much of it by rote himself.

From the outset, Gray demonstrates his mastery of his domain—from the assertion of his own status (FRS and professional post on the title page, and the third person Author, with a capital A, of Belgravia, thanking others for assistance in the *Preface*) to the last word of text, he declares the field his own. The voice of the professional anatomical scientist fills the book: no personal pronouns, no doubt. Austere, all-knowing and authoritative. There is neither enthusiasm nor excitement, simply a steady stream of informative description, which starts at the beginning with the bones, and ends without faltering, with surgical anatomy.

Reading Gray's prose, page after page, one does grasp what an enormous task it would have been to create an entire book full of text on this massive subject. Yet the subject, for all its extent and detail, is also finite. There are only so many bones, only one brain, two eyes with lids to them, one diaphragm, and there was less known in the 1850s than now. Quain might have three volumes, but with strict brevity Gray could boil it down to one. Anatomy (like botany, as Anne Secord observes) is 'a classificatory observational science',[14] so a book which purports to give a student a good grounding in anatomy has to be a detailed and prolonged exhaustive three-dimensional inventory of the human body; there is no way round it. The key for Gray was to make every word tell, to keep it well organized, in small manageable chunks, well signposted, and to give variety to the student's interest with a diversity of accessible data.

The organization of the text of *Gray's* may in itself provide an insight as to how the writing of the book was accomplished—planned rather like a military campaign, I imagine. The dissecting season ran from October to March, six months, half a year, over two years, so 365 days. The book was designed as a teaching as well as a learning text, so would have been structured by the anatomy course 'Anatomy Descriptive and Surgical' which Gray had studied, and taught, for several years. The finished volume had 363 of Carter's drawings, and the unused spine engraving in the Butterworth and Heath proofs makes that almost one a day for a year. For Gray, those 364 illustrated topics had to have text written in a year: if each topic had two pages of text that would be say, 730 pages, and each topic could be seen from a regular series of aspects: definition, description, form/shape/structure, position/relations, actions. Two pages a day would take a year to write. We know from Carter's diary that the original time span planned for the work was fifteen months, so Gray looks to have built-in three months' slack, but it was nevertheless a punishing schedule.

There is certainly a recognition in *Gray's* of the limits of students' concentration. Examining the book, one sees that the policy of the breaking up of subject matter into digestible gobbets visibly works on almost every page, with varying sized typography, diagrams, tables, a short paragraph for each point (often only a couple of lines long), the regular use of italics for the main subject of each, and of course the illustrations and their captions. The presentation is logical, simple, linear, and clear; and the turgidity into which the subject can lapse is avoided by the policy of variety. The *Contents List* at the front makes clear the logicality of the multiple topics the book covers, each split down into smaller headings, often three or four sub-topics to a page, and easily findable.

The book itself is a three-dimensional solid which can be grasped and held in a moment. But there inheres that elusive fourth dimension: time. The contents need time to be grasped and held, just as they needed time to compose. So the object held has also a temporal quality through which it resembles a lecture series, laid out in linear fashion for the self-education of the reader. The *Lancet* reviewer had recognized the full teaching course the book contained when he warned against book-learning alone.

Gray is at his best when he speaks direct to the student about practical matters, and especially when he explains how to do surgical anatomy. His brisk directions for dissecting the area concerned in inguinal hernia, for example, are a model:

> For the dissection of the parts concerned in inguinal hernia, a male subject, free from fat, should always be selected. The body should be placed in the prone position, the abdomen and pelvis raised by means of blocks placed beneath them, and the lower extremities rotated outwards, so as to make the parts as tense as possible. If the abdominal walls are flaccid, the cavity of the abdomen should be inflated by an aperture through the umbilicus. An incision should be made along the middle line, from the umbilicus to the pubes, and continued along the front of the scrotum; and a second incision, from the anterior superior spine of the ilium to just below the umbilicus. These incisions should divide the integument; and the triangular-shaped flap included between them should be reflected downwards and outwards, when the superficial fascia will be exposed.[15]

Many readers may not perhaps recognize the technical terms here, but effectively what Gray is saying is to make an L-shaped cut from the front crest of the hipbone across to the navel and down to the pubic area, and to fold back the skinflap at an angle across the groin, to rest on the upper thigh. If you can imagine the dead body and the cut as he speaks, you are doing what all anatomy students do—imagining/remembering the geography of the area, and what the cut reveals. After opening up this area in the reader's imagination, Gray dives in again to explain what is in there, so there is a process of varying focus, too, sometimes on the body's surface, sometimes a deeper excavation of its innermost strata and workings. Part of the reason, perhaps, that Gray intersperses regional anatomy throughout the book is the varying focus it allows, adding to the interest, the motivation, and at the same time providing a dimensional mesh which marries the various bodily systems to the regions of the body.

Gray chose to feature the most common operations of the day: fractures and amputations, the correction of club feet, hernias and lithotomies, ligatures, simple operations on the eye, and tracheotomy. His emphasis on surgery served to reorient the priorities of those studying anatomy by the use of his book. A description alone—this is how it is—is less memorable than a description plus an appeal to reason—this is how it is and you need to know this *because.* The prioritizing function of surgery helped make *Gray's* fresh. In *Gray's,* surgery was not just an end-of-the-third-volume-afterthought, as in Quain, but the very reason for doing anatomy. Quain's placing of surgery at the end of anatomical teaching was perhaps his way of discouraging dangerous foolhardiness, but Gray's insistent integration of the *use* of anatomy was a teaching technique which kept the audience awake; a way of keeping the vital importance of the anatomical knowledge in mind throughout the learning process, and of preparing the student to visualize its future use. It treated students as adults, as future surgeons.

The importance of words in the anatomical tradition was high. An old Hogarth engraving of the dissection of a criminal at the Old Bailey shows a reader reading out loud from a book, a man with a stick pointing at what was being described, and a cutter working to dictation, doing what he hears he must, before a select audience.[16] By the mid-eighteenth century, these were ceremonial events, rehearsing an obsolete manner of dissecting dating back to the middle ages. By the same date, elsewhere in London, bodies were being snatched from graves for hands-on dissection, where practitioners learned their anatomy by doing it themselves, watching, listening, reading, cutting, looking, learning. By Gray's era, as we have seen, the bodies were ready-provided, the words had proliferated, pictures were available, but the cutting and looking still needed to be done for the learning to take place. The directional and descriptive details were crucial, because if the anatomy was misunderstood, real lives and limbs might be lost. The responsibility shouldered by Gray, honing the words down to a minimum, is therefore great.

But Gray's writing didn't spring like Athena from the head of Zeus. There were innumerable precedents in the 1850s for verbal descriptions of human anatomical structures, to which he had access. As we have seen, a considerable number of teaching texts published during the previous fifty years had no illustrations whatever, or were illustrated only minimally, so the particularity of verbal description in the teaching of anatomy was a very well developed scientific specialism. Gray had many precedents to work from. It seems clear from the *Lancet* review, that originality was not regarded as the most important point about *Gray's* text at the time; what was regarded as significant about the book was that it brought together the best aspects of a number of works, including continental ones. It may be that Gray's 'professional' voice served a similar function to the copyist's hand in Gray's *Spleen* essay—that it served to smooth over and unify a variety of contributing voices.

Gray's voice in *Anatomy Descriptive and Surgical* is one of studied neutrality. This kind of voice seems to be an acquisition of the professional scientist of his day, and has parallels with what the literary scholar Audrey Jaffe has analysed concerning the 'Omniscient Narrator' in the Victorian novel. She characterizes this non-being as belonging to 'a series of cultural phenomena through which ... knowledge itself—is coded as white, male, and middle class'. Jaffe does not address the voices of Victorian non-fiction, but she does allow for the possibility of congruence in other fields, and it does seem curious that although Gray is not an omniscient narrator (little other than the dissection process is narrated in *Gray's*, rather, it is all individually itemized, inventoried) he does seem to want to appear omniscient, all encompassing, with the ability to 'manage a vast potentially unmanageable amount of information'.[17]

The voice in *Gray's* is disembodied, its guidance round the human body has that stamp of authority which, as Jaffe says, resides in a tension between 'a voice that implies presence and the lack of any character to attach it to'. Gray himself remains quite opaque in the pages of his own book, revealing nothing personally of himself beyond his status and ownership of the entire domain at the outset, and his desire to remain opaque. Boz dissects his subjects, and so does Gray, but to him their characters and biographies are of no consequence, so long as they end up on his slab. His interest resides elsewhere: his role is to know and reveal more about his subjects than they ever knew about themselves.

The omniscient enumerator, or itemizer, in *Gray's*, shares something else with the omniscient narrator of Victorian fiction, which is what Jaffe describes as 'mobility', and his subjects truly epitomize what she sees as the fixity of fictional characters. The agility of the role assumed by Dickens when narrating his tales—looking through walls, into locked cells, quiet bedrooms—has its parallel in Gray's ability to cut where he likes, look where he likes, roll the body over, and excavate where he likes, down to the bone. *Gray's* is rooted firmly in the physicality of the dead body, but it accesses too the living imagination, bypassing natural disgust and repugnance, subverting Victorian modesty, embarrassment, or prurience, displaying its author's mastery of the physical world of the dead and the living, and by means of emotionless language passes on an imaginative grasp of the human corpse at its most intimate and vulnerable which somehow also encompasses it—and by implication the author and the reader—as an abstract living entity.

Gray's metaphors are functional and mechanical, architectural, structural, botanic, and geographic—we have grooves, furrows and ridges, fissures and apertures, pivots, levers, branches, and arches. At the same time, the directional and morphological clarification of attachments and movements, flows and functions also conveys a balletic quality of great delicacy, detailing specificities so carefully that the precision of the body's most minute and slender mechanisms are appreciated in all their intricacy and frailty. How much of this is actually Gray, and how much derives from the observations of his predecessors it is impossible to know. One would indeed need the computer to help inform us. Nevertheless, as a new work in the field of anatomy, *Gray's* lays before the reader a wide and imperative inventory of such extraordinary confidence and sweep that there can be no doubting its force.

When, back in 1852, the 21-year-old Carter recorded in his Paris diary:

> design in head for making drawings of arteries &c.

he was describing what Eugene Ferguson has memorably described as the workings of his own 'visualizing faculty', the 'nonverbal thought and nonverbal reasoning of the designer, who thinks with pictures'.[18,19] Carter was certainly a picture-thinker, but his psyche was formed from the united love of both father and mother: his father the artist, his mother the bearer of the Word.

Both word and image in *Gray's Anatomy* placed a strong emphasis upon the practical, upon exploring and doing for oneself, an interlacing of intention which reveals the two young anatomists' concord in their practical desire to share their anatomical know-how with others younger than themselves. Carter's illustrations complement Gray's prose in the most obvious and painstaking way. Obvious because as Gray describes, Carter is there to show, and the placing of most of the images on the pages is arranged as appropriately as Wertheimer and Parker between them could make it. Painstaking, because it is obvious to all that have eyes to see, that Carter had spent hours and hours

---

The arteries at the base of the brain, from the first edition of *Anatomy Descriptive and Surgical*. A good example of the simple presentation of the complex so typical of Carter's work. The image overhangs the letterpress area by 8mm.

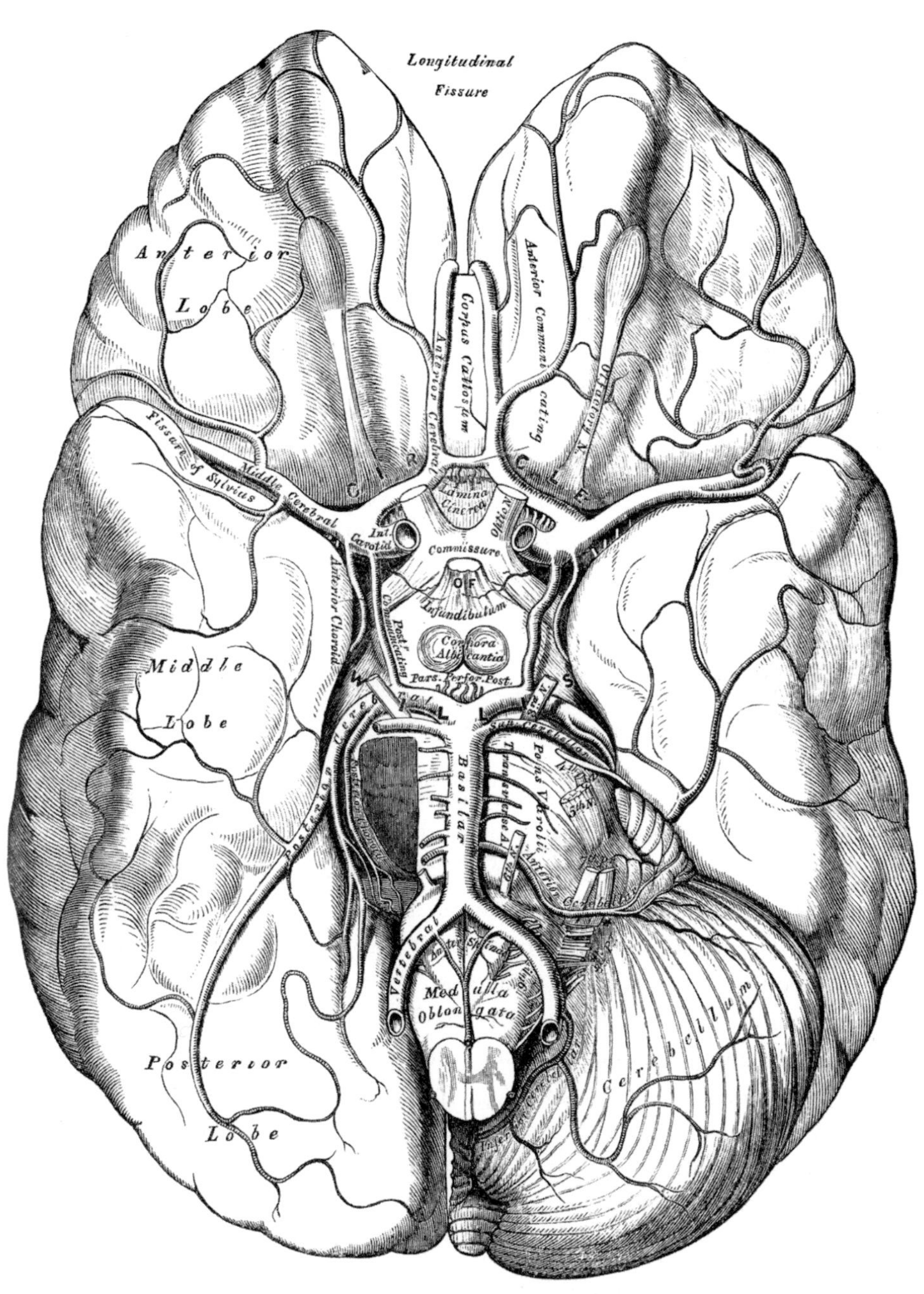
Longitudinal
Fissure
Anterior
Lobe
Corpus Callosum
Anterior Cerebral
Anterior Communicating
Fissure of Sylvius
Middle Cerebral
Lamina Cinerea
Int. Carotid
Commissure
Infundibulum
Anterior Choroid
Post. Communicating
Corpora Albicantia
Pars. Perfor. Post.
Middle
Lobe
Basilar
Pons Varolii
Vertebral
Medulla Oblongata
Posterior
Lobe
Cerebellum

of careful effort on these images, doing his utmost to make them as good as they could be.

The book is a sort of anatomical look and say. One of the most distinctive qualities in Carter's illustrations is that they are a look and say all on their own. They are definitely visual, but also verbal. When you open an illustrated page in *Gray's*, especially when the image is a goodly size, it's hard to look first at the text. The eye is drawn to the illustrations, with their physical demonstration of what Gray's words endeavour to describe in the adjacent letterpress. Carter's illustrations lay out the body, *showing* behind, above, below, concave, smooth, passes through, arises from, and lies along; and at the same time *telling* trapezius, clavicle, sternum, hyoid, sterno-thyroid. And as you look, you're already down deep in the structures, discerning their names and shapes for yourself as you go. Even without being conscious of thinking, the eye takes in the structure and its name at the same time.

Contemporaries like the *Lancet* reviewer, who immediately recognized the quality of Carter's images, could see that they were designed to assist the student know what they were seeing, and memorize name and structure. Carter's attentiveness is not only to the anatomy, and its accurate portrayal and labelling, but also to the student's needs, the student's thought-process, the learning process. It was only six or seven years earlier that Carter himself had discovered that he liked dissecting, but needed a guide.[20] Carter had created the guide he himself had sought, now, for students he might never meet.

While the words of the letterpress activated the mind's eye, Carter's images probed other routes into the memory, activating the eye's mind. We know now that the excellence of his images in part resides in the fact that they activate different parts of the brain at once, working to store and trigger visual *and* verbal memory, helping the reactivation of memory by mutual support, as it were, with word and image. Embedded as his illustrations are in appropriate positions in Gray's letterpress text, there is a further reinforcement in the letterpress alone, and iteration goes on during the course of verifying the letterpress against the drawing. So checking between illustrations and text during dissecting would help any student reader get well on the way to remembering the details. It's called driving it home.[21]

Just as we saw the way in which the text has a temporal dimension of being a series of lectures, Carter's drawings offer themselves up as if they were the blackboard drawings that go along with the lecture, built up during the presentation. This is what is left on the blackboard at the end of the lecture.

Or, closer still perhaps, these are entire body parts slowly and painstakingly dissected over time, with all the elements of anatomical structure one seeks for, already lying exposed to view. But Carter's paper prosections are not merely presentations of the parts, ready-made dissections, as it were. The anatomical structures in his dissections are not passive, nor dead meat: they reveal their own identities. Some images in the anatomical tradition show animated corpses lifting their own skin, or raising flaps to show internal organs to the viewer, trying to engage the viewer eye to eye, in a curious form of collusion between the living and the dead. This is not at all Carter's way. Like Gray, his focus is not upon the person of the corpse, but their anatomy. By presenting the names of anatomical structures upon the structures themselves, Carter's work ensures that the very act of looking involves the act of reading along the morphology of each structure, actively involving the viewer-reader in identifying the part and its lie.

In other illustrated textbooks of the 1850s for medical students, as we have seen, the illustrations were small, and their structures often difficult to make out by comparison to the real human body on the dissecting-room table. Matters were seriously exacerbated by the use of proxy labelling (a, b, c or 1, 2, 3) or arrowed proxy labelling (a, b, c or 1, 2, 3 arranged around the structure at the end of an arrow, often in a nimbus) which involved the eye in irregular ricocheting journeys round the page, between the proxy letter, the structure, and the legend, the last usually in a footnote set in small type, but often even harder to find through being integrated with the letterpress text. The hapless eye wastes its effort in navigation, and finds itself astray, instead of seeing what needs to be seen, and learning what needs to be learned.

The naming of the parts in anatomical illustration is a fundamental and ubiquitous problem. A deep and serious conflict has been fought out over the centuries between presenting the parts visually, and their labelling. Very often the conflict has been exacerbated by problems of size, and the practical printing difficulty of placing legible lettering on or near structures. But that cannot have been the only reason why body parts have traditionally gone unlabelled in anatomical illustrations. There are very large anatomical atlases whose size would not have precluded it. Proxy labelling is a convention adopted consistently throughout the long history of anatomical illustration, even before the illustrations by John Stephen of Calcar in Vesalius's *Fabrica* of 1543, which seems to stem from a desire to foreground and valorize the aesthetic of the visual, by subordinating to it the information transmissible by means of type and word.[22, 23]

The illustrations in *Gray's Anatomy* have been denigrated by those whose predilection seems to favour traditional proxy labelling. But to accuse Carter of 'greyness', and to compare his images to engineering drawings, as the art historian Martin Kemp has done, not only seems to miss their artistry, but is also historically imprecise: engineering drawings traditionally used proxy labels too.[24] Not to see their artistry is to miss their art. But Kemp is not altogether wrong to notice affinities between Carter's anatomical illustrations and technical drawing. The illustrations in the great eighteenth-century French project of the *Encyclopédie*, as Daniel Brewer has observed, are a 'visual counterpart to utilitarian rationalism', 'not just savoir, but savoir faire'. The *Encyclopédie* illustrations, Brewer says:

> disassemble the finished machine or building, dividing it into its parts, and orders them into a series, the logic of which leads from part to whole, from tool to its use, from raw material to finished product ... The gaze of the reader–viewer retraces ... the stages in the object's transformation from matter to object, from thing to property.[25]

This might indeed be one way to describe Carter's deliberately pared down outline drawings of the human body. His simple schemata for demonstrating dissection procedure offer themselves in such a way—their dark lines signifying cuts, and turned back corners of skin showing from where to peel back, and in which direction. These are 'how to' drawings indeed, but contrariwise to Brewer's, transforming person to thing, object to matter: these are images of dismemberment in process. The end product here is never shown. The property involved, however, is portable—the book carries it, and helps transfer it. The reassembly—as in all dissection—takes place in the student's mind. The kinship with technical drawing resides in the fact that in both, conventional aesthetics are happily subordinated to the carriage of information, or rather, that their aesthetics encompass both image and word: the simplicity of unadorned practicality is a form of beauty in itself.[26]

To bring words into the heart of the visual field in anatomical illustration was an unconventional move in a printed anatomical work.[27] It can be seen both as a revival of a much older tradition, and as a direct engagement with contemporary Victorian graphic techniques. Prior to John Stephen's magnificent illustrations to Vesalius's *Fabrica*, woodcut images of human anatomy did feature words on structures, so Carter's reintroduction of the word to the anatomical image can be seen as a revival of mediaeval anatomical educational practice. Carter's work exhibits, too, the visual influence of contemporary graphic devices vibrant in the mid-Victorian era: above all geographical and geological mapping, the enormously popular 'outline style' championed by the Pre-Raphaelites (and used, for example, in numerous images in the Art Union's *Catalogue* of the Great Exhibition) and the animated lettering of caricature and advertising, which Gerard Curtis has memorably called 'visual words'.[28]

For a person as visually senstive as Carter, the graphic vocabulary of the 1850s was unavoidable. Posters on walls, advertisements on omnibuses, sandwich-board men covered in advertisements: the liveliness of the graphic arts, especially the interwoven words and images of advertising, were among the most evident characteristics of London street-life. A similar confluence of words and images permeated mid-Victorian caricature. A cartoon in *Diogenes* in 1854, for example, entitled 'The Jug of the Nightingale' featured a bird with a bonnet flying past the cliffs of Dover, bearing a Victorian jug, on which was written 'Fomentations', 'Embrocations', and 'Gruel', a visual play on words/image for Florence Nightingale's expedition to the Crimea. Another of the same year, 'Publication Extraordinary', showed a spectacular spoof 'Illustrated London Shirt', a wearable newspaper, shown printed all over, sleeves and collar included, with news: 'Latest Intelligence' on the lower belly, army and navy news by the shoulder seams, arts down the centre, law and police news on the tails, Poet's corner under the left arm, an anatomical figure composed of blood vessels advertising Morrison's Pills on one collar, and a phrenological diagram of a human head on the other. The phrenological head is itself a noticeable forbear of the tautological presentation of word on human image, found in Carter's anatomical imagery. Considering how ubiquitous these kinds of hieroglyphics were, it is really rather extraordinary that large wood engravings featuring directly labelled anatomical structures had not appeared before the mid-1850s in student texts.[29]

Carter was mercifully free of the baggage of an art school education, untrammelled by the doctrinaire insistence on forcing nature to conform to the classical profile and the statuesque pose, of finding reality disappointing.[30] Such conventions emerge most curiously in the anatomical tradition—from Vesalius to Quain—because anatomical artists were often 'classically' trained, or because the ideal body was held to derive from classical statuary. Carter is free of such prejudices—though he was surely aware of them. He saw, like the Regency anatomist John Bell, 'a continual struggle between the anatomist and the painter, one striving for elegance of form, the other insisting on accuracy of representation'.[31] Bell had complained that anatomical illustration rarely offered consonance between image and reality, 'close comparison which the student seeks, and misses, with disappointment which is continually renewed'.[32]

John Bell had derided what he called 'plan' anatomy, likening it to anatomizing a statue: 'it is a figure which the student can never compare with the body as it lies before him for dissection.'[33] Carter may have sympathized with Bell's opinions, but he did not follow his practice, which was to create an extreme form of dissection-room realism, *winceworthy* in its brutality. No doubt, the arty-fication of dissection was not to Bell's taste, but his reaction against it went too far. His dead are remnants of the human, meaty, mangled lumps cut to dangling shreds. Barely recognizable, human body parts lie awkwardly, in positions of unwarrantable intimacy and pain, hooks and chains claw and hold human skin, ropes suspend human joints, decapitated heads have faces with mouths agape, look agonized.

Bell's dissecting room procedure is without pity, or the slightest concession to human dignity. The closest equivalents in the European art tradition are the terrible images of Goya's *Disasters of War*.[34]

Such unhappy imagery would never have appealed to Carter, to whom the body was something altogether more cherished. Carter's reaction to his mother's death, in the midst of his work for *Gray's Anatomy* at Kinnerton Street, raised an issue that I did not deal with at that time, but shall try to do so here, a question that seldom receives discussion or elucidation in works on anatomy or its illustration: how does the anatomist perceive the dead body they dissect? Carter's diary is silent concerning his own—or anybody else's—feelings or thoughts towards the human beings whose bodies found their way to Kinnerton Street. This is something never discussed verbally at any level, personal, religious, or existential. But there are moments when he does give glimpses of his thoughts in words, which reinforce what is evident from his images. During his Studentship at the Royal College of Surgeons in 1853–4, Carter occasionally wrote down his reflections about his daily activities, revealing his emotional and devotional involvement with the objects of his studies, which (as we saw in Chapter 2) derived from every corner of the natural kingdom, from fossils and exotic creatures, to foetal abnormalities and other human curiosities.

> trust am getting nearer to God—all my ideas tend to Him & strive hard to understand Him—rejoice in his works and trust ever to regard them as His handiwork & look through them to Him. ... have systematically cultivated love of natural objects—so much so that [I am] almost sentimental at times

> ... I know not however how far severe study may be carried—with perfect innocuity to higher interests—of that a man's own conscience is probably the only guide—when rectified and enlightened from above. [concerning a devout man's studies]—all his acts are acts of worship inasmuch as they are prosecuted in a right spirit & for good ends, and are good, even in their operation, as well as conception.[35]

Carter is not as declamatory as the French anatomist Leon Jean-Baptiste Cruveillier, who regarded a work of anatomy as 'the most beautiful hymn which man can chant in honour of his Creator', and 'every acquisition of knowledge is a conquest achieved for the relief of suffering humanity',[36] but Carter did regard the close study of nature as an exploration of God's works, and ethical research in medicine as an act of worship. After his mother's death, Carter's reactions to her body were written out simply:

> the poor weak frame still retained its warmth when Lily saw it. Great was the shock of all. She was laid out in the same room [she had made the shroud herself several years ago—all was ready], when we saw her the vital spark was alone wanting to realize her usual calm expression. ... the earthly tenement of my dear Mother's soul rests now in the upper part of the new burial ground, adjoining the church.[37]

While the words Carter used to record his feelings are conventional, almost formal, it is also evident from the way he uses them that he genuinely thought of the soul as separable from the body, as a *vital spark of heavenly flame*, as the popular funeral hymn had it.[38] The body (here an object of pity) was its earthly dwelling-place, rather as it had been portrayed in Parker senior's book, *The House I Live In*, or as it would be in the words of another Parker author, the poet Coventry Patmore:

> Little sequester'd pleasure-house
> For God and for His Spouse;
> Elaborately, yea, past conceiving, fair,
> Since, from the graced decorum of the hair,
> Ev'n to the tingling, sweet
> Soles of the simple, earth-confiding feet,
> And from the inmost heart
> Outwards unto the thin
> Silk curtains of the skin,
> Every least part
> Astonish'd hears
> And sweet replies to some like portion of the spheres.[39]

This view of the body is one which is without prudery, honest, open-eyed, and matter-of-fact about the spiritual inherency of its corporeality. A little later in his life Carter delivered an interesting paper on the way in which the formation of bone in the human foot matches the stresses of weight-bearing and pressure, which he prefixed with the hope that his medical readers:

> will partake of the interest which the subject itself affords to us as medical men, and also of the pleasure naturally felt by every educated person in tracing the evidences of design universally present in animated nature.[40]

By 'design', Carter clearly meant his readers to understand that he was talking about what Svetlana Alpers calls locating 'God in the details of his creation', which is, I feel, a key characteristic of Carter's work.[41] Huxley might have howled out loud in derision, but there can be no doubt from Carter's words that his view of the human body was deeply respectful, a view that is expressed equally well in his images.

It would be true to say that Carter did not wholly embrace either realism or the notion of the ideal in his anatomical images. What he has done is aim midway. He portrays bones and organs, hands and feet as individually as they were created, but de-particularizes them: he simplifies out in his drawings all the features of a specimen which it might not readily share with others. This simplification process allows the images to partake of a greater universality than if they were completely realistic renderings, and lends them a dimension of heightened presence not unlike the notion of archetypes so championed by Richard Owen.[42] Carter's body parts are real human body parts, carefully delineated, but they also carry an archetypal quality which a slavish documentary realism could not endow. They show the anatomical structures of everyman, and of everywoman.

From the number of acknowledgements he made in his own List of Illustrations in *Gray's Anatomy*, it is evident that Carter had searched widely, and studied carefully, among anatomical illustrators. But what we see in his acknowledgments are the anatomists he used, not those he rejected. An example of what he rejected is evidenced in his illustration of the uterus and its appendages, which occupies about a quarter-page, and shows the womb, fallopian tubes, and ovaries separate from the body, without geographical context other than the vagina. It is healthy and bloodless, shown diagrammatically, labelled directly on the important structures, and appears with no external shadow.

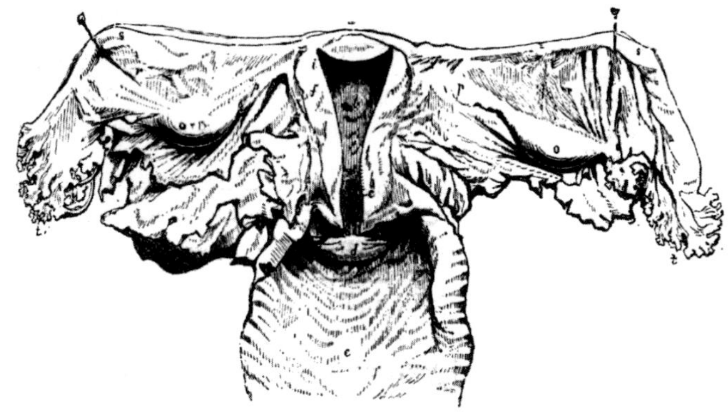

Carter's illustration offers a direct contrast to the same subject in Quain, which he had probably seen a thousand times during learning and teaching.[43,44] Quain's version is in realistic mode, showing a real woman's womb separated from her body, the key parts enveloped in a ragged curtain of flesh. The entire specimen is nailed to a board in two places. The board has been propped up vertically, so gravity exerts its force on the entire specimen. It is not easy to look at, since it is, essentially, a crucifixion. Quain's version exemplifies the static sadism which anatomical images often re-present, which is absent in *Gray's Anatomy*. Carter's image is a diagram of the life-giving womb, it does not—as in Quain's case—convey the impression that a woman has died for us to see it.

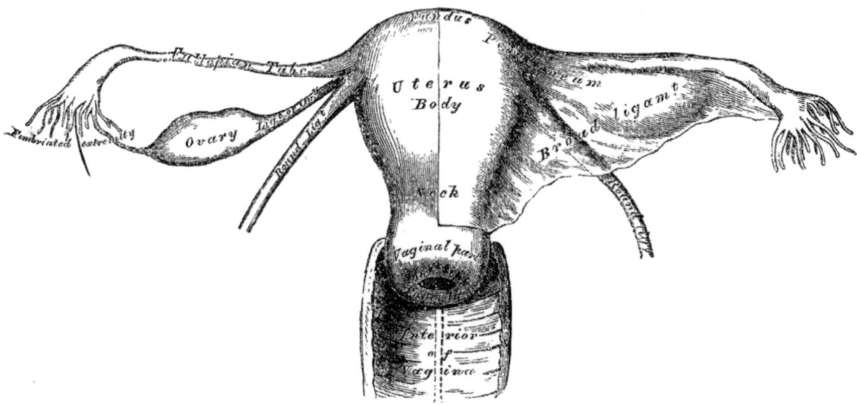

Quain and Carter compared: the Uterus from Jones Quain's *Elements of Anatomy* (1856) and Carter's from *Anatomy Descriptive and Surgical*. From the author's collection.

Carter's dissections are products of their era, and something we must remember when we look at his images, is that *Gray's* was the first major anatomical textbook to be created after the arrival of chloroform. These are painless dissections.

On a single occasion at St George's, Carter fainted during an operation. He was embarrassed to have done so, explaining in his diary that he had not breakfasted that day. He also mentioned that the operation had been an amputation without chloroform, upon a child. The subtext to Carter's faint is his tenderness of heart, and his own sense of powerlessness to prevent the child's agony. In the book he had the opportunity to administer chloroform figuratively to his subjects, and if he did so in an endeavour to prevent fainting among the audience, his audience was probably most grateful for it.

By comparison with other anatomical books, neither the subjects nor the viewers experience pain in Carter's rendering: his images minimize the flinch, and maximize the interest for the viewer. Moreover they show their human subjects with eyes and mouths closed, as if asleep, chloroformed, or as if they had been respectfully laid out by a caring person—which of course, they had.[45]

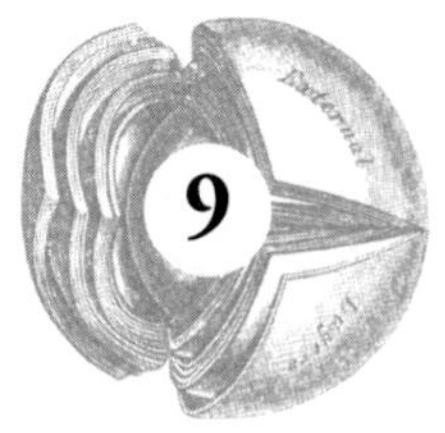

CALAMITY

1860–1861

Professor Henry Vandyke Carter, of Grant Medical College, Bombay, continued making occasional notes in his diary. In mid-October 1858, he recorded the arrival of his copy of *Gray's Anatomy.* Parker had been good enough to parcel it up and send it to him via the overland mail. Carter doesn't mention a covering letter, though there probably was one. As usual, his note is succinct:

> The Book is out and looks well—had a copy sent me by [Parker] but feel that except under some ruling mind shall do nothing self—too superstitious—analyse life on too small a scale.[1]

Carter was still dissatisfied with himself, and still seeking Providence. He had arrived in Bombay the previous March, after a journey he had thoroughly enjoyed, and was sent to serve up-country. But he was recalled almost immediately and installed at Grant College: 'The very thing I had wished for.'[2] The first College prize-giving he witnessed had prompted him to remark respectfully in his diary: 'sharp fellows, these Parsees and Hindus.'[3]

Carter was twenty-seven. He wanted to marry, but thought he lacked the pluck to find a wife. By October 1858, when the book arrived—despite the change of scene and role—he was suffering his old lassitude. He thought the state of affairs at the medical school resembled the situation at St George's, and feared he would do little in India: 'may not stay here long—such a fool!'[4] Time was to show him otherwise.

Back at St George's, Gray had waited well into the new year for the review from the *Medical Times and Gazette.* The journal had taken *Anatomy Descriptive and Surgical* so seriously that it had cleared four pages for the review, which appeared at last in early March, 1859.[5] But Gray would have grasped in moments that the long-awaited piece was not long on admiration: quite the reverse, in fact: any author receiving such an onslaught would become ashen with discomfort. Despite its justice in some respects—one cannot help but feel for him, and hope he was at home when he first read it.

Like the positive reviews in the other journals, this one was not signed. Some commentators have dismissed its content entirely, on the grounds that it was rooted in personal animosity or jealousy. Charles Mayo Goss attributes it to jealously, or to personal partisanship; Irvine Loudon calls it both savage and ridiculous.[6] But even if we reach the same conclusions, the review deserves to be taken seriously. Its criticism was both detailed and copious, and provides insight concerning aspects of the book some contemporaries regarded as irreparable flaws. First, then, a condensed version of the review itself is presented here, then follows a discussion of the main grounds of its censure.[7]

> It is a serious thing to review a book like this of Mr Gray's. One sits down to the task with the oppressive feeling of sadness which comes over a man when he has seen a wrong done, when he finds the occasion of such wrong has been unnecessarily sought for, and that the ill deed is after all, ill done. It is grievous not so much as to do what is fitting under such circumstances, as to come in contact with such a necessity; especially when it compels us to turn reproachfully upon one whose first essay deserved and received our full commendation.
>
> Mr Gray has published a book that was not wanted, and which, at any rate ought not to have been dedicated to Sir Benjamin Brodie. It is low and unscientific in tone; and it has been compiled, for the most part, in a manner inconsistent with the professions of honesty which we find in the preface. It is not even up to the mark of existing vade mecums. Mr Gray has worked under a false estimate of his duties as a teacher. A more unphilosophical amalgam of anatomical details and crude Surgery we never met with.
>
> [Gray] is careful enough to borrow from Wilson's Vade Mecum 'Plan of the Relations of Arteries' simply altering the square or oblong centre-piece into a round. But where was the need of this? Elementary books we must have; but let them be the first rounds of a scaling ladder, not the useless ledge of a treadmill.
>
> Is it honest to compile, without citation of originals, in such a manner as to lead [to the] inference that the facts and conclusions so positively stated, are

> Mr Gray's own? [The reviewer accuses Gray of systematically keeping the names of] our great anatomists & physicians out of sight [there follow the names of some of those on whose shoulders the reviewer thinks Gray stands].
>
> [Gray] furnishes us with a great many statements of facts, unattested, simply adopted. [The reviewer cites the example of the coverage of bone development in *Anatomy Descriptive and Surgical*:] If we did not know better, we might be induced to conclude Mr Gray the first, and a most profound, investigator of this obscure province of developmental anatomy. Confronting, however, his 'simple method' abstracts with the corresponding parts of the textbook of Quain and Sharpey, it becomes manifest that he is entirely indebted, not only to others for all his material, but, as well, to them for the preparation of it.

The reviewer suggested Gray owed an unacknowledged debt to *Elements of Anatomy*, and accused Gray of 'ablating [the] results of Quain's labours'. Numerous instances of borrowing were listed, and the reviewer provided parallel texts within the review to demonstrate the close similarities between Gray's text and that of *Quain's*. The reviewer continued:

> Our difficulty has not been to find such passages, but to know which to select from many equally startling. Those who take the trouble to compare the plan of the book with that of Quain, and will examine the two books together, chapter by chapter, section by section, paragraph by paragraph, sentence by sentence, parenthesis by parenthesis, and oftentimes word by word, must admit that the interests of the Profession demand a full exposure.

Gray's Anatomy, the reviewer said, was an example of 'debased compilation and unscrupulous assumption', it was 'not honest'. He asked:

> How should the student avoid the idolatry of Mr Gray, when his broad sheets interpose between them and the source of his inspiration? [The reviewer says the omission of preceding anatomical investigators/discoverers is a] naked treatment of the subject [which] may give strength of memory … but … never strength of mind.

The review closed with two parting shots, both intended to wound. The first impugned Gray's intellectual standing: 'Mr Gray is neither an Owen nor a Huxley, and he cannot help it.' The second attacked his methods of work:

> Curiously enough, the same plan of reduction from the text-book of Quain and Sharpey is followed in the account of the spleen, where, at least we might have expected some degree of originality. Such a coincidence would almost lead us to suspect that Mr Gray must have employed some literary day-labourer to compile it for him, and that he himself has been duped. May we hold this charitable thought?

Reading this review would have been a most deeply discomfiting experience for Gray himself, who had waited so long, and probably with suppressed excitement and trepidation, for the journal's verdict. Appearing in the 'establishment' medical periodical, it would have been seen by everyone that mattered to him. The journal had been founded by Roderick Macleod, a George's man: it was a child of his own institution. The *Medical Times* had declared Gray's work superfluous, derivative, unscientific, discourteous, and dishonest. Gray himself was publicly proclaimed an intellectual lightweight, of an unoriginal cast of mind, a person who purloined and perhaps procured others' efforts, and took the credit for himself.

At the medical school and elsewhere, the review would have been the subject of gossip, probably to Gray's intense mortification. It would have been difficult to laugh off, without imputing some ill-motive to the reviewer, which is indeed how Gray and his contemporaries seem to have dealt with it. At St George's it was regarded as the malicious effusion of a disgruntled rival author from another medical school, anxious to damage the competition. And it must be said, there *is* something in the reviewer's tone to justify that idea; especially when pronouncing Gray's book superfluous.[8]

Yet, even Gray would have had to admit, if only to himself, that while the *Medical Times* review certainly overstated the case, this was not just a turf war: there were elements of truth in its imputations. The reviewer's key allegation was that *Gray's* was a compilation, an abstract or paraphrase of *Quain's*: a work of ablation, or surgical reduction. Reading through the parallel texts provided in the review, it is clear that the passages from *Quain* are wordier and contain more detail in every instance, and Gray's summaries are well done, succinct, clear, and to the point. *Quain's* words have not always been 'lifted' verbatim. But, by the simple expedient of giving each passage from both books in full in the *Medical Times*, the reviewer demonstrated to contemporaries that there was indeed, in places, something noticeably derivative about Gray's book.[9]

Perhaps the simplest demonstration of the accusation against Gray concerns the weight of the brain. *Quain's* tabulated data on the topic occupied half a page, and had additional commentary. Gray summarized both the table and *Quain's* commentary in one paragraph. The summary is an admirable lesson in concision, but the evidence of borrowing is clear to see: the numerical figures of weight are essentially the same, and the brain weights of famous named individuals—Cuvier, Abercrombie, and Dupuytren—all appear in the same passage of *Quain's*.[10]

Prose descriptions of the embryological formation of the hyoid bone (a bone in the underjaw, which helps anchor the tongue, near the pharynx) manifest the debt to *Quain's* equally well:

> *Quain's*: 'There are five points of ossification for the os hyoides—one for each of its parts. Nuclei appear in the body and the great cornua towards the close of foetal life, but soonest in the cornua. Those which belong to the small cornua make their appearance some months after birth.' [51 words][11]
>
> *Gray's*: 'Development: By five centres; one for the body and one for each cornu. Ossification commences in the body and greater cornua towards the end of foetal life, those for the cornua first appearing. Ossification of the lesser cornua commences some months after birth.' [43 words][12]

The second passage is a succinct summary of what was known. But every anatomist of a cohort older than Gray would have been aware that the material on the formation of bone in the embryo and the infant did not appear in early editions of Jones Quain's book, but had been added only in the previous decade by his brother Richard Quain, also an anatomist and surgeon at University College Medical School.[13]

Gray might have justified himself by arguing that by the 1850s, anatomical descriptions were difficult to write without *some* elements of formula and repetition, and he would surely have been right. In a description of how the anatomical structures of swallowing actually work, for example, one would have to follow the process directionally, from the mouth to gullet, via tongue and palate; and since by 1850 all the structures had recognized technical names, and their actions had been worked out, described, and named, subsequent writers' descriptions of the same process might well bear similarities. The facts were largely known, and Gray was making them easily available, cutting through the dross. He was not engaged in an elaborate work of medical etiquette, but creating a student workbook.

Thirty years after this controversy, in 1888, GD Pollock, who had been Assistant Surgeon at St George's in the 1850s, delivered a Presidential Address to the Royal Medical and Chirurgical Society of London. Pollock had taught Henry Gray and knew him quite well.[14] Now, as President of the 'Med & Chir', Pollock was charged with the duty of discoursing a little about the lives and achievements of those members of the Society who had died during the previous year. Among them in 1888 was numbered Richard Quain. In his address, Pollock praised Richard Quain's additions to Jones Quain's book, *Elements of Anatomy*:

> its careful investigation of the question of development of the bones adds greatly to the interest of the study of the human skeleton. It is doubtful whether any better work, for accuracy in detail or clearer style of description, has ever passed through the medical publisher's hands.[15]

Pollock's comment on Richard Quain's anatomical claim to fame in the matter of bone development demonstrates that to contemporaries, the topic belonged to Quain. The implied criticism of *Gray's Anatomy* in Pollock's address, coming as it did from a long-established George's man, would probably have been noticeable to them.

The *Medical Times* reviewer also suggested Gray's debts were not to *Quain's* alone. The order in which the major elements of human anatomy were dealt with—bones, joints, muscles, circulation, lymphatics, the nervous system, then organs: of sense, digestion, respiration, urinary organ, generation, and finally, surgical anatomy—was the same as in *Quain's*. But the arrangement was not unique to Quain: it is similar in Erasmus Wilson's *Human Anatomy*, and indeed it was quite possibly the shape of many medical school anatomy courses where Quain or Wilson were course-books, including those at St George's on which Gray had learned his anatomy, and had later taught. The *Medical Times* reviewer highlighted that Gray had adopted without acknowledgement Wilson's schematic maps of the arteries, with the sole difference that whereas the artery in Wilson's graphic is represented as a rectangle, Gray's has it as a circle (see opposite). A straightforward visual comparison demonstrates the truth of the allegation.[16]

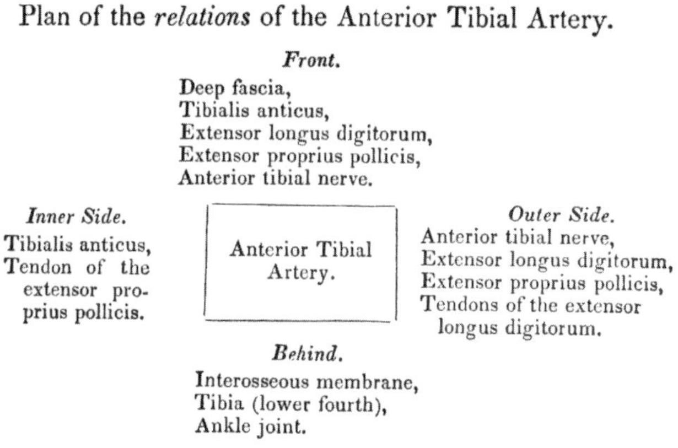

Plan of the *relations* of the Anterior Tibial Artery.

Front.
Deep fascia,
Tibialis anticus,
Extensor longus digitorum,
Extensor proprius pollicis,
Anterior tibial nerve.

Inner Side.
Tibialis anticus,
Tendon of the extensor proprius pollicis.

Anterior Tibial Artery.

Outer Side.
Anterior tibial nerve,
Extensor longus digitorum,
Extensor proprius pollicis,
Tendons of the extensor longus digitorum.

Behind.
Interosseous membrane,
Tibia (lower fourth),
Ankle joint.

Gray and Wilson compared: blood vessel diagrams from Erasmus Wilson

The fact that Knox also used schematic presentations of arteries like Wilson's had perhaps allowed Gray to feel they did not belong to anybody.

The reviewer's assertion that Gray had assumed credit for all previous anatomists' work is not altogether reasonable. The title page announcement that the anatomy had been verified from fresh dissections did not amount to laying a claim to knowing everything in the book from first-hand discovery. Gray made no claim to originality in *Anatomy Descriptive & Surgical.* But his authorial tone did give the impression of being all-knowing: Gray presented himself as lord of the domain, but without showing a genealogy of title.

The lack of credit to others was really the telling point: Gray did not own up to his debts in the straightforward way Carter had done. The *Medical Times*'s fundamental objection was Gray's failure to thank, or even to indicate, his sources. Gray liked to work fast, under stress, and his impatience with having to recognize every debt—intellectual and presentational—had clearly been too much for him to entertain. In a similar case in the world of geology, Martin Rudwick quotes those whose work was used without credit as saying: 'a single sentence, a mere parenthesis (if to the point) would have satisfied us'.[17] Jones Quain's original edition in 1828 had done exactly that, mentioning his duty 'willingly undertaken' to express 'a debt of obligation to those works [he had been] in the habit of consulting'.[18] A symbolic nod to where Gray had obtained material would have been sufficient. Not to give that was a public discourtesy, which revealed to those within the discipline both the nature of Henry Gray's activity, and his attitude towards those on whose shoulders he stood.

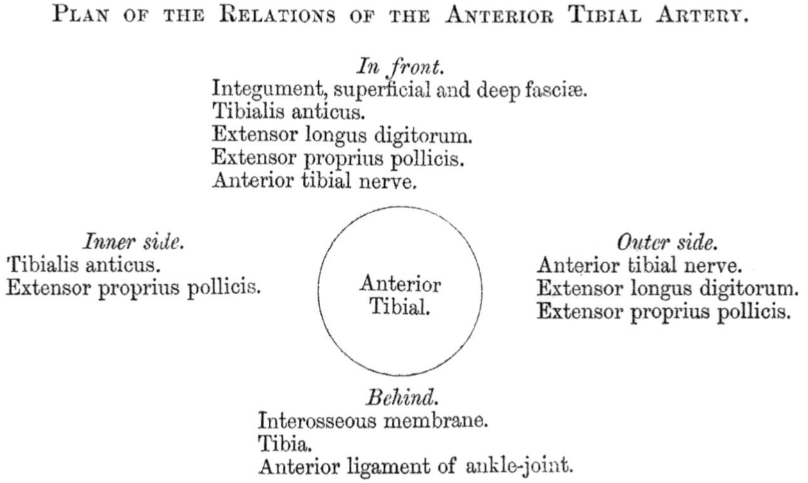

PLAN OF THE RELATIONS OF THE ANTERIOR TIBIAL ARTERY.

In front.
Integument, superficial and deep fasciæ.
Tibialis anticus.
Extensor longus digitorum.
Extensor proprius pollicis.
Anterior tibial nerve.

Inner side.
Tibialis anticus.
Extensor proprius pollicis.

Anterior Tibial.

Outer side.
Anterior tibial nerve.
Extensor longus digitorum.
Extensor proprius pollicis.

Behind.
Interosseous membrane.
Tibia.
Anterior ligament of ankle-joint.

and from *Gray's Anatomy.* From the author's collection.

There is a substantial element in *Gray's Anatomy* of something the reviewer in *The Lancet* had noticed with approval, which is the bringing together within one volume the best aspects of a number of existing works. Henry Gray's text is perhaps better understood not as an original work at all, but as an anatomical anthology unified by one voice. The book's text clearly owed most to *Quain's*, but it was better organized and much more easily navigable, presenting everything more simply, and demonstrating an aversion to cross-referring, one of *Quain's* worst aspects. *Gray's* owed features to other works, especially to Wilson's *Anatomy*, not just the diagram above, but the clarity of the book's design, the less cluttered larger pages, the way the text is broken up typographically, and the italicization of key subject words for headings, especially in its American edition.[19] Doubtless there were other debts. Without his having observed traditional courtesies, it is not simple to perceive what of the text was original to Gray himself. It seems it was this that the reviewer found so objectionable.

Having seen Gray's earlier dealings with Carter, we know that a reluctance to share credit was an element of Gray's character, particularly perhaps where his debt was greatest. It would have cost nothing to thank his sources, but it was evidently too much *for him.* Gray had apparently learned little from Carter's remonstrance about recognition at the outset of the work on the book. Had he taken it on board, such devastating criticism as appeared in the *Medical Times* would not have been levelled at Gray in print.

The surmise by the *Medical Times* reviewer that Gray might have employed other people to do his writing seems either inspired, or well-informed. We know that employing others really was of a piece with Gray's methods of work. Carter had witnessed Gray's working practice early in 1856, and there can be no reason to doubt him.[20] One cannot help wondering whether the reviewer knew Crisp, or had read or heard news of his protests after the publication of *The Spleen.*

As for the dedication, quite why the reviewer felt Gray should not have dedicated the book to Sir Benjamin Brodie is unclear, but he seems to have thought it unseemly, regarding Gray as toadying to a genuinely great man, misusing the great man's name to promote a shabby book. But as we shall see, Sir Benjamin Brodie himself did not regard the dedication or the book in the same light at all. He remained one of Gray's most steadfast supporters.[21]

Most Victorian book reviewers did not identify themselves, and they invariably remain anonymous even today. However, as Martin Rudwick has observed apropos of book reviews in geology, informed guesses or rumour meant that insiders often had a good idea who might be responsible.[22] So far, little evidence has been found of contemporaries discussing the matter in any depth. It may be that Gray's sudden early death inhibited the act of recording for posterity the informal gossip the controversy generated at the time. A later observer has suggested that the buzz at St George's was that the review had been written out of jealousy by Richard Quain and William Sharpey, editors of *Quain's*, and others have felt the attack on *Gray's* was openly partisan.[23,24]

The field of likely contenders was small. The voice is apparently that of an older person, and looks to be either someone with a grievance against Gray, or someone genuinely offended by his book. Only one main contender emerges in the first category: Edwards Crisp. But I think he can be ruled out as the reviewer, because had he been responsible, the tone would have been resentful rather than regretful, and the method of attack less measured. The reference to Huxley rules out James Paget, who might otherwise have been a tenuous possibility in the second category, but to whom a taunt of such a nature would have been unthinkable. Erasmus Wilson seems an improbable choice, since the review made nothing of the debt owed by *Gray's* to the illustrations of Luther Holden, who was Wilson's colleague at St Bartholomew's. Another possibility is GV Ellis, who, when he agreed to join Sharpey as co-editor of *Elements of Anatomy*, had taken on Richard Quain's subject responsibilities, including the formation of bone. Ellis would have been familiar with the pre-existing text, and might well have taken umbrage on behalf of his predecessor.

Sharpey, too, would have known Jones Quain's work intimately, would have known his old teacher's phrasing, and would have been sensitive to the extent of Gray's use of Quain's work. He would have recognized Richard Quain's additions, and of course the material he himself had added to the work.[25] Sharpey's involvement in the review is a most likely possibility even more particularly because he was Secretary to the Royal Society, and would not only have witnessed Gray, Huxley, and Owen in performance, but would have known of Brodie's backing of the young man from the presidential chair. Sharpey may well have witnessed the circumstances surrounding the Society's grant of £100 to Gray, and probably had to deal with Edwards Crisp's letter of remonstrance to the Society's Council in 1857, concerning the many hands involved in Gray's prize essay on the spleen, and the unfair advantage Gray had obtained towards winning the Astley Cooper Prize, as a result of that funding.[26]

The reviewer's past respect for Gray's *Spleen*, and the comment about Huxley, is a pairing which presents a key element in any attribution of the review to Sharpey, with or without help from his colleagues Richard Quain and/or GV Ellis. Sharpey is known to have had a high estimate of Huxley, having awarded him a gold medal in anatomy and physiology for his outstanding performance in his first MB examination at the University of London in 1845.[27] It had probably been Sharpey who, in the coverage of the human spleen in the 1856 edition of *Quain's*, had cited Henry Gray's book *The Spleen*, and mentioned a difference of opinion between Gray and Huxley concerning the formation of the Malpighian corpuscles of the spleen. Gray had asserted the existence of a distinct capsule, but from researches in comparative anatomy, Huxley (with his colleague George Busk) had said there was none. The matter had been left in doubt in the 1856 edition of *Quain's*, making it clear that there was still work to be done.[28] Sharpey, being editor of *Quain's*, would have been an obvious person for the editor of the *Medical Times* to ask to review *Anatomy Descriptive and Surgical*. He may have agreed out of curiosity, and subsequently perhaps found himself pondering Crisp's allegations, particularly had he originally perceived Crisp as a bit of a crank. The constellation of arguments—the assertion of *Quain's* importance to *Gray's*, earlier respect for *Spleen*, the reference to Huxley and Owen, and the mention of Gray's employing other people—together suggest that Sharpey either wrote or had an active hand in the review.

But, when one reads Sharpey's prose in his letters, he comes over as a man of great courtesy and good manners, not at all as aggressive as in the reviewer's voice in the *Medical Times*, though he did have a running feud with *The Lancet*, whose review had been generous to Gray.[29] Richard Quain was known to have a reputation for being aggressive, to have a fighting edge to him, and occasionally Sharpey would have preferred him to hold his tongue.[30] On this occasion, it seems, rumour may indeed have been correct. Gray's offence was perceived to be so great, that these two men may well have collaborated on the review.

The *Medical Times* review allows us to see partly what was so new about *Gray's Anatomy*, and why it was appreciated, as well as why it was denigrated in some quarters. It felt like a new start. The book was created by a young man to whom the minutiae of the history of anatomy was so much clutter, a man impatient of traditional anatomical courtesies. Gray saw his own task as summarizing anatomy for students, rather than chronicling the history of anatomical esoterica, and his book therefore treated anatomical history as separable from anatomy; his anatomy is largely shorn of its sources, dehistoricized.[31]

Gray's impatience with the past is closely parallel to opinions expressed by the surgeon Robert Liston, in the preface of his book *Practical Surgery* (1846). Liston had said that history is important, but not in a book about practical surgery. Liston himself was highly impatient with prolixity and debate:

> The reader of this work would not be much wiser, ... or more capable of understanding the management of difficult cases, were the author to enter into a long detail of what this or the other of the moderns have recommended, or what he has found by experience in practice to be useless or hurtful.[32]

Liston presented his views 'concisely stated and without conflicting opinions', saying without more ado that he had based his book upon his own practical experience.

Gray wanted to be a top surgeon, like Liston: not a top anatomist. For him, the dissecting room was a necessary stage on his trajectory, no more. We have seen that for his prize essay and book, Gray had delegated the work on the history of ideas about the spleen to Timothy Holmes, and that Gray had demarcated it from the rest of the essay so he could get on with reporting his own researches. *The Spleen* had been a staging post for Gray, who had now applied Liston's sentiments to Anatomy. The trouble was, the content of Gray's experience of anatomy seems to have been derived in part from studying and teaching with *Quain's*, whose wordy volumes, small print, and paltry illustrations were, for him, part of the problem ... and, it seems, also part of the solution. Gray's titlepage announced his book's contents were rooted in and verified by the practical activity of dissection, but the book had still deeper unacknowledged roots too.

Gray had produced an admirably lean and succinct anatomy book, which many people found liberatory. But in failing to recognize the labours of others, Gray found himself standing on the real human toes of other anatomists, to whom the discipline of Anatomy was a painstaking process of incremental discovery, with its own modes of accretional proceeding, and its own disciplinary etiquette. That the book generated friction within the discipline is hardly surprising, because it had *not* made new, but reworked the old, and presented it in a new and simple way. Its author's discourtesy was objectionable, but the book had raised the important question what Anatomy was really for.

The *Medical Times* published no protest or other comment about the review, and Gray himself remained silent. But before March 1859 was out, a vigorous defence of *Gray's Anatomy* appeared in *The Lancet.*

The writer was George B Halford, Lecturer in Anatomy at the rival medical school to St George's in Grosvenor Place. He emphasized that—in contrast to the *Medical Times* and its reviewer—he was not writing anonymously. The *Medical Times* piece, in his view, brought shame upon the whole profession. Urging readers to compare the book with the review, and to give Gray his due, Halford said Gray must answer for himself the charge of dishonesty, but that his book was

> especially adapted to the medical student who has to complete his study of anatomy by the end of the second winter session—the minimum required by the Royal College of Surgeons. It is illustrated in such a manner as to command unqualified praise, and I feel sure that those conversant with the teaching of anatomy and the requirements of the student will say that Mr Gray has done good work by showing us the style of book needed at the present day.
>
> By all means let the student possess his Quain, and work with his Ellis, Holden, or *Dublin Dissector*, but preach as you will, the majority of men will do neither: hitherto they have got on better with Wilson, and henceforth I predict they will get on faster and better with Gray.[33]

Halford's letter was a splendid endorsement of the book by a rival teacher. For Gray the generosity of its support probably felt better than any advertisement. The controversy no doubt drew attention to the book, and such a letter addressed the book's strengths directly, as well as the gossip. But Gray could not regard it as a complete vindication: Halford had been careful to differentiate between the intellectual and educational achievement embodied in the book, and the personal morality of its author. Gray might have been discourteous, or not honest (not *candid*, in Carter's formulation), and for that he must answer; but the book was nevertheless superb.

Sales were good as soon as the book was out, and it became swiftly evident that a new edition should be planned. Gray was probably preparing his answer at the time. In the second edition, which appeared the following year, 1860, he inserted a new paragraph, set in small type, at the end of the osteology section. The passage acknowledged the 'valuable aid' Gray had 'derived from the perusal' of the work of twenty or so authors. Just after half way through, singled out for its minute description of bone development, was Sharpey and Ellis's edition of *Quain*. The acknowledgement was belated, but it was there.[34]

Edwards Crisp looks to have gained temerity from the *Medical Times* review, and later in 1859 (apropos of a difficulty relating to the administration of an Irish medical prize) went public in *The Lancet* with the substance of his allegations concerning Gray and the illegality of his Astley Cooper Prize award.[35] Nothing seems to have come of the bad publicity, but after the triumph of the book's publication, by the end of 1859, Gray may have felt a little careworn.

The best thing about 1859 for Gray had been the appearance of the new edition of *Anatomy Descriptive and Surgical* in the United States of America. Blanchard and Lea, of Philadelphia, had asked their printer Sherman and Son, to re-typeset the whole of Gray's text entirely afresh. They used an American typeface, which took up almost the same page space as the one used in London by Wertheimer, but was lighter in character, leaving the page looking brighter.[36] While keeping the book's dimensions pretty much the same, the Philadelphia publisher widened the letterpress area on each page by an extra ¼ inch (7 mm). By this slight alteration, the printer managed to minimize the appearance of forcing, accommodating the majority of the illustrations within the printed territory of the letterpress.[37] A more extensive index was added, and an American anatomist oversaw the entire edition, by which process a number of small errors that had crept into the first English edition were eliminated before the first American one reached the bookstores.

Carter was mired in controversy, too, and endeavouring to deal with it. In his Bombay lodgings he had met up with a 'ladylike', 'lively and agreeable' young widow whom he knew by the name of Harriet Bushell. After a brief passionately sexy courtship, Carter and she were married, and only afterwards did Carter discover that she had lied to him and to the minister conducting the marriage service. He discovered, too late, that Harriet was clever at running up debts, and that they were not at all compatible. Their life together was unbearable.

Among Carter's papers there survives a cutting concerning a divorce case in the Cape Colony (South Africa) which is where Harriet Bushell had been living before her arrival in Bombay. The case is that of Argent *versus* Argent. The wife in the case was originally named Amelia Adams, married to a soldier in the Cape Mounted Rifles: their marriage had been dissolved 1852, at the Cape, on grounds of the wife's adultery.[38] Carter seems to have obtained the cutting through lawyers much later, and preserved it because it revealed the original identity of Harriet Bushell, and the fact of her divorce. She had arrived in India following the lover in the case, or perhaps a later one, who had cast her off, with a child.

Sadly for Carter, although she had lied in presenting herself as a widow, and was living under a false name, the marriage was held to be legitimate because she had been legally divorced. Carter's efforts to extricate himself from the relationship did not succeed. Harriet had become pregnant almost immediately, and Carter helped deliver the baby in September 1860, a beautiful little daughter he named Eliza Lily after his mother and sister. He adored the child, who he said, 'seemed like Providence smiling on our sin'.[39]

Because of Harriet Bushell's deceit, Carter himself considered the marriage invalid. 'Certainly', he wrote later, 'she most egregiously deceived me and deserves no more regard.' To obtain a divorce was impossible in India at the time, and to try for an annulment would have made the child illegitimate. 'She' (Carter preferred to avoid naming her) was 'dearly resolved to stick on as long as there is money to be had'.[40] Her banishment, on a regular allowance of £150 a year from Carter's Indian salary, deprived Carter of the child, the only pure love he had known since the death of his mother.

Henry Vandyke Carter was devastated by the entire experience. The events were very public, and involved him in scandal and notoriety across the small Bombay expatriate community. He was aware that he was regarded as 'a notorious—fool, at the least'.[41] He disposed of furniture she had acquired, paid off the debts she had run up, and resumed his bachelor life with a broken heart.[42] His diary between May 1859 and November 1861—the period covering his unhappy marriage—is missing. Eight pages have been ripped out.

But, in a curious way this ghastly series of events was providential for Carter. He came through it by finding refuge in incessant work, and his work was among the poor of Bombay.

In early November 1860, number 445 Strand was visited by an undertaker, who went upstairs to measure the corpse of John William Parker for his coffin. Young Parker was only 40 years old, and had probably succumbed at last to the same disease as his mother and older brother: tuberculosis.[43]

Parker had been cared for in his last illness by James Anthony Froude. Young Parker, characterized as 'a great soul in a pygmy body' had been frail for quite some time. Froude, newly a widower himself, was said to have nursed him like a brother until the moment of his death. The hearse and its plumes wound its way through north London to Highgate Cemetery a few days later. The funeral service was conducted by FD Maurice, and RC Trench, the Dean of Westminster. Gray and Holmes, as well as Kingsley, may have been among the many other authors and contributors to *Fraser's* likely to have been there, since Parker was much loved. Baron Pollock noted sharing a coach with 'Theodore Martin, Helps and Froude, who has nursed Parker most tenderly during his last illness'.[44]

We do not know what the Parkers, father or son, had made of the review in the *Medical Times*. It would probably have made pretty mortifying reading for them. 1860 had been far more difficult. Notoriety had attached itself to a Parker book published under the inoffensive title of *Essays and Reviews*.[45] It contained a clutch of theological essays by seven Anglican churchmen, sympathetic to the notion that traditional Anglican theology could not avoid the implications of modern-day studies in the sciences, and in biblical history.[46] Interest in the disparity between biblical and geological time was already widespread among the Victorian public: how could Adam and Eve have brought death into the world if fossils were as old as the hills, and the hills were as old as the geologists were saying? Was the Bible the actual word of God, or, of human making, and therefore amenable to study? Churchmen had so far kept a lid on their own conversation. *Essays and Reviews* lifted it, questioned biblical infallibility, and polarized contemporary attitudes to scripture. The book caused an enormous stir, and sold like hot cakes.[47]

When read today, *Essays and Reviews* seems quite harmless, but at the time it seems to have hit a cultural nerve, and the hullabaloo was so extreme in some quarters that its authors were denounced as heretics, and accused of promoting a movement to subvert Christianity. The Archbishop of Canterbury believed no graver matter could be imagined since the Reformation, while at the 1860 meeting of the British Association for the Advancement of Science, the controversy drowned out discussion of Charles Darwin's long-awaited book, the *Origin of Species*. The lives of all the contributors (collectively characterized as 'Seven against Christ') were affected by the book's public reception, and Parker senior may have feared losing friends at the SPCK as a result of his involvement in its publication.[48]

Young Parker had been stricken with the illness that killed him during the late summer of 1860. His father may have had some warning of his son's impending illness, or may have been seeking to cut down his own workload, because that year he had taken his long-serving assistant, Mr Bourn, into partnership, the company becoming Parker, Son and Bourn.[49] Parker senior himself was getting on in years—in 1860 he was 68—and had bought a house in the country near Farnham, some distance to the south west of London. Having moved away from Charing Cross with his wife Ellen and their growing family, he may have been hoping to work only part-time, as in his Cambridge days. Ever since his childhood, Parker senior had lived a long hard-working life in the midst of London's bustle and din. Now he wanted to unwind in a new life, the quieter comforts of which he had earned.

The first edition of *Gray's* was almost sold out, and the second edition was probably at the printers at the time young Parker fell ill. This was to be Henry Gray's definitive edition, and it had some adjustments from Parker, too. It was given thirty-four additional pages, and although Gray made additions, there was some space for the text to feel less crowded, so the whole book feels a little more relaxed. Gray added to the title page his new Fellowship of the Royal College of Surgeons, which he had achieved by examination in May 1860, and corrected some minor typographic errors. He also added some extra material on 'development'—lifelong changes in the body beginning in the embryo—and enlarged the sections on neurology and generation. Gray's *Preface* to the second edition was extremely brief:

> In preparing a second edition of my 'DESCRIPTIVE AND SURGICAL ANATOMY,' I trust that I have corrected any inaccuracies contained in the previous one; every page has been carefully revised; much additional matter has been added to the text; and several new Illustrations, executed with great care and fidelity by Dr Westmacott, have been inserted.

Gray's new illustrator seems to have made an initial effort to keep the new images in the spirit of Carter's, but he did not quite succeed in the task: his line is timid by comparison, and the engraving looks to have been done by a different hand.[50] The scale of the new illustrations was governed by the need to avoid forcing, so altogether the new illustrations are noticeably smaller and less bold in conception than Carter's. Their inferiority must surely have been evident to Gray. In an endeavour to keep the appearance of the book close to that of the first edition, the Parkers had returned to Butterworth and Heath for the engraving, and to Wertheimer for the printing, but the second edition nevertheless lacks the unity of the first. Most of the added illustrations are not really in keeping, and mark the earliest development of *Gray's* as the visual salmagundi it later became.

The new edition had probably been planned for the October start of the dissecting season in 1860, but because of young Parker's fatal illness, it did not reach the shops until early 1861.[51] *The Lancet* greeted it with a short notice in early February:

> This edition is much improved, and contains several new illustrations by Dr Westmacott. This volume is a complete companion to the dissecting room, and saves the necessity of the student possessing a variety of 'Manuals'.[52]

Gray's new *Preface* was dated December 1860, so it was written after his publisher's death. Notably, it made no reference to the loss of young Parker.[53]

James Anthony Froude wrote an affectionate and respectful obituary for his friend in the *Gentleman's Magazine*, in which he called him an 'uncommon man', and said that young Parker had 'a peculiarity about him':

> he was one of those rare persons to whom 'success' in the mercantile sense of the word was by no means the first object. He carried into business the strongest conceptions of duty and responsibility … [and] looked on his position as an opportunity of doing good in the largest sense in which he understood the word …. He did not wish to be known as the publisher who had made the largest fortune in the trade, but as one who had added most to the enduring literature of England.[54]

These were not just kind comments about a friend, though they were that. Froude's words were plainly true, and his confidence in the achievement of Parker's intentions has been confirmed with the passage of time. A later observer has commented that depite his sadly premature death, young Parker 'significantly encouraged Liberal and Broad Church causes'.[55] Passing over for our purposes all the other remarkable and influential books which emerged from 445 Strand, had the Parkers not planned to invest heavily and to carry a deficit on the first edition of *Gray's Anatomy*, and had they not actually borne additional unexpected losses, the book may not have seen the light of day, certainly not in the form we know it. The book they fostered and brought to market has assisted generations of doctors in learning their human anatomy, which cannot be anything other than a great good in Froude's terms. As a publisher, Parker genuinely did add something enduring to the literature of England, something vigorous, enduring to this day.

The torn-out pages of Carter's diary covered more than the era of his disastrous first marriage. During the period between May 1859 and November 1861 several other important events occurred, his opinion of which it would have been good to read, however briefly. Young Parker's death was one of these events. Another was an important scientific discovery made with the help of his new microscope by Carter himself.

Grant College Bombay shared a large site with a much larger institution, now known as the 'JJ' Hospital.[56] It is still there, buzzing with activity, having overbuilt much of what in Carter's day was intervening open land. As Professor of Anatomy and Physiology at the College, Carter had responsibility for dealing with specimens sent over by the Hospital's surgeons for the medical school's Museum. Carter was highly competent to deal with such things, and the work was probably how Carter was drawn to make his discovery. Examining a foul and hideous-looking swelling on a deformed amputated foot fresh from the operating theatre in early October 1859, Carter decided to take a closer look, and found himself confronting the fruits of what he discerned to be a fungus.

During his studentship at the Royal College, Carter had been intrigued by the development of the branching filaments and spores of a fungus growth inside a famous cedar in the old Physic Garden at Chelsea, which had silently brought about the demise of a noble tree.[57] [58] He was witnessing the fruits of the same kind of insidious activity now, but in a human limb, belonging to a Hindu patient named Pandoo, from near Poona. Fungal diseases were known to affect the human body, but this was the discovery of what is now known as a deep mycosis. Before Carter's discovery the disease had a local name, 'Madura Foot', but had been thought of as a tubercular hypertrophy.

The growth and its disabling effect on patients had been described, but its pathology was unexplained. Never before had anyone attributed this malady to a fungus.

Through the desperate days of his miserable marriage, Carter worked at the subject when he could—dissecting and drawing, taking microscopic samples, preserving the growths, trying to grow the fungus, endeavouring to understand the pathological process of the fungal activity, and trying to identify the organism responsible.[59] A senior doctor at the medical school took a strong rather predatory interest in Carter's discovery, but just prior to the torn-out pages, Carter wrote anxiously: 'there is little of the Gray and Carter principle in me', meaning he was not about to abnegate himself in order to advance yet another senior's career.[60] He parcelled up a sample preparation of the fungus in situ, and sent it off to the Pathological Society in London with a letter describing his discovery. He lost no time in researching the subject in depth in the medical school library, and began to write up his findings.

By the time Carter delivered his paper to the Medical and Physical Society of Bombay towards the end of 1860, he had examined material from several further cases and had discerned at least two varieties of the fungus, one black, the other with white patches. He had sought cooperation from a range of doctors working in the field, and their responses and specimens had assisted his researches. He had combed the literature, and by his own microscopic investigations had distinguished the disease by its pathology from leprosy, elephantiasis, and other diseases of the foot, including malignancy. His paper gave credit to others for their help and observations, but made clear that the discovery of the fungal origin of the disease was his own. He recommended amputation—at least four inches clear above signs of fungal activity—to prevent its further spread. Carter had also formulated a new name for the disease, '*mycetoma*', a tumour caused by a fungus. This is the name it still bears today.[61]

At the end of the medical school year in May 1861 it became clear that the two most senior surgeons at St George's would retire, and that their own assistants would apply for the vacant positions. This was the time-honoured way of making high appointments at St George's. There had been no new appointments at the Hospital for some time, and the excitement generated by the prospective rearrangement of the upper ranks was felt by staff and students

View of the Dissecting Room
of St George's Hospital. Taken 27 March 1860
To Henry Gray Esq. F.R.S.
as a respectful remembrance from

Photograph of the Dissecting Room at St George's Hospital Medical School, Kinnerton Street. Joseph Langhorn, 1860.

This remarkable picture, taken in the late morning of 27th March 1860 by Joseph Langhorn, shows Henry Gray and others in the dissecting room at Kinnerton Street, belonging to St George's Hospital Medical School. Gray is in black in the left foreground sitting by the corpse's feet, looking serious. The view is posed.

Langhorn himself is said to be the balding older gentleman standing at the extreme left of the picture, with eyes looking downwards. He was a mature student who entered medical school at about the age of 60, and said to be kindness itself. Despite difficulties with memorizing the amount of information required, Langhorn obtained his diploma, and practiced midwifery in the nearby district of Brompton. Pickering Pick, who wrote about this photo in the 1890s remembered meeting Langhorn carrying hot beef tea to his impoverished patients.

The photograph offers a significant contrast to the macabre jocularity found in Victorian dissecting room photography from the United States examined by James Terry.* Such photography in the USA was an underground convention, Terry has argued, associated with the transgression of taboos, dramatizing and emphasizing the gulf between all those who never dissected, as against those few who had. This view of the Dissecting Room at Kinnerton Street is a sober portrait of over thirty anatomists (teachers, students and possibly technicians) grouped in their own natural environment: a purpose-built, top-lit barrel-vaulted room, hung with anatomical drawings. There is no sign at all here of hilarity or mirth. Only one student (at the very back right) is smiling, and the atmosphere of the entire room is quiet, still, serious, and earnest. The dissecting room skeleton is almost out of sight at the very back. Apart from a skull posed in the foreground, and a few long bones, also evidently posed, only the lower limbs of one human body is visible. The end of March was traditionally the end of the dissecting season, so this is unsurprising, but it is also worthy of comment that the rest of the body is not exposed to the camera.

The students wear overalls resembling artists' smocks, the teachers, smart day clothes. Gray, in the foreground, third from left, wears a dark suit, elegant wing collar and neckcloth. He sits informally, one hand on his knee, the other in his trowser pocket or tucked on his hip. The endeavour is sober, serious, thoughtful, respectable.

Some of the anatomical diagrams hung on the walls may have been by Henry Vandyke Carter, but they look rather too old for that. His work may have been for the adjoining Anatomical Museum, or for the Lecture Theatre in the same building. Carter does not appear in the photograph, because at the time it was taken he had already travelled to India, and was teaching at Grant Medical College in Bombay, where he was Professor of Anatomy and Curator of the Anatomical Museum.

The first American edition of *Gray's Anatomy* had already appeared when this photo was taken, and the second English edition was in its proof stages, probably due to appear that October. Publication was delayed when the publisher JW Parker junior fell ill during the summer. Gray himself died unexpectedly the following June, 1861.

* Terry, JS: Dissecting Room Portraits. *History of Photography*, 1983, 7 (2): 96–98.

of the entire institution, especially when it became known that genuine competition for the various posts in the lowerarchy would be facilitated by the larger shifts further up.

A paragraph announcing the situation appeared in *The Lancet* of 15th June 1861:

> MEDICAL VACANCIES
>
> Mr Caesar Hawkins and Mr Edward Cutler have resigned as Surgeons, and Mr Prescott Gardner Hewett and Mr George D. Pollock as Assistant-Surgeons, to St. George's Hospital, thus causing four vacancies in the medical staff of the Hospital.

Gray would have known long in advance that this was the opportunity for which he had been waiting, and had already thrown his hat into the ring. Appointments at St George's were made by the Governors, and Gray's referees' letters were ready for nomination day on 6th June, as were those of other applicants for the two vacant posts of Assistant Surgeon, Timothy Holmes and Athol Johnson, both older men than Gray. Gray was extremely well-placed to get one of these jobs, but there was clearly going to be a public competition.

On the day of the nominations, Gray arrived at St George's looking unwell. By the time news of the vacancies was published in *The Lancet,* he was dead: in fact news of his death appeared on the facing page.[62]

> DEATH OF MR. HENRY GRAY, F.R.S. The profession will regret to hear that this estimable gentleman expired on Thursday morning, at his residence in Wilton-street, at the early age of thirty six.[63] The deceased, who was Lecturer in Anatomy at St. George's Hospital was nominated on the 6th inst., for the Assistant-Surgeoncy of that Hospital, on which occasion it was observed he looked ill. On the following day symptoms of fever came on, terminating in confluent small-pox, and in one short week this rapidly-rising surgeon passed from us. Mr Gray was well known by his valuable contributions to the advancement of science. In 1849 he obtained the Collegial Triennial Prize from the Council of the Royal College of Surgeons of England for his Essay, 'On the Anatomy and Physiology of the Nerves of the Human Eye'. In 1853 he obtained the Astley Cooper Prize of three hundred guineas for his Dissertation, 'On the Structure and Function of the Human Spleen'; in 1858 he published his 'Descriptive and Surgical Anatomy', and at the time of his death was engaged in preparing another edition of this excellent manual. The deceased was a Fellow of the College of Surgeons by examination, and Surgeon to the St George's and St James's Dispensary.

Henry Gray's memorial card, showing his real age.

People ignorant of the circumstances surrounding Gray's death might wrongly have assumed from *The Lancet*'s last line, that Gray had caught smallpox from his work amongst the poor. In fact he had been nursing his little nephew, Charles Gray, in Belgravia, and had taken the disease from him. Gray had certainly been vaccinated as a child: his death certificate records 'Confluent Smallpox, 7th day. Vaccinated in Childhood, Certified.' He and his family had probably thought vaccination sufficient to protect him. He may have taken on the child's care to protect other family members from its ravages, only to take it himself. Because of cases similar to his, there had been a debate in the medical press concerning whether or not is was advisable to re-vaccinate. A new, more virulent strain of smallpox had recently been circulating, with which Gray's childhood immunity clearly could not cope.[64,65]

In reporting the vacancies, and then Gray's death, *The British Medical Journal* took it for granted that Gray would have obtained one of the posts at St George's:

> Mr Gray, whose election to one of the vacancies would have been certain is snatched away by death; and Mr Athol Johnson … a favourite candidate retires from the scene. Mr Holmes, therefore … alone remains … It is not for us to

> reason upon this sad event, but it is impossible not to feel, at such a moment, of how little account are all our great endeavours for professional advancement. It may well teach us all to pause for an instant amidst the busy struggle of life! On the very threshold of his long cherished wishes, just as he was about to realise the object of his ambition and reap the well earned fruit of his labours, death marks our brother as his own.[66]

To the grand old man of St George's, Sir Benjamin Brodie, Gray's death was calamitous. He wrote to his friend Charles Hawkins:

> I am most grieved about poor Gray. His death, just as he was on the point of obtaining the reward of his labours, is a sad event indeed. If you have any means of doing so, I wish you would express to his mother how truly I sympathise with her in her affliction. Gray is a great loss to the Hospital and the School.
>
> Who is there to take his place?[67]

The question was an important one, especially as Athol Johnson had withdrawn his name from the list of applicants. Timothy Holmes was appointed without opposition to the post of Assistant Surgeon, and the remaining position was filled when an older George's man, Henry Lee, who had been working as a surgeon at King's College Hospital since 1847, was persuaded to apply.[68]

Readers of the medical journals, especially at St George's, would have been bemused by the coverage of Gray's sudden death in the *Medical Times and Gazette*, which made no reference whatever to its lacerating review of Gray's book a couple of years before, and instead recorded emolliently:

> ...the feeling of general regret expressed at the great loss the Profession sustains by the death of Mr Gray. We do not remember another instance in which so young a surgeon has attained such varied honours ... the works he leaves behind him [serve] as encouragements to others to follow the path by which he gained the respect of his contemporaries, and an enduring place in the Roll of Medical Worthies.[69]

How Carter reacted to the news of Henry Gray's death when it reached him in Bombay a month or so later is not known. Had he written down his feelings about the shock of the news, the record would have been in the lost pages of his diary.[70] Carter had commented on the premature deaths of many individuals since he had begun his diary, but there had been no one with whom he had had such a close working relationship, or known for so long.

Carter had grown up knowing Gray, from his own first student days in the dissecting room, back in 1848. Gray had always been a brisk and hard-working older presence at St George's, always there whenever Carter returned to the place. Carter probably had difficulty imagining the dissecting room at Kinnerton Street without him in it, or somewhere nearby, and might well have wondered with Brodie, who could take Gray's place. Carter would also have known Gray's mother, as he had doubtless met her on many occasions at Wilton Street, and would perhaps have pondered her shocking experience and subsequent awful loneliness in the quiet house, having lost her active, busy son. Carter had witnessed smallpox at the Hospital, he knew it was a ghastly way to die.[71] He would surely have felt a profound pity for Gray, and probably brooded over how he had caught the terrible disease, the distress, the waste, and the tragedy of his death.

It is very likely that Carter sent Gray's mother a letter of condolence, and perhaps also wrote to Holmes, whom he probably expected to be in the running for one of the vacancies. He and Holmes remained in touch. Invariably, in other cases of early or unexpected death, Carter would note the death and perhaps say a word or two about the person, often ponder incredulously why he himself had been spared, and then, gird himself up to renewed activity. If this was the pattern in 1861, Gray's death may have been a 'great steadying shock' to him: helping keep firm his resolve to work steadily thereafter.[72]

After the torn-out pages, Carter wrote very little. He mentioned grief in a letter to his sister Lily, but he was referring to his own failure to make a happy family life, and his sorrowful prospect of lifelong loneliness.[73] Yet despite his continued sense that he was under the burden of 'a crowd of intractable defeats', and his consciousness of an 'entire absence of Christian spirit', Carter thanked God that he was able to relieve his mind with work. 'The time is the most endurable when it is absorbed thus', he wrote to Lily:

> Such a quiet obscure and monotonous life would not suit many, but I have been schooled to it, beginning with those dreary hard-working London days; in India comparative prosperity nearly brought me to ruin; thank God it is not quite so.[74]

In his final diary entry Carter still felt destitute of divine aid. He noted 'working incessantly the only relief ... and as to England—have resolved to indulge in no more reminiscences'.[75] Aware that his own predilection for introspection inclined towards indulgence, and that he himself was his own greatest tormentor, the fact that his diary ceased here suggests Carter decided to renounce it along with reminiscing.[76] He was keenly aware that Gray's achievements were related to his freedom from introspection; indeed, this had been one of Gray's character traits that had most interested Carter early on in their relationship: 'what does, does well! think—*do* it.'[77]

Perhaps Carter's significant new determination to live life forwards, to do and be in the world, was at last provoked by his calamitous marriage, and reinforced by Henry Gray's terrible and unexpected death. Carter shared his determination with Lily some time later: 'Since I commenced study in earnest', he wrote, ' "frivolities" appear to me unworthy.'[78] Carter probably harnessed his intensified activity to further mycetoma work, and to some new enquiries towards which his researches now beckoned.

Carter had already begun his life's work.

FUTURITY

After 1861

With the Hospital's move away from Hyde Park Corner, out of the bustle of what is now central London, the dissecting room at St George's moved too, to Tooting, several miles south. Today, it is a huge basement room, longer and wider than the old one at Kinnerton Street, but with a lower flat ceiling, and bathed in brilliant artificial light. There are no unpleasant odours, either of decay or chemicals, which is what people always expect from descriptions of dissecting rooms. Refrigeration and preservatives between them ensure that the dissecting season runs all year round, and the room is highly productive. The stainless steel trolleys on which the wrapped bodies lie line the walls like hospital beds in a great cool ward. The students, in their white coats, sit on tall stools beside their corpses, as in the old days, assiduously bent forwards over them, working away at the structures of the human tenement. There is an atmosphere of diligent quietness and concentration, order, decency, and endeavour.

By comparison with the great old room in which Gray and Carter probably met, and worked on the book, this new dissecting room feels far less bleak. The light is better, and even if the quiet working atmosphere at Kinnerton Street was similar at times, and the people doing the dissecting equally young and serious, more than half of them today are women.[1] The biggest change, however, is not the stainless steel or the artificial light, nor the presence of young women anatomists, but the fact that the bodies are there by consent.

Today British anatomy rooms are supplied by the kindness of donors, which means that, thankfully, the relations between the dead and the living are quite different to what they were in mid-Victorian dissecting rooms. Right up until the Second World War, body donation in the UK was a rare thing, but since the establishment of the National Health Service, bodily giving has increased to such an extent that the old means of procurement have faded from view. All the bodies on the tables at St George's today are there by gift, donated by members of the public.

In a confidential survey of future whole-body donors for *The Lancet*, respondents explained they wanted to assist medical education, and to benefit Joe Public.[2] When asked what they hoped would be the outcome of their donation, one replied that 'some young budding doctor/surgeon will gain knowledge and go sailing through exams', and another expressed the hope that:

> someone, somewhere, a student, a surgeon, perhaps a group working on a problem, may improve their skill or find an answer to a problem in their study of the body that once contained—and served—the person that was me.

The knowledge that the bodies are there by consent dissolves some of the transgressive hard-heartedness which resonated in the old dissecting room. It is difficult now to imagine how it may have felt to carry out the ritual pillaging of the dead house, tell untruths to enquirers, and dismember people whom one knew had no wish to be dissected. For many of those who trained in the great room at Kinnerton Street, the guilt of the transgression may have been a painful thing to carry. The act of dissecting seems altogether more benign now that the bodies are gifts, and the dissecting willed. An alliance of this kind between dissector and dissected in a joint endeavour is something Victorian medical students experienced only in dream.[3] So different is the ethos of the anatomical world today, that medical schools now run annual ceremonies of thanksgiving to honour those who donate their bodies for dissection, and to thank their families for seeing donors' wishes done.

Butterworth and Heath continued engraving in the Strand, working on all sorts of creative projects for books and magazines, including engraving for Sir John Tenniel, the famous illustrator of *Alice in Wonderland*.[4] But, like colleagues and competitors in the rest of the wood-engraving business, their fine skills were engulfed by the march of new technology. First photography did away with the engraver's draughtsman, and then photographic lithography and etching—known as photo-litho and process engraving—made the laborious carving involved in wood engraving redundant. Once you could photo-sensitize a metal plate, you could transfer an image and etch or mechanically engrave it. Boxwood went out. Half-tone—the process which broke images up into dots for printing—lasted to within living memory. Pixels are its current incarnation. How Mr Butterworth's company survived in the Strand until his descendants cleared out the office in 1923 is anybody's guess, as the bottom fell out of the wood-engraving market in the 1880s.[5]

John Wertheimer survived the dangerous glitch in his finances and, perhaps because he had been mentored through it by his own largest creditors, learned a great deal by it. He went on to become extremely successful, developing his specialist book and periodical printing business through the nineteenth century, both in English and other languages. In 1917, the company advertised itself as printers, lithographers, manufacturing stationers, and account-book makers. In the mid-twentieth century the company was selected to print the British Union Catalogue of Periodicals (1955–6) a complex printing feat indeed, as anyone who has used it will appreciate. More recently, the firm John Wertheimer founded has grown into a major international company.[6]

After his son's death, Parker senior retired from publishing. He had lost heart, and sold out to Longman. He retired to his house near Farnham, and lived there for another ten years with his wife and their family, amid the fields. But he did not turn his back on Charing Cross before doing some essential things. Right across the road from the Pepperpots, gangs of labourers were busy tearing down the grand old building of the Hungerford Market, demolishing its many levels right down to the River, to make way for the new railway station planned for Charing Cross. The noise and dust and destruction laid waste the view from the shop as Parker and his long-time assistant/partner took a long inventory of their stock.

Old Mr Parker, and old Mr Bourn, together counted up their assets: the number of volumes they could put up for sale, and the intellectual property they owned at half profits or full copyright, their stereotype copies of wood engravings were checked and wrapped up carefully, their stores in the cellars at 445 Strand, in the shop, and in their warehouse, were all assessed, the lease for 445 Strand brought out, the cashbook totals totted up and the books in hand at other booksellers... then the royalties due to their various authors, payments due to suppliers like Spalding and Hodge, and other debts.

It was during this process that Gray's book *The Spleen* was finally dealt with—all 797 copies still lying on the shelf were offered for sale at a fraction of the original price of fifteen shillings. Out of the thousand printed in the days of young Parker's optimism, only 203 had been shifted, some of which had been review copies, author's copies, and gifts. Even at a heavy discount, there was only a single customer. George Henry Lewes, the companion and lover of Marian Evans (George Eliot) bought fifteen copies at four shillings each, perhaps as gifts for like-minded friends. The remainder Parker sent for waste, which is why copies are so scarce today.[7]

The death of young Parker, and then of Gray himself, had been a potentially serious problem for the successful survival of *Gray's Anatomy.* But old Mr Parker knew it was a valuable thing, both intrinsically and commercially, and ensured its safe survival by asking Timothy Holmes to edit the third edition. Holmes was the ideal person: trusted colleague of Gray and Carter at St George's, the man who had done so much work on the first edition, and who had got on so well with Parker's son.

Possibly during the first proofing process, when Gray had delegated everything to Holmes so he could get away on the Sutherland yacht, young Parker had hatched an idea for an entirely new project, with Holmes in charge. It turned into a series of four fat volumes called *A System of Surgery*, edited by Holmes, containing an array of fine essays commissioned from the pick of the upcoming young medical profession in mid-Victorian London, writing on subjects of their own choosing.[8] It was an unusual venture. Some brilliant doctors were encouraged to share their expertise, and given the space of an essay to discuss a field in which they had a personal interest. Several of Holmes's contributors remain well-known even today—John Simon, James Paget, Henry Gray, Ernest Hart, Jonathan Hutchinson, Brown Séquard, and Joseph Lister. Their prominence offers a good measure of Holmes's acuity in asking them to write, and equally, the respect in which he was held in their agreeing to do so. This commission from Parker gave Holmes a recognition which had hitherto evaded him, and allowed him to return a compliment to Gray, by commissioning him to do something in a book with Holmes's name on the spine. The series probably stood Holmes in good stead when the Assistant Surgeoncies came up at St George's in 1861.

By 1862, if not before, *Gray's Anatomy* had become the standard work, not just for students, but for teachers of anatomy. A delightful letter from an anatomy teacher published in *The Lancet* was severely critical of the top examiners at the Royal College of Surgeons for being old-fashioned and getting their anatomy wrong, having marked down one of his own students as being in error, when in fact they were incorrect themselves. The letter-writer cited three published works to support the accuracy of his own teaching: first Gray, then Quain & Sharpey, then Ellis, demonstrating that his mental arrangement of these authorities was neither by date or alphabet, but by choice.[9]

Wisely, Messrs Longman respected Parker's decision to recruit Holmes. He subsequently remained Editor of *Gray's Anatomy* until 1880. Having seen the ninth edition through the press, Holmes relinquished the project in which he had been involved since almost the beginning, and had helped steer towards its long-term survival. Had the hiatus after Henry Gray's death been less well managed, the book might have lost its status as the best anatomy book on the market, and could have gone the way of all its contemporaries, first into intellectual limbo, and then publishing oblivion. Holmes kept it institutionally grounded at St George's, kept it abreast of current anatomy and surgery, and retained the book's visual character as best he could, gradually adding to Carter's illustrations as Gray had done.

It is a very great pity that Timothy Holmes never published a memoir of Gray or Carter, or indeed of young Parker, whom he clearly liked very much. Had he written what he knew, there might have been little need for a book like this one, since he knew so very much. As a historian, I often speak to doctors' groups, urging them to record their lives and experiences for future historians. Too few realize that what they see as everyday and ordinary will in a hundred years' time be perceived as extraordinary, compellingly interesting, and specific to them, and that if not recorded now it will die with them. I fear that this may be what happened when Timothy Holmes died in 1907. Had he recorded his life, especially that interesting period at St George's alongside Carter and Gray, Holmes would have been able to tell us almost the entire inside story of *Gray's Anatomy*. He knew St George's like the back of his hand—every ward, every member of staff, and the same at Kinnerton Street. Holmes might have told about the shop at 445 Strand and the talk around Parker's table, father and son, but most especially about Gray and Carter, and the process of creating *Gray's Anatomy*. Holmes might have commented upon how Gray got on with the nobility, and whether he came back from the Sutherlands an insufferable snob or humbled by the experience. The 1858 title page suggests the latter unlikely, but he may have altered. Gray's reaction to the good and bad reviews would have been interesting to hear, as would the talk buzzing about the corridors at Kinnerton Street, or at the Royal Medical and Chirurgical Society; and the sorrow at the funerals of both young Parker and Gray. Had he wished to do so, Holmes could have detailed the body procurement system at St George's, and perhaps could have told us something about those whose faces appear in Carter's famous illustrations. Holmes would have known Gray's mother, might have met John Wertheimer, may have visited Butterworth and Heath. He may even have known the story behind the title page proofs.

Had Holmes chosen to do so, he could have saved me years of work: no book like this would have gone begging to be written, and many things which remain mysterious might have been explained. It may be, of course, that there is a manuscript in his neat hand sitting somewhere in an attic, a modest manuscript history of *Gray's* in an archive that I've missed quarrying: I do hope so. But equally, Holmes may never have recorded what he knew. Perhaps he felt he knew too much, or, not being able to say all good things, chose not to say any bad. Perhaps his was a silence of discretion. Holmes kept his silence for nearly fifty years, and in that time, people who might have spoken also kept quiet. The silence might be explained away by detecting in Holmes a stiff English manner and an inability to express feeling. He is said to have been a terror to students: with a harsh voice, and only a single eye (the other lost to an infection accidentally acquired during an operation) and his manner 'carefully cultivated to hide any interest he might feel'. But those who knew him discerned that his 'lack of sympathy was entirely assumed', and that 'he was the friend and trusted adviser of all who sought his help.' Holmes was capable of deep feeling, and whatever his manner, he did express a public lament for young Parker, twice, in different volumes of his *System of Surgery*.[10]

But late in life, speaking apropos of the death of Henry Lee, who had been appointed along with himself to the vacant posts at St George's in 1861, Holmes allowed himself to touch for an instant on his old friend Gray, whose death, he said, had

> brought to a sudden close a career of perhaps the brightest promise of all in the ranks of the junior surgeons of London at that date.[11]

Holmes looks to have pondered many times what Gray might have made of his life had he lived. Yet he chose never to publish an obituary of his friend in *Gray's Anatomy*, and when he wrote a biography of Sir Benjamin Brodie, Gray did not feature in it. Holmes remained at St George's all his working life, a stalwart of the institution, becoming its Treasurer after he retired from the wards. He outlived Carter by a decade, so that cannot have been the reason.

Holmes's editorial silence was preserved by others for a decade after his death in 1907: no reference to the life or character of the author of *Gray's Anatomy* appeared in its pages, and no tribute to him, until after the First

World War, when the memorialization of heroic young men was more general. Carter went entirely unhonoured in its pages until much more recently.

The book itself has survived by dynastic succession, mentor to junior, for most of its long life. The man to whom Timothy Holmes handed on the editorial baton was one of his own students, Pickering Pick. A young man of twenty at St George's at the time Henry Gray died, Pick was a long-term Editor of *Gray's* (1883–1905) and also Inspector of Anatomy, so he was sure to have known a lot from personal observation, and had probably heard stories about Gray from George's men, and perhaps from other anatomists, too.[12]

In the 1890s, Pick wrote an article in the *St George's Hospital Gazette* on the large Langhorn photograph of the dissecting room at Kinnerton Street, reproduced on page 248 in this book. It is an interesting piece, in which Pick takes the reader on a tour of the room, telling stories and speaking with great warmth about many of the individuals in the photograph, almost all of whom he seems to have known by name. But curiously, despite his detailed knowledge, and the sharpness of his memory, Pick found nothing to say beyond the obvious about Gray, sitting there right at the front of the photograph, by the corpse's feet. He merely asked the rhetorical question:

> And what shall I say of Henry Gray?

What Pick chose to say was minimal, and evasive: that Gray worked hard, that the great ambition of his life had been to become a Surgeon at St George's, and that there was something very tragic about his end. He related not a single anecdote, or memorable phrase or deed of Gray's. Pick conveyed no real sense of Gray's personality, and related nothing kind or witty about him, or said about him by anyone else.

It seems strange that Pick could bring himself to say no more than he did. It was common for Victorian writers to be reticent about living individuals, and Pickering was silent about people he knew were still alive; but he behaved in the same way concerning Gray, dead for thirty-five years, almost as if there was some superstitious reason why he might not share any scrap beyond common knowledge. Was this because he felt the *book* still lived, or was this the reticence of knowing too much? I have puzzled about this silence, which is not confined to Pick. It may have been because the younger man was deferential towards Holmes and his steadily maintained silence.

One is left with the feeling that Pick had plenty to say, but whatever it was, it was not warm. He may have been inhibited by Holmes himself; or perhaps like Holmes, preserved silence through a reluctance to speak ill of the dead, or from loyalty to St George's, or to the founder of the institution that was *Gray's Anatomy*.[13]

Pickering Pick remained in post at *Gray's* until the early twentieth century. The next editor, Robert Howden, saw the book through from the fifteenth to the twenty-third editions, and then TB Johnston was in post to 1958. Over time the book has grown in size, and has become less a student work than a major work of reference, an indispensable encyclopaedia of anatomy, whose illustrations are extraordinarily various, colourful, and informative. Since the 1960s a succession of editors have been assisted by a gradually growing team, until for the most recent edition, the Editor-in-Chief, Professor Susan Standring, the book's first woman editor, was coordinating a team of eighty-five editors and other contributors, and twelve illustrators.[14]

The field of knowledge the book contains is vast. The later appointment of editors whose main interest was anatomy, rather than surgery, meant that Henry Gray's original intention of a book for students, and of keeping surgery at the forefront of the book's preoccupations, fell by the wayside to some extent, and the detailed accretions of anatomical discovery included in its pages grew like a pathological proliferative process. For the 2005 edition, Susan Standring, determined to ensure the great book's continuing relevance to surgery, took the historic decision to reorganize the entire volume, and to depart from the division by systems—the nervous system, the lymphatic system, and so on—to bodily *regions*—limbs, the abdomen, the chest—which is how surgeons need their anatomy. The current edition, now published by Elsevier, is over 1600 pages long, but also features a compact disc of the contents, and access to updates on the Internet.[15]

So *Gray's* remains topical, fresh, relevant, and accessible by technology undreamed of in Gray and Carter's day. But, if the Editors of *Gray's Anatomy* were to be able to stand in line, it would take only seven handshakes to get back to Gray and Carter.[16]

Teaching anatomy in Bombay, Carter had to write up his annual report for the official record of the Grant Medical College's achievements for the medical school year 1859–60, recording the statistics of students taught and lectures delivered. After all the details of the heavy workload, Carter gave a sketch of the course itself, by saying:

> the order of subjects in descriptive anatomy has been generally conformed to that of the text-book, *Quain's*, or (which is becoming a favourite book with the students) *Gray's Anatomy*, and without vanity I may say that the figures in the latter work are calculated to greatly assist students.

To this, Grant College's Acting Principal, Professor Peet, added a footnote:

> Dr Carter is the author of the beautiful plates by which *Gray's Anatomy* is illustrated.

In this unassuming Indian official document, then, we can see that Carter was at last able to enjoy some satisfaction from his own achievement on the book.[17] Not only was his Principal full of admiration for the beauty of his work, but Carter was using it to teach. Best of all, his illustrations were found to be helpful, and the book was a favourite among his students, the new cohort of those intelligent Hindus and Parsees he had so admired at the first prize-giving.

Holmes remained in touch with Carter in India, and indeed presented scientific papers on his behalf in London.[18] Carter remained an employee of the Indian Medical Service in the Bombay Presidency for the rest of his working life, ending up as Principal of Grant Medical College, and First Physician at the JJ Hospital. During his working years in India, in addition to his teaching and clinical responsibilities, Carter became an indefatigable clinical researcher of the indigenous diseases of India, especially those affecting the very poorest levels of Indian society. Those same eyes that saw and those same hands that drew the beautiful micrographs for Gray's *Spleen* essay, and dissected and drew the illustrations for *Anatomy Descriptive and Surgical*, also focused upon the terrible malady of leprosy, the widespread killer famine or relapsing fever, cholera, deforming diseases such as mycetoma, elephantiasis, and lymphscrotum, and those other major killers, tuberculosis and malaria.

The breadth of his interests show that Carter remained at heart a generalist, but the depth of understanding and expertise he carried with him place him also as a specialist in a number of fields. Remarkably, Carter was the first researcher in the entire subcontinent to confirm the existence in India of the organisms which cause leprosy, described by the Norwegian researcher Armauer Hansen; tuberculosis, described in Germany by Robert Koch; and of malaria, famously discovered by Alphonse Laveran of France, and described by William Osler of Canada.[19]

Carter remained recognizably the same good-hearted shy man, however. He named the lectureship in physiology he endowed at Grant Medical College after someone else.[20] In a letter to a London friend in 1887, Carter wrote:

> next year at this time God willing my time will be up—the 29th March—and then for such repose as may come. The [Royal] College [of Surgeons] is almost like a home—at least a house of call—was I there before you 'joined' ? Always too reserved Uno? I regret not accepting your kind invitation to call—until it became too late. I have longed to say this many times.
>
> I am in sole charge here of a nice looking hospital (native General) and have just the sort of work I like, wanting only a congenial soul or two fond of our science—if it can be called such, too perplexing as it still is. The love of medicine as an object of man's study I hope never to lose … [but] since suffering from Relapsing Fever my eyesight cannot be overtasked. India's Pathology is an almost virgin field.[21]

After thirty years in India (1858–1888) Carter returned to England, and settled in Scarborough, near his sister Lily and her family. His first wife having died, Carter was free at last to marry again, and he found late happiness with Mary Ellen Robison, whom he married in 1890. They had two young children before he died of tuberculosis at the age of 65, in 1897. They are buried together in Scarborough cemetery. By the time you read this book, a blue plaque honouring his life's work will mark his old home, in Scarborough's Belgrave Square.[22,23]

Henry Vandyke Carter, with his wife Mary Ellen and their young son, photographed on the Esplanade at Scarborough, in 1892.

APPENDICES

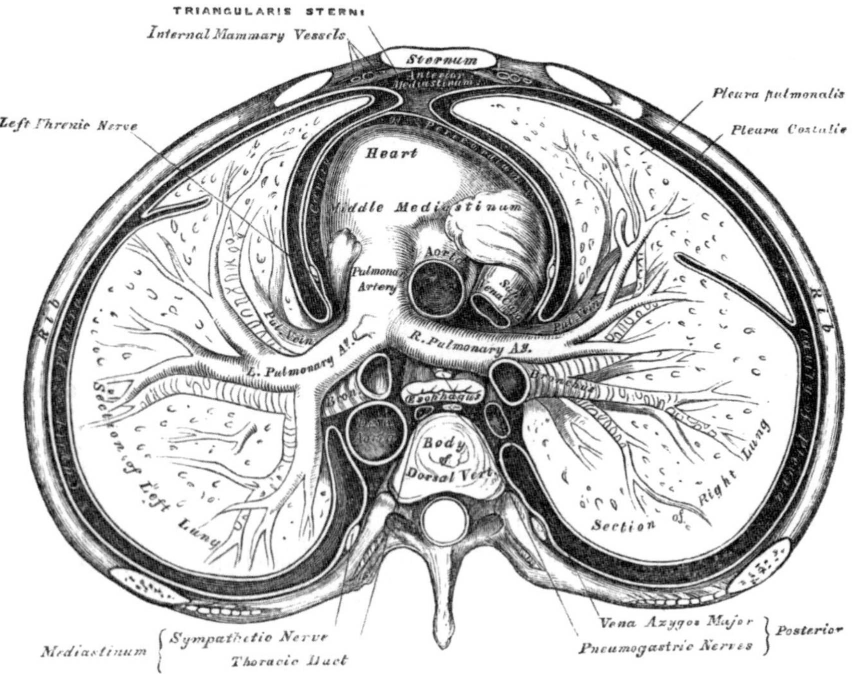

TRIANGULARIS STERNI
Internal Mammary Vessels
Sternum
Anterior Mediastinum
Pleura pulmonalis
Pleura Costalis
Left Phrenic Nerve
Heart
Middle Mediastinum
Aorta
Pulmonary Artery
Pul. Vein
R. Pulmonary Aa.
L. Pulmonary Aa.
Oesophagus
Body of Dorsal Vert.
Rib
Rib
Section of Left Lung
Section of Right Lung
Vena Azygos Major
Pneumogastric Nerves
Posterior
Mediastinum
Sympathetic Nerve
Thoracic Duct

APPENDIX 1: *Gray's Anatomy* Editions 1858–2008

	Date	Author/Editor/s *TITLE/Details*	Publisher
1st	1858	Henry Gray	JW Parker & Son
		ANATOMY DESCRIPTIVE & SURGICAL	
		The drawings by Henry Vandyke Carter. The dissections jointly by the author & Dr Carter.	
2nd	1860	Henry Gray	JW Parker & Son
3rd	1864	T Holmes	Longmans
4th	1866	T Holmes	Longmans
5th	1869	T Holmes	Longmans
6th	1872	T Holmes	Longmans
7th	1875	T Holmes	Longmans
8th	1877	T Holmes	Longmans
9th	1880	T Holmes	Longmans
10th	1883	TP Pick	Longmans
11th	1887	TP Pick	Longmans
12th	1890	TP Pick	Longmans
13th	1893	TP Pick	Longmans
14th	1897	TP Pick	Longmans
15th	1901	TP Pick & R Howden	Longmans
16th	1905	TP Pick & R Howden	Longmans
17th	1909	Robert Howden	Longmans
		ANATOMY DESCRIPTIVE & APPLIED	
		Notes on applied anatomy by AJ Jex-Blake & W Fedde Fedden.	
18th	1913	Robert Howden	Longmans
19th	1916	Robert Howden	Longmans
20th	1918	Robert Howden	Longmans
		First edition to feature a photograph and obituary of Henry Gray.	
21st	1920	Robert Howden	Longmans
		Notes on applied anatomy by AJ Jex-Blake & John Clay	
22nd	1923	Robert Howden	Longmans
		Notes on applied anatomy by John Clay & John D Lickley	
23rd	1926	Robert Howden	Longmans
24th	1930	TB Johnston	Longmans
25th	1932	TB Johnston	Longmans
26th	1935	TB Johnston	Longmans
27th	1938	TB Johnston & J Whillis	Longmans
		GRAY'S ANATOMY	
28th	1942	TB Johnston & J Whillis	Longmans
29th	1946	TB Johnston & J Whillis	Longmans
30th	1949	TB Johnston & J Whillis	Longmans
31st	1954	TB Johnston & J Whillis	Longmans
32nd	1958	TB Johnston & DV & F Davies	Longmans
33rd	1962	DV Davies & F Davies	Longmans
34th	1967	DV Davies & RE Coupland	Longmans
35th	1973	R Warwick & PL Williams	Longmans
		Separate neurology volume: Functional Neuroanatomy of Man.	
36th	1980	PL Williams & Warwick R	Churchill Livingstone
37th	1989	P L Williams	Churchill Livingstone
38th	1995	P L Williams & editorial board	Churchill Livingstone
		Historical introduction by Dr Tilli Tansey	
39th	2005	Susan Standring & editorial board	Elsevier
		Historical introduction by Dr Ruth Richardson	
40th	2008	Susan Standring & editorial board	Elsevier
		150th anniversary edition	

APPENDIX 2: The Wertheimer Parchment

Names (deciphered from signatures)	Amount owed in £sterling	Likely or possible identity*
John Cooper for Self & Partner	514+	[?Grocer, Finsbury <u>or</u> Analytical Chemist, Blackfriars? *1846*]
John Hopkinson for Self & Cope	187+	[Printing Press manf, Finsbury, *1846*]
Edward Dewick for Self & Partners	807+	[Wholesale Stationers, Herring, Dewick & Hardy, Walbrook, *1846*]
KM Sanderson for Self & Partners	602+	*untraced*
John Hodge, for Self & Partners	832+	[Paper manf., Spalding & Hodge, Drury Lane, 1846]
George Astle, for Self & Partners	99+	[Bookbinder, Crystal Palace Library *Hodson's 1855*] [*Mother & Sons, Bookbinders, Kelly, 1846*] [*Astle & Co, Bookbinders, 80 Coleman St, 1852 Kelly, POD.*]
Joseph Collinson	49+	[Wholesale Stationer, 3 Great Trinity Lane, *1846*]
for Partners & Self, John Mathewes	202+	[?Mathews, Wholesale Stationers, High Holborn, *1846*]
George Fagg, for Self & Partner, HW Caslon & Co	1,322+	[Typefounders, Caslon & Fagg, Chiswell St, *1852.*]
Charles Knight, for Self & Partner	227+	[Publisher, 90 Fleet St, *1852*]
Horton Harrild, for Partner & Self	148+	[Printing Press manf & supplies, 1846, 1852/ Litho printer, *1852*]
James Figgins, for Self & Partner	504+	[Typefounders, Vincent & James Figgins, Smithfield, *1846, 1855*]
GP Nicholls, for Self & Partner	72+	*untraced* [*?George Nicholls, printer, Leicester Sq, 1845*]
Wm McMurray [in pencil]	97+	[Bookbinder, 2 Lillypot Lane, Noble St *1846, 1855*]
FH Edwards, for Self & Partner [in pencil]	—	[Ink manf., *Edwards & Hopkins, ink, dye & Chemical Manf, 1846*]
Robert Bryansdon or Bryanston	197+	*untraced*
James King	211+	[Engraver, Blackfriars, *1846*]
for Self and Partners, ?D Ward Chapman	543+	*untraced* [*illegible*] [*?ironmonger, 1846*]
for London Discount Co. Limited, Edward J Woodhouse	295+	[bank]
Spencer, Budden, for Self & Partner	113+	[E.India & Wine Merchants of Fenchurch St [London Gazette, 1863]
Charles West	113+	*untraced;* [*Chas West Chemist, Chas West Physician, or RCWest engraver?*] [*William West was a specialist lithographic printer, in Hatton Garden.*]
The City Bank, J White Manager	135+	[bank]
By authority of James English, William Vallance	28+	[Vallance = solicitor]
pp Harrison and Co, A Bence or Benn	31+	[Harrison, printer, St Martin's Lane, *early partner of JWParker senior*].

* References are to London Directories between 1845 and 1865, mainly to Kelly Post Office Directories.

? = uncertain: there is clearly plenty of scope for further research here. If any reader can offer clarification I would be most glad to hear of it.

APPENDIX 3

A Word About Victorian Money

Pounds, shillings, & pence, & wages

In the Victorian era, the British currency was the £ pound sterling.

Each pound was divided into 20 shillings.

Each shilling contained 12 pence.

Each penny contained 4 farthings.

Unskilled male weekly wages in the 1850s ranged between 10 and 20 shillings, depending on the nature of the work. Women usually earned much less.

'Relative sums of money' website: http://privatewww.essex.ac.uk/~alan/family/N-money.html#1264 accessed June 2008.

A Word About Victorian Measurements

Feet & inches

In the Victorian era, the standard linear measurement was the inch, which is roughly 2 ½ centimetres.

Twelve inches made one foot (roughly 30 centimetres).

Three feet (36 inches) made a yard (roughly 90 centimetres).

Inches were generally divided into 8ths, 10ths, 12th, or 16ths.

'x' height is the height of a lower case letter x.

NOTES
&
REFERENCES

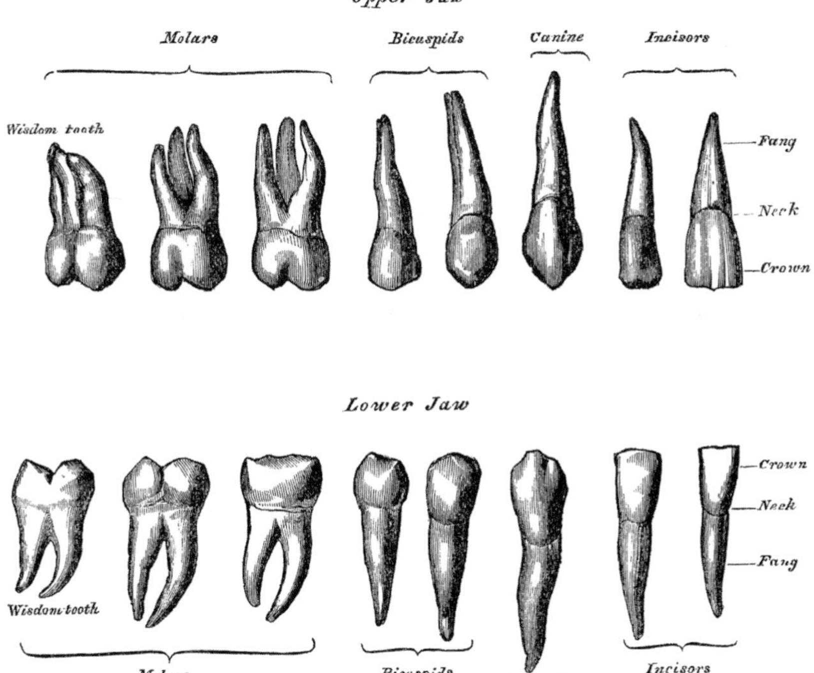

BIBLIOGRAPHICAL NOTE

Carter Papers—refers to the collection of material relating to Henry Vandyke Carter, held at The Wellcome Library, London.

Longman Archive—refers to material relating to Messrs Longman, held at the University of Reading Archives.

Nicol Archive—refers to the private archive of Gray material of Mr KE Nicol.

INTRODUCTION

1. The copy of the book John Fawthrop MRCS inscribed was a first edition, dated 1858. It is a hundred and fifty years old, and has been rebound. It feels elderly, more than second hand, not at all in pristine condition. It sits on the open shelf in the Wellcome Library, in London, that hospitable and blessed place every historian or reader with a love of medical history appreciates for the many riches it holds. Any reader may consult the volume, and find Mr. Fawthrop's inscription. After he qualified, John Fawthrop settled in a country district near Halifax. Two other Fawthrops listed in the district in the 1861 Medical Directory (perhaps his father and brother) were apothecaries.

CHAPTER 1: WORDS

Mr Gray of Belgravia

1. My grateful thanks to an anonymous referee at Oxford University Press for the Roget simile.
2. In the novel *Arrowsmith,* by Sinclair Lewis (New York, Harcourt Brace, 1925: 4). My thanks to Chris Hamlin for this splendid reference.
3. Most of the copies of the first edition I have managed to locate show evidence of having been passed down or sold on more than one occasion. Copies of most Victorian editions of the book, whether in private hands or libraries, have usually been well used.
4. Carter Papers, Wellcome Western MSS 5818: 25.11.1855.
5. Hudson, D: *Munby: Man of Two Worlds.* London, Murray, 1972: 50.
6. *The Times,* 6.10.1830.
7. Wilkins used iron girders in place of wooden floor beams, which made the building capable of bearing the extra weight of an added floor when new wards were added in the late 1850s. Wilkins's building is still in use. See also Blomfield, J: *St George's 1733–1933.* London, Medici, 1933; and the history of the other medical school nearby: James, RR: *The School of Anatomy & Medicine adjoining St George's Hospital 1830–1863.* London, printed for the author, 1928.
8. Anon: Good Results of the Amputations at St George's Hospital. *The Lancet,* 11.8.1860: 136. Pyaemia is what we would now call septicaemia. There was no cure, and it was generally fatal.
9. Landon, LE: St George's Hospital, Hyde-Park Corner. In *Poems,* London, Routledge, n.d (*c.*1875): 293.

10. See MacNally, AS: *A Biography of Sir Benjamin Ward Richardson.* London, Harvey & Blythe, 1950. Gould, T & Uttley, D: *A Short History of St George's Hospital.* London, Athlone Press, 1997: 44–5.
11. [Wardrop, J]: Intercepted Letters. *The Lancet,* 4.7.1835: 456–7.
12. *Oxford Dictionary of National Biography* entry for Brodie. Holmes dates the school's opening to 1836, but I suspect in error. See Holmes, T: *Masters of Medicine: Sir Benjamin Collins Brodie.* London, Fisher Unwin 1898: 117–18. Opening of Kinnerton Street see Gould, T & Uttley, D: *A Short History of St George's Hospital.* London, Athlone Press, 1997: 7. There seems to be some confusion about the site of the Kinnerton Street School, and its date of opening. The rented premises seem to have been a temporary expedient, while the Hospital and the purpose-built medical school were being constructed. It was for the *running* of the new building for which Brodie had put up the original capital that he later asked the Hospital for help, so as a speculation, it didn't seem to have yielded him a great profit. Brodie should perhaps be seen as a benefactor rather than a speculator. Gould and Uttley seem confused as to when the new building was opened, saying 1844 on p.8 and 1849 on p.74. *The Lancet* refers to Brodie's school as a 'new building' in 1835, and covers the event. *The Times* mentions its opening in 1840, when it was referred to as the Anatomical Theatre. See *The Times,* 5.5.1840: 6.
13. Barton, N: *The Lost Rivers of London.* London, Historical Publications, 1992. For a stunning description of the slum housing which lay along its southerly banks, see *Household Words* 1854: 398–402.
14. *The Times,* 24.6.1839: 3.
15. *Charles Booth's Descriptive Map of London Poverty, 1889.* London, London Topographical Society, 1984.
16. My knowledge of this building is gleaned from a variety of sources and from surmize based on careful observation, indoors and out, of the building as it survives. My description may not be 100% correct in every detail, but gives a good impression nevertheless, of this early purpose-built medical school.
17. Noad gained a doctorate in physics at Geissen. *Oxford Dictionary of National Biography* entry for Noad.
18. Chaplin, S: John Hunter and the Anatomy of a Museum. *History Today,* 2005, 55: 19–25.
19. The connection to William Mews was still there in 1888—see *The A to Z of Victorian London.* London Topographical Society, 136, 1987 which reproduces George W Bacon's 1888 Ordnance Atlas of London: Sheet 22.
20. The building is now called Bradbrook House.
21. Gould, T & Uttley, D: *A Short History of St George's Hospital.* London, Athlone Press, 1997: 75.
22. The researches of Mr KE Nicol have located some materials which were preserved by the Gray family after Gray's death. His discovery gives hope that other things may have survived.
23. Irvin Loudon's *Oxford Dictionary of National Biography* entry portrays Gray as a heroic figure: 'it is a measure of Gray's single-minded devotion to anatomy and authorship that *Gray's Anatomy* remains even today, not only an important book of reference but as virtually a household phrase.'
24. Gould, T & Uttley, D: *A Short History of St George's Hospital.* London, Athlone Press, 1997: 73–5.

25. 1st March 1819. Transcript of document in Nicol Archive.
26. Nicol, KE: *Henry Gray of St George's Hospital: A Chronology.* Published by the author, 2002. See also *Oxford Dictionary of National Biography* entry for Henry Gray. The Hospital had been named for the local parish patron saint.
27. Personal communication, Archivist, Westminster School, 2005.
28. Goss, CM: *A Brief Account of Henry Gray.* Lea and Febiger, Philadelphia, 1959: 16–17.
29. Gray used this motto twice on prize essay submissions.
30. Gray was born in 1827. He was awarded the St George's Hospital Medical School Junior Prize Certificate on 24th April 1843, at the end of the medical school winter season, which had begun the previous October. St George's Hospital Prize Register, starting 1837. See also Nicol, KE: *Henry Gray of St George's Hospital: A Chronology.* Published by the author, 2002.
31. Personal communication, Dee Cook, Archivist, Society of Apothecaries.
32. Gray, H & Carter, HV: *Anatomy Descriptive and Surgical.* London, JW Parker & Son, 1858: Dedication. The reasons for Brodie's hostility to apothecary training are discussed by his biographer. See Holmes, T: *Masters of Medicine: Sir Benjamin Collins Brodie.* London, Fisher Unwin, 1898: 28–9.
33. Gray may not have been aware of the public ridicule poured on another Brodie enthusiast in the 1830s, known in *The Lancet* as 'Dedication Hope'. See 'Clique at St George's', *The Lancet,* 27.6.1834: 425–7. A spoof dedication appeared at the time in *The Lancet* thanking Brodie 'for the excellence of the Hospital arrangements which have afforded [Hope] an invaluable sphere for the prosecution of researches on the large scale'. *The Lancet* 15.11.1834: 256.
34. *Oxford Dictionary of National Biography* entry for Brodie.
35. None of the materials examined in the archives at St George's offers clues to why Sir Benjamin Brodie took such an interest in the younger man. It is possible that Henry Gray was related to the distinguished naturalist Dr JE Gray of the Natural History Museum, and the Zoological Society, or the famous family of seedsmen of Pall Mall. Another Gray dedicated a book on artificial limbs to Brodie. See Gray, F: *Automatic Mechanism.* London, Renshaw, 1855. 'Dedicated by special permission to Sir Benjamin Brodie Bart. with upwards of 100 Illustrative cases and correspondence by F Gray Esq., Operator to the Marquis of Anglesey.' A Sir Henry Gray served George IV. As I have noted Gray is not an uncommon name—once one starts looking they are everywhere. The reason may not, of course, have been familial.
36. Gould, T & Uttley, D: *A Short History of St George's Hospital.* London, Athlone Press, 1997: 44. *The Lancet* in 1834 was buzzing with criticism of the 'disagreeably notorious' and corrupt state of the hospital.
37. It is stating the obvious to observe that better-off students therefore found it easier to ascend the hospital heirarchy.
38. The career of Henry Lee, is an instance in point. Timothy Holmes might have had a similar fate, had Gray lived.
39. Nicol, KE: *Henry Gray of St George's Hospital: A Chronology.* Published by the author, 2002.
40. St George's Hospital Medical School Register shows that Carter was registered on 27th May, five days after his birthday. St George's Archives, Tooting.
41. It is possible, though not probable, that Carter provided Gray with a few of the illustrations for his Triennial prize essay, as there are one or two images in the second

half of the essay which look as though they could have been from his keen young hand. See Gray, H: *An Essay on the Origin and Distribution of the Nerves of the Human Eye and its Appendages.* Royal College of Surgeons of England, Triennial Prize Essay 1848. Manuscript. RCS Library, London: 172, 198, 256. Gray gave no credits for the illustrations, not even to himself. It is known that Gray employed at least one other artist, so these may not in fact be Carter's.

42. Gray submitted the essay on 13th December 1848 according to Nicol, KE: *Henry Gray of St George's Hospital: A Chronology.* Published by the author, 2002. Moles were used by von Baer, who in 1828 published his discovery of mammalian ova and spermatozoa. They were also cheap and easy to obtain. See KE von Baer: *De Ovi mammlaium et Hominis Genesi.* (1828). Trans. O'Malley, CD. Cambridge, Mass. History of Medicine Society, 1956.
43. The winner of the second prize was Alfred Poland of Guy's, who was also awarded an equal 50 guineas, so a very close second. These were large sums.
44. Royal College of Surgeons, London, 31.7.1849. Nicol, KE: *Henry Gray of St George's Hospital: A Chronology.* Published by the author, 2002. Edward Stanley was President of the RCS at the time.
45. Todd, RB & Bowman, W: *Physiological Anatomy and Physiology of Man,* published in parts by JW Parker & Son 1843–1856. Sir William Bowman is known to posterity as a preeminent ophthalmologist. He was the first to describe the rods and cones in the eye. He was the first President of the Ophthalmological Society, and the figure of a bowman/archer on its historic crest (prior to incorporation as a Royal College) was a reference to him. See *Oxford Dictionary of National Biography* entry for Bowman. How Gray and Bowman knew one another is an interesting question: possibly through Brodie, a friend of Bowman, or through the Royal Medical and Chirurgical Society of which both were members. This sort of networking was crucial to Gray's career-building.
46. Gray was present at the event, on 31st January 1850. Nicol, KE: *Henry Gray of St George's Hospital: A Chronology.* Published by the author, 2002.
47. Gray, H: On the development of the Retina and the Optic Nerve and of the Membranous Labyrinth and Auditory Nerve. *Phil. Trans.*, 1850 (1): 189–200. Received 18th October 1849; read 31st January 1850.
48. Royal Society records: EC/1852/04/291.
49. Dr Diana Manuel, the biographer of one of Gray's supporters, Marshall Hall, has observed that in her view, the youth of a successful applicant to the Royal Society can be understood as a surrogate measure of insider status. Many scientists of greater importance than Gray were invited to join the Society only much later in life, or excluded altogether. Dr Manuel, personal communication, January 2008. Nathaniel Ward, for example, inventor of the Wardian case, was made a Fellow on the same day as Gray. He was in his seventies.
50. Cooper Prize Archives Administrative records. Manuscript in the Archives of King's College, London: Pasted slip, folio 45. Cooper died in 1841.
51. Gray, H: On the development of the ductless glands of the chick. *Phil. Trans.* 1852: 295–309.
52. Gray, H: *On the Structure and Use of the Spleen.* 1853, Manuscript in the Cooper Prize Archives, King's College, London.
53. The list is exhaustive, and includes Aristotle, Plato and Galen via Vesalius and Leuwenhoek, Malpighi, Hewson, Dupuytren and Cuvier, to Home, Hodgkin,

Virchow, Kölliker and Funke, among many others, including Müller, Giesker, Henle, Purkinje and Schwann. Concerning Hewson, Gray cited the work of Gerlach, Schaffner and Bennett.

54. Gray, H: *On the Structure and Use of the Spleen.* 1853, Manuscript in the Cooper Prize Archives, Archives of Kings College, London: 307.
55. For the rumour re Kölliker see Carter Papers, Wellcome Western MSS 5816: 21.10.1850.
56. Gray, H: *The Structure and Use of the Spleen.* London, JW Parker & Son, 1854. The British Museum Reading Room copy of Gray's book is stamped as having been received on 7.6.1854. The printer was Savill & Edwards, of Chandos Street.
57. Goss, CM: *A Brief Account of Henry Gray.* Lea and Febiger, Philadelphia, 1959: 3–21. For the allegations about Müller, see p. 18.
58. Carter Papers, Wellcome Western MSS 5816: 14.6.1850. For microscope-worship among younger students, see Galton, F: *Memories of my Life.* London, Methuen, 1908: 41.
59. Richardson, R: Microscopical Conversaziones. *The Lancet,* 2001, *358*: 2004. For the contemporary feeling of excitement see Galton, F: *Memories of my Life.* London, Methuen, 1908: 41.
60. Carter Papers, Wellcome Western MSS 5819: 11.12.1854.
61. Between 14th June and 5th August 1850, Carter mentions painting/drawing/engraving spleens for Gray on at least nine separate occasions; between 20th September and 2nd December, he mentions dissecting/drawing/working with Gray on chicks on eight occasions. Carter Papers, Wellcome Western MSS 5816: 1850. Gray's illustrator for the published chick paper, however, was someone else, so it seems Gray may have had some other publication in mind, or that these were perhaps for teaching. It is possible that the beautiful illustrations for Gray's paper on the retina, which are unsigned, were by Carter. So far I have been unable to trace whether, though they are dated 1850, the date may reflect presentation, rather than publication, and Gray may have had time to submit the work Carter was doing for him in the autumn of 1850 for the printed volume of *Philosophical Transactions.*
62. Further research may yield clues as to where these illustrations may have appeared.
63. Brodie, BC: (Advertisement), To the Governors of St George's Hospital. *The Times,* 7.11.1834: 4. See also *The Lancet,* 21.6.1834: 477–9; 5.7.1834; 12.7.1834.
64. Boott, F., Bigelow, HJ., Liston, R: Surgical operations performed during insensibility. *The Lancet,* 2.1.1847: 172–9. Simpson, JY: On a new anaesthetic agent, more efficient than chloroform. *The Lancet,* 20.11.1847: 549–50. Both papers are available on-line via ScienceDirect and in Richardson, R: *Vintage Papers from The Lancet.* London, Elsevier, 2006: 19, 75–8, 81–2.
65. Snow, J: *The Casebooks of Dr John Snow* (Edited by Richard H. Ellis). London, Wellcome, 1994. For St George's, 624; for St George's men, see the Index of Surgeons and other Medical Men: 570–6.
66. *Ibid.*: Felton case: 308.
67. *Ibid.*: 271.
68. *The Lancet,* 14.5.1853: 453 (editorial).
69. Nicol, KE: *Henry Gray of St George's Hospital: A Chronology.* Published by the author, 2002.
70. Uninteresting patients were sent home, or to the nearby workhouse infirmary.
71. As a contemporary colleague later observed, 'Practice among the poor is the highway

to become rich.' Ogle papers, Royal College of Physicians Archive MSS 2420. *c.*1871. Manuscript margin comment. Ogle was at St George's with Gray.

72. *The Lancet*, 24.6.1854: 669–70.
73. For Brodie and *Fraser's* see Holmes, T: *Masters of Medicine: Sir Benjamin Collins Brodie*. London, Fisher Unwin 1898: 162. The involvement is dated to 1861, after these events, but the contact is suggestive. My knowledge of Gray's friendship with George Kingsley is due to Mr K Nicol. Transcript of document in Nicol Archive.
74. Cooper, BB: *The Life of Sir Astley Paston Cooper Bart.* London, JW Parker & Son, 1843.
75. *Ibid*: 2: 478.
76. The Royal Society was housed at Somerset House at the other end of the Strand.
77. *The Lancet*, 24.6.1854: 669–70.
78. No record of them remains in the Astley Cooper Prize Archive.
79. Administrative records of the Astley Cooper Prize, Guy's Hospital archives at King's College, London, 17.6.1850: fol.9.
80. Two of Carter's best spleen drawings reappeared (credited to Gray) in Todd and Bowman's final volume, published soon afterwards by JW Parker & Son, probably from the wood engraving, or a metal stereotype copy. Todd, RB & Bowman, W: *The Physiological Anatomy and Physiology of Man.* JW Parker & Son, 1856: 513, 514.
81. Desmond, A: *Huxley: the Devil's Disciple.* Michael Joseph, 1994: fig.14, facing p. 230.
82. Kölliker, A [Trans. Huxley, T & Busk, G]: *A Manual of Human Histology.* London, Sydenham Society, 1854.
83. Desmond, A: *Huxley: The Devil's Disciple.* Michael Joseph, 1994: 183.
84. The annual salary at the RI reveals the largesse of the Astley Cooper Prize purse of £300.
85. Gray's letter of application (though sadly, not the testimonials) has survived in the RI Archives. Gray asked for the testimonials to be returned to him. Transcript of Gray's letters in Nicol Archive.
86. *Ibid.*
87. Desmond, A: *Huxley: The Devil's Disciple.* Michael Joseph, 1994: 195, 199, 227. Wharton Jones may have recommended Huxley, as they knew one another from Charing Cross Hospital days; Huxley also had excellent testimonials, no doubt, and would probably have been brilliant at interview.

CHAPTER 2: IMAGES

Dr Carter of Scarborough

1. Some of the text of this chapter is based upon my essay on HV Carter in Bell, GH, Credland, A, & Richardson, R: *HB Carter & Sons: Victorian Watercolour Drawing and the Art of Illustration.* Pickering, Blackthorn Press, 2006: 32–67. See also Bayliss, A: Dr Henry Vandyke Carter (1831–1897) Yorkshire Illustrator of 'Gray's Anatomy'. *Yorkshire History Quarterly* 1998, 4: 69–71; Roberts, S: Henry Gray and Henry Vandyke Carter: creators of a famous textbook. *Journal of Medical Biography*, 2000, 8: 206–12.
2. Bell, GH: Henry Barlow Carter & Sons. In Bell, GH, Credland, A, & Richardson, R: HB Carter & Sons: *Victorian Watercolour Drawing and the Art of Illustration.* Pickering, Blackthorn Press, 2006: 1–31.
3. I owe this insight to Professor Gordon Bell, the authority on HB Carter.

4. Carter Papers, Wellcome Western MSS 5819: 3.7.1853.
5. Carter's diary entries are quoted here by permission of the Wellcome Library and Carter family descendant Sarah Potts, for whose kindness I am most grateful.
6. I have battled with this difficulty, and have decided not to penalize Carter because Gray's private writings are missing. The picture received of Gray is therefore Carter's: one could not hope for a closer view.
7. Carter Papers, Wellcome Western MSS 5810: 21.2.1851 (Letter to his sister Lily).
8. Hippolyte Taine, quoted in Gillian Tindall: *The Journey of Martin Nadaud.* London, Chatto, 1999: 186.
9. Carter Papers, Wellcome Western MSS 5818: 3.2.1858.
10. Carter Papers, Wellcome Western MSS 5809/4. Letter to his Mother from HVC 27.4.1851. Carter signed the letter with his initials, but reconsidering, crossed them out and wrote 'Harry' instead.
11. *A Catalogue of the Choice and Valuable Library of Books, including a Select Library of School Books … of the late JD Sollitt Esq., to sell by Auction, at The Grammar School, North Church Side, Hull, on Monday, July 6th, 1868.* Hull, Office of the *Hull Independent,* 1868. I am grateful to Arthur Credland for bringing this document to my attention.
12. Credland, A: The Carters and the Hull Art Scene in the Nineteenth Century. In Bell, G; Credland, A, and Richardson, R: *HB Carter and Sons: Victorian Watercolour Drawing and the Art of Illustration.* Pickering, Blackthorn Press, 2006: 76–7.
13. Henry Clark Barlow was an MD and a member of the Royal Geographical Society. The mature Carter found him rather a pathetic figure, commenting that he never finished anything, but when he was younger he may have seemed more impressive, older than he by 25 years, and a great traveller. See also *Ibid.*: 4, 29, 32, 98.
14. Bell, G (ed.): *Theakston's Scarborough Guide.* [Originally published Scarborough, 1841] 2nd edition, Pickering, Blackthorn, 2002: 141. The Guide was dedicated to the elder Travis, so he and Carter's father may have been friends.
15. Carter Papers, Wellcome Western MSS 5819: 3.7.1853. In 1853 Carter looked back at his apprenticeship, and wrote: 'entered the Surgery of a general practitioner in Scarbro' [Travis & Dunn] where were 3 fellows—got picked [on] among them too, being too sedate tender hearted—would not join in loose conversation or run after servants &c or smoke & was seriously disposed too—& not pushing—being home-nourished … had the good opinion of the head of the firm…: one of the assistants was unsteady—one always 'chaffing'—and one half-jealous, and proud—hence was not very comfortable, and learning nothing, came to town. Made no friends there, being considered as proud (as p'raps was) by the last who was the youngest:—rather soft by the 'chaffer' and p'raps thought more of by the 'sot'. (2 last since dead!)'
16. Society of Apothecaries, London. *Apprenticeship Registers.* HV Carter's certificate, 7.10.1852.
17. *St George's Hospital Medical School. Pupils' Register.* HV Carter's entry 1848–52.
18. Carter Papers, Wellcome Western MSS 5816: 11.2.1849.
19. Carter Papers, Wellcome Western MSS 5819: 3.7. 1853.
20. Carter Papers, Wellcome Western MSS 5810/2: 26.3.1848; 5816: 13.5.1849.
21. Carter Papers, Wellcome Western MSS 5819: 3.7. 1853.
22. Carter Papers, Wellcome Western MSS 5816: 7.7.1849.
23. *Ibid.*: 7.7.1849; 11.7.1849.
24. *Ibid.*: 24.6.1850. The Oxford English Dictionary dates the modern meaning of the

word to 1848. 'B&G' might mean Brodie and Gray, but the B could stand for Frank Buckland (a fellow student) or for someone else.

25. Carter Papers, Wellcome Western MSS 5816: 14.6.1850.
26. *Ibid.*: 14.6.1850. One hopes he was able subsequently to re-negotiate terms.
27. Carter Papers, Wellcome Western MSS 5817: 5.7.1852. Three pounds sterling in 1852 was sixty shillings, so even at the other artist's rate, this represents many hours of work.
28. Carter Papers, Wellcome Western MSS 5810: undated [2nd May] 1851.
29. *Ibid.*: 8.10.1852.
30. Carter Papers, Wellcome Western MSS 5817: *passim.*
31. *Ibid.*: 9.11.1852.
32. *Ibid.*: 8.11.1852.
33. Carter Papers, Wellcome Western MSS 5818: date uncertain (early January 1853). The anatomist Joseph Swan was thinking along more materialist lines at the same time: his book *The Brain in Relation to the Mind* was published by Longman in 1854.
34. Carter Papers, Wellcome Western MSS 5818: date uncertain (early January 1853).
35. *Ibid.*: date uncertain (early January 1853).
36. Richard Monkton Milnes quoted in FM Brookfield: *The Cambridge Apostles.* London, Pitman, 1906: 226.
37. Huxley moaned about string pullers and influence, but when push came to shove pulled strings himself. See Desmond, A: *Huxley.* London, Michael Joseph, 1994: 169.
38. *Ibid.*: 53–146 *passim.*
39. Carter Papers, Wellcome Western MSS 5810: 8.10.1852.
40. Carter Papers, Wellcome Western MSS 5818: 1.1.1853.
41. Blumfield, J, Ransome, GH, F.H., music by Chadborn, CH: *Nine Medical Songs.* London, St George's Hospital. n.d.
42. Carter Papers, Wellcome Western MSS 5819: 3.4.1853.
43. Carter Papers, Wellcome Western MSS 5818. The reference to Owen here is to Richard Owen at the Royal College of Surgeons, and suggests they were on good terms before Carter won his Studentship. There is a slight indication in the diary that Carter had tried and failed for a previous RCS Anatomical Studentship the previous summer, before he went off to Paris. This meeting with Owen suggests that he had probably received encouragement to try again.
44. Carter Papers, Wellcome Western Mss. 5818: 16–18.3.1853.
45. *Ibid.*: week ending 22.12.1855.
46. RCS Museum Committee Minutes, 14.6.1853.
47. *Ibid.*: Report to Council, 2.5.1855.
48. The job description of the studentship holder officially included the following tasks: Dissections of rare animals; Dissections of animals for Physiological series; Dissections of Pathological cases and specimens; Making Preparations for the Museum; re-putting-up and remounting specimens; Assisting in the making of the Catalogues; Assisting in the making of Diagrams for Lectures & anatomical drawings; Explaining the Museum to visitors. RCS: Museum Committee Minutes, 17.3.1855.
49. Carter, HV: RCS Anatomy Studentship Journal 7.7.1854. Royal College of Surgeons Archive.
50. The creature died on the 6th July 1854, it arrived at the RCS on the 7th. Carter was involved in drawing details of its anatomy for much of the rest of that summer.

Owen, R: On the Anatomy of the Great Anteater (Myrmecophaga Jubata, Linn.) [verbal report] July 25th, 1854. Published in *Proc. Zool. Soc, London,* Part XXII, 1854: 154–7, and plates. It took Owen another eight years to write up the finished report and to publish Carter's drawings of the post-mortem of this unusual creature. This fuller report is to be found in: Owen, R: On the Anatomy of the Great Anteater (Myrmecophaga Jubata, Linn.) Communicated at a scientific meeting of the Zoological Society, July 25th 1854. *Trans. Zool. Soc.,* 1856, 4(4) 117–35; 4 (5) 179–81, and plates. The parts in which the illustrations were published appeared in 1856 (4) and 1858 (5), but the full volume did not appear until later (London, Longman, 1862). The first plates of 1856 were misattributed to 'EVC', and remained so until recently in the Zoological Society's indexes. It is not known if Carter knew about this error. The mistake was corrected in the second batch of 1858, when Owen praised Carter's drawings. His comments were doubtless welcome, but too tardy to assist Carter's career.

51. Carter, HV: RCS Studentship MSS Diary 29.5.1854. Royal College of Surgeons Archive.
52. The list has been condensed from a longer version. RCS Museum Committee Minutes: 7.2.1855. Royal College of Surgeons Archive.
53. Wellcome Western MSS 5819: 18.3.1855. Condensed from original, paragraph breaks added, punctuation/spelling erratic in original. Carter does not rate Henry Gray here with scientists like Owen, Quekett, or BW Carpenter.
54. Carter's RCS diary makes clear the level of work required. In June 1855 for example, after he had left, the Museum Committee was informed that Carter, a porter, and a junior student had created '99 large lecture diagrams'.
55. In 1849 the entire staff at the Royal College were listed as: 3 curators, 3 students, 6 porters, 1 charwoman. The third curator in 1849 had been William Clift, who died, and was not replaced. Looking at the number of galleries, and bottles and skeletons, my own pity is greatest for the poor charwoman, who was probably paid even less than the students. In 1857, after Richard Owen had also left, Quekett suggested the abolition of the studentships, and asked for 3 permanent assistants at £200 per annum each.
56. Carter, HV: RCS Studentship MSS Diary, May 1855. Royal College of Surgeons Archive.
57. [Hunt, FK]: What there is in the roof of the College of Surgeons. *Household Words*, 1850 (1):464–7.
58. Carter Papers, Wellcome Western MSS 5819: 20.11.1853. Carter failed along with 40% of other applicants. The difficulty of the examination immediately gave it a high reputation. See *The Lancet*, 63.12.1862: 538. Willson, FMG: *Our Minerva.* London, Athlone, 1995: 179.
59. Carter Papers, Wellcome Western MSS 5818: 10.6.1855–24.6.1855.
60. Carter Papers, Wellcome Western MSS 5820/4. Letter from Eliza S Carter to her sons. 13.7.1855 Spellings original.
61. Carter Papers, Wellcome Western MSS 5818: 5.8.1855.
62. *Ibid.*: 11.11.1855.
63. See Chapter 6 here, and *J.Path.Soc.Lond.* 1855–6, *7*: 242–4; plates II, XII. Carter is listed as a member at the front of the volume.

CHAPTER 3: ENTERPRISE

JW Parker & Son of West Strand

1. For Dr Johnson see Desmond, A: *Huxley*. London, Michael Joseph, 1994: 23; Sala, GA: *Twice Around the Clock*, London, Houston & Wright, 1859: x.
2. Its eastern arm extended to Paternoster Row, Cheapside and beyond.
3. The current street numbering runs rather confusingly: 442, 443, 448. The best view of the West Strand Pepperpots building is in *Tallis's Street Views of London*. (1839) London, London Topographical Society, 2002. Reproduced here on page 63.
4. His obituary in *The Bookseller* says a naval officer, the *Oxford Dictionary of National Biography* an army officer. Parker published books which show an interest in both services.
5. Langley, L: The Life and Death of *The Harmonicon*: An Analysis. *Research Chronicle of the Royal Musical Association*. 1989, *22*: 137–63.
6. By c.1840 Clowes was employing nearly 150 men and 30 boys, almost 80 staff more than his nearest competitor, Luke Hansard. See Howe, E: *The London Compositor*. London, Bibliographical Society & Oxford University Press, 1945: 298–302. See also Clair, C: *A History of Printing in England*. London, Cassell, 1965: 227–9; and Clowes, WB: *Family Business*. London, Clowes & Son, 1953: passim.
7. Records of the Sun Fire Office, Guildhall Library Archives, City of London.
8. *Ibid.*
9. Apple, MA: *Teachers and Texts*. London, Routledge, 1986: chapter 4 passim.
10. *Popular Poems selected by 'EP'*. London, JW Parker, 1837. The two imprints feel like a Victorian artisan's version of the book-jackets Vanessa Bell designed for her sister Virginia Woolf's books.
11. I like to think that as a boy John W Parker Senior received some of his visual education from an ancestor of mine, a William Richardson who had a large shop on the corner of Villiers Street and the Strand (the site is now *air* in front of Charing Cross Station Hotel, opposite the Pepperpots building) and whose windows were brim full of prints of all kinds.
12. British Library. Add.MSS 41267A f.154. 19.1.1839.
13. McKitterick, D: *A History of Cambridge University Press*. Cambridge, Cambridge University Press, 1998, *2*: chapter on Parker, 328–51 passim. See also Leanne Langley's fine article, The Life and Death of the Harmonicon: An Analysis. *Research Chronicle of the Royal Musical Association*. 1989, *22*: 137–63. The lovely book *A Stickful of Nonpareil* by George Scurfield and Edward Ardizzone which dates from 1956 gives a genuine feeling about the worklife of the CUP printshop in the twentieth century, and it doesn't take much to imagine back from it, because Ardizzone's illustrations are so evocative. See Scurfield, G and Ardizzone, E: *A Stickful of Nonpareil*. Cambridge, privately printed 1956.
14. Parker's handwritten letter, dated 7 December 1842, is in the Archives (Add.MSS A 65/29) at Trinity College Cambridge, to whom I am grateful for having been allowed to study it, and to quote from it here. Most of its text appears in McKitterick, D: *A History of Cambridge University Press*. Cambridge, Cambridge University Press, 1998, *2*: 332–3. It is worth noting that Parker had been University Printer for six years when this letter was written, and that the printing trade was at the bottom of a trough in trade in 1842.

15. Trinity College Archives hold a cache of letters together with the one quoted above, which shows the Whewell–Parker relationship continued after Parker had left Cambridge University Press. See also Frisch, M & Schaffer, S: *William Whewell.* Oxford, Oxford University Press, 1991: 213. The Archive at the University of St Andrews holds correspondence between Parker and JD Forbes, which demonstrates contact between Whewell, Forbes, and the Parkers in 1847–1848. Whewell also supported Parker & Son's campaign on the Bookselling Question in 1852, see below.
16. The Parker Society was named after Archbishop Matthew Parker. McKitterick, quoting Mark Pattison, says the controversy ended when JH Newman converted to Rome. McKitterick, D: *A History of Cambridge University Press.* Cambridge, Cambridge University Press, 1998, *2*: 340.
17. Illustrated *Ibid.*: Fig. 31: 376.
18. *Ibid.*: 336.
19. *Ibid.*: 328–51.
20. Anon: *Words by a Working Man about Education.* London, JW Parker & Son, 1852. Haight, G: *George Eliot and John Chapman.* New Haven, Yale, 1969: 51–4; Barnes, JE: *Free Trade in Books.* Oxford, Oxford University Press, 1964; See also Ashton, R: *142 Strand.* London, Chatto, 2006: 146–51; McKitterick, D: *A History of Cambridge University Press.* Cambridge, Cambridge University Press, 1998, *2*: 344.
21. Concerning Darwin, see *The Makers, Sellers, and Buyers of Books.* London, JW Parker & Son, 1852: 9. For JS Mill on book prices, see *The Opinions of Certain Authors on The Bookselling Question.* London: JW Parker and Son, 1852: 47. For a succinct overview of the entire matter see Plant, M: *The English Book Trade.* London, Allen & Unwin, 1974: 448–61.
22. Thomas Carlyle to Lady Bulwer Lytton 7.1.1851; Thomas Carlyle to Delia Bacon 14.12.1856. Accessed via Carlyletters online: CLO <http://carlyletters.org> accessed 6 Sept 2007.
23. Many thanks are due to the Archivists at King's College London, and at King's College School, for their detective work in confirming these details. Parker's entry in the *Oxford Dictionary of National Biography* states in error that Parker went to King's College Cambridge. The best treatment I have yet found of the cultural cluster around literary Christian Socialism is in Cazamian, L: *The Social Novel in England 1830–1850.* London, Routledge, 1973 (originally published 1903): chapter 8 passim.
24. *The Magazine of Popular Science* was edited under the direction of The Society for the Illustration and Encouragement of the Practical Arts, which ran a museum in the Lowther Arcade, just along the Strand. Its work looks to be related to the Society of Arts nearby, and emphasizes Parker's strong interest in art and science. For FD Maurice see: Maurice, F: *The Life of F.D.Maurice.* London, Macmillan, 1884: 461–3, 471–5, 480–3. Maurice lost his post at King's because he doubted the eternity of hell fire.
25. The first two lines, and a third which does not match, were attributed on the Internet in 2008 to Archbishop Trench.
26. I thank Mr Christian Forsdyke for the translation. This text was a constant, which appeared used on a number of variant emblems which changed over time. The Parkers also used authors' and institutions' own emblems on title pages in some instances, especially for staff members at King's College, London.
27. Hudson, D: *Munby, Man of Two Worlds.* London, Abacus, 1971: 27, 114. In 1852 Clough described young Parker to the American poet Emerson, as 'The Editor of *Fraser* … and indeed, the Acting Publisher' at JW Parker and Son. The contributor to *Notes*

and Queries who submitted the correspondence for publication, David Bonnell Green of Boston University, mentions many other literary figures in their circle, including TL Peacock, Edward FitzGerald, Matthew Arnold, Coventry Patmore, and George Meredith. *Notes and Queries*, January 1963: 24–6. For Kingsley and Thackeray, see Ellis, SM: *Henry Kingsley*. London, Grant Richards, 1931: 30.

28. Pollock, Sir Frederick: *Personal Remembrances of Sir F Pollock*. London: Macmillan, 1887, 2: 71–2, 79, 86. These evenings fell on a Thursday and two Wednesdays, so it is not easy to tell whether they were a regular fixture. Pollock was the Lord Chief Justice. Frustratingly, Pollock says absolutely nothing about what was said.
29. In his fine history of the book, Adrian Johns quotes the reminiscences of the medical publisher John Churchill concerning his shop in about 1800 as having been the daily resort of medical men, with newspapers provided, and visitors meeting in friendly intercourse. See Johns, A: *The Nature of the Book*. Chicago, Chicago University Press, 1998: 121. Thomas Wakley regularly had convivial gatherings at his offices in the Strand when the Victorian *Lancet* went to press. See Richardson, R: *Vintage Papers from The Lancet*. Elsevier, 2006: 7.
30. 'Those old times in Johnny Parker's room are wae to think on—so many dead and gone.' Dunn, WH: *James Anthony Froude*. Oxford, Oxford University Press, 1963: 175, 349–50. See also *Oxford Dictionary of National Biography* entry for Froude.
31. Dunn, WH: *James Anthony Froude*. Oxford, Oxford University Press, 1963: 279. See also *Poole's, Waterloo & Wellesley Indexes*.
32. Quoted from Hayek, FA: *John Stuart Mill & Harriet Taylor*. London, Routledge, 1951: 140–1, 143. The original letters are in Yale University Library, to whom I am grateful for permission to quote from them.
33. See Trench's entry in the *Oxford Dictionary of National Biography*.
34. See advertisements at the back of the British Library copy of *Companion to the new Rifle Musket*. JW Parker & Son, London, 1855. They were not typical Parker productions. They appear to have been semi-official publications, as they were issued from Horseguards and sold from a dedicated shop called the Military Library, in Whitehall. Among their identified authors were eminent army figures such as Lord Frederick FitzClarence.
35. Concerning fish to eat: The book was dedicated to Thomas Watson, FRCP, whose book *Lectures on The Principles and Practice of Physic* was also published by the Parkers. Concerning *Essays and Reviews*: Ellis, I. *Seven Against Christ*. Leiden, Brill, 1980. *Essays and Reviews* is discussed more fully in Chapter 9.
36. A number of Parker medical books were not included, among which may be mentioned William Bowman's *Thoughts for the Medical Student* (1851); WA Miller's *The Importance of Chemistry to Medicine* (1845) which remained in print until the 1880s; Rutter's work on *Human Electricity*, and Brande's massive doorstopping textbook of chemistry. The inclusion of poetry and other arts-based books in their medical advertisement lists was typical of the Parkers. See Holmes, T: *A System of Surgery*. London, JW Parker & Son, 1860, I: bound in advertisement pages.
37. I have consulted widely on the history of mid-Victorian publishing, and have found the following the most valuable sources: Ball, D: *Victorian Publishers' Book-bindings*. London, Library Association, 1985; Carey, A: *The History of a Book*. London, Cassell, 1873; Carter, J: *Binding Variants in English Publishing, 1820–1900*. London, Constable, 1932; Carter, J: *Publishers' Cloth, 1820–1900*. London, Constable, 1938; Gaskell, P: *A New Introduction to Bibliography*. New Castle Delaware, Oak Knoll Press, 1995;

Glaister, G: *Encyclopedia of The Book.* Cleveland, World Publishing, 1960; Jamieson, E: *English Embossed Bindings, 1825–1850.* Cambridge, Cambridge University Press, 1972; King, EMB: *Victorian Decorated Trade Bindings 1830–1880: A Descriptive Bibliography.* London, British Library & Oak Knoll Press, 2003; McKerrow, RD: *Introduction to Bibliography for Literary Students.* Oxford, Oxford University Press, 1959; McLean, R: *Victorian Publishers' Bookbindings in Paper.* London, Gordon Fraser, 1973; McLean, R: *Victorian Publishers' Bookbindings in Cloth and Leather.* London, Gordon Fraser, 1974; Plant, M: *The English Book Trade.* London, Allen & Unwin, 1974; Secord, J: *Victorian Sensation.* Chicago, Chicago University Press, 2000. Robin Alston, Des McTernan, Edmund King, and Malcolm Marjoram of the British Library have been extremely kind and helpful.

38. The paperback nature of the parts is evident from the survival of the part wrappers preserved within the library binding in the copy currently held by the British Library, at St Pancras. British Library shelfmark 7406 cc 13.
39. In Frederick Tennyson's *Days and Hours,* London JW Parker & Son, 1854; British Library copy, a remnant of Burn sticker; the second example, in my own copy of *Lances of Lynwood.*
40. Machell, Mrs [Margaret]: *Poems and Translations.* London, JW Parker & Son, 1856. British Library copy, shelfmark 11649.c.33.
41. *Diogenes* was only in circulation between 1853 and 1855, and went out of business in 1855. I thank Brian Lake of Jarndyce, bookseller of Museum Street, London, and his staff, for the brilliant sleuthing which led to the identification of the cartoon.
42. One was signed with the monogram 'GD'. A search of Rodney Engen's splendid *Dictionary of Victorian Wood Engravers,* Cambridge, Chadwyck Healey, 1985, yields a choice between George Dalziel, or George Dorrington. Because the Dalziels illustrated another Parker book (see my discussion of *The Lances of Lynwood* in the text) I lean towards the former, but I do not wish to exclude Dorrington at this stage, as he worked close by the Strand. The monogram however is distinctive, so the problem should be resolvable.
43. The Dalziels were also engraving for DG Rossetti in 1855. See Vaughan, W: *German Romanticism and English Art.* Paul Mellon/Yale University Press, New Haven & London, 1979: 175. The copy in my own collection has a bookbinder's ticket by Burn, of Hatton Garden, manuscript fillet on the gatherings, newsprint on the spine, and sisal strings in place of tapes.
44. There were two processes, the woodcut went along the flank of the wood, wood engraving, developed by Thomas Bewick, went across the cut grain of the wood. See Chapter 6.
45. *The Times,* 1.10.1853: 8.
46. John Wertheimer (1799–1883) was an Englishman of Jewish descent, whose father was a schoolmaster. There are no records relating to the years in which he founded, established, and ran the company, and so far I have not managed to find any family papers. I live in hope, however. Twyman, J: *An Investigation into the history of Williams Lea and Co, 1820–1984.* London, 1984. Typescript, lacks pagination. Twyman quotes an obituary for John Wertheimer from the *British & Colonial Printer,* 10.1.1884, which said he was known as 'one of the most painstaking and exact reproducers of Oriental texts'. Another obituary, from the *London, Provincial and Colonial Press News* 10.1.1884 said: 'He was without doubt the greatest linguist of his time amongst printers, and his fame for foreign printing is well known.' This document was kindly loaned to me

by Tony Williams, Life President of Williams Lea, in 2007. After Todd and Bowman had completed its issue in parts, Parker issued it as two volumes. It is these that were printed by Wertheimer. Todd, RB and Bowman, W: *The Physiological Anatomy of Man.* London, JW Parker and Son, 1856. This edition was dedicated to Sir Benjamin Brodie.

CHAPTER 4: IDEA

Person or Persons Unknown

1. Girtin, TC (ed): *The House I Live In.* London, Parker, 1837.
2. Birkett, J: *The Diseases of the Breast and their Treatment.* [Jacksonian prize, 1848 awarded by Council of the RCS] London, Longman, 1850.
3. The 7th edition was advertised in *The Times* 12.6.1854 at 2s. 6d. per copy. The 8th edition appeared in 1855.
4. Parker Ledgers. Longman Archive: University of Reading.
5. Holmes, T: *System of Surgery.* London, Longman, 1864, *4*: ii.
6. Kölliker, A [Trans. Huxley, T & Busk, G]: *A Manual of Human Histology.* London, Sydenham Society, 1854.
7. Mantell, G: *Medals of Creation.* London, Bohn, 1844: title page. See also *Oxford Dictionary of National Biography* entry for Gideon Mantell.
8. Cheselden, W: *Osteographia.* London, Richardson, 1733. I warmly thank Monique Kornell, who allowed me to look at the copy she was studying at the Wellcome Library.
9. Hunter, W: *The Anatomy of the Human Gravid Uterus.* Birmingham Baskerville, London, Murray, 1774. The illustrations were drawn by John van Rymsdyk, and engraved F.S. Ravenet.
10. My discussion of Hunter and Smellie is based on Richardson, R: Human Remains. Essay in Arnold, K & Olsen, D*: Medicine Man: the Forgotten Museum of Henry Wellcome.* British Museum Press, 2003: 316–45. Hunter was Smellie's pupil.
11. Charles Kingsley's novel *Alton Locke* contains a fine portrait of a generous bookseller.
12. Smellie, W: *A Sett of Anatomical Tables.* London, Smellie, 1754. Illustrated by Jan van Rymsdyk.
13. Thornton, JL: *Jan van Rymsdyk.* Cambridge, Oleander, 1982.
14. Richardson, R: Chamberlen's Forceps. *The Lancet,* 2001, *358*: 1279.
15. *Anatomical Plates of Midwifery with Concise Explanations selected and reduced from Smellie's Large Tables Principally Intended for the use of Students.* London, E Cox & Son, 1823. Further editions appeared in 1833 and 1835. The engraver is as yet unidentified.
16. Carter Papers, Wellcome Western MSS 5818: 25.11.1855.
17. Quain, J & Wilson, WJE: [*Anatomical Plates of the human body*]. London, Taylor and Walton, 1836–42. The portfolio work was reissued in two Royal Folio (large) volumes in 1844, in half bound morocco leather, gilt tops, at £12. Coloured £20.
18. The text of an advertisement reads: 'Quain's Anatomy. *Elements of Anatomy* 2 vols 8vo 400 woodcuts £2 cloth edited by Dr Sharpey and Mr Quain 5th edition London, Walton & Maberly, 28 Upper Gower Street.' See *The Times* 7.12.1853: 13. This was still the 1848 edition.
19. For example, Renshaw sold MW Hille's works on *Anatomy, Regional Anatomy* and *Physiology* which went through numerous editions 1840–1880.

20. Soemmering, ST: *Traite d'osteologie*. Paris, 1843.
21. Todd, RB & Bowman, W: *Physiological Anatomy and Physiology of Man*. London, JW Parker & Son, 1843–57. Miller, WA: *Elements of Chemistry, Theoretical and Practical*. London, JW Parker & Son, 1856. Watson, T: *Lectures on the Principles and Practice of Physic*. London, JW Parker, [several editions] 1843, 1848, 1857.
22. Todd, RB & Bowman, W: *Physiological Anatomy and Physiology of Man*. London, JW Parker & Son. Published in five parts: 1843, 1845, 1847, 1852, 1857. Part 5 (1856–7): 511–4, figs 247–9.
23. Miller, WA: *Elements of Chemistry* JW Parker & Son, London, 1855, Vol. I: advertisement bound in the back of British Library copy, for: Tomes, J: *Dental Physiology*. John W Parker & Son, 1855.
24. Kölliker, A [Trans. Huxley, T & Busk, G]: *A Manual of Human Histology*. London, Sydenham Society, 1854.
25. Facsimile engravers worked in such a way as to produce as close a version of an original drawing as possible; drawings in which lines are drawn in black were reproduced with black lines. It was a careful and difficult technique to acquire real fidelity to an artist's style and intention, and was time-consuming, especially where the artist used cross-hatching for shadows, as each diamond-shape lozenge between the crossed lines had to be individually sculpted out. Carter tends to avoid cross-hatching for shadows and shaping, as he had experience of wood engraving. White line was a technique used by followers of Bewick, and 'artistic' engravers, for the cutting of white lettering or detail out of darker areas. It is a wood engraving technique, not generally a technique of drawing.
26. There was an enormous range of engraving talent in central London at this time. See Engen, RK: *A Dictionary of Victorian Wood Engravers*. Cambridge, Chadwyck Healey, 1985.
27. The previous 5th edition of Quain, 1848 was still on sale in 1853 in two volumes at 40 shillings the set. *The Times* 7.12.1853: 13.

CHAPTER 5: RAW MATERIAL

The Friendless Poor of London.

1. Barlee, E: *Friendless and Helpless*. London, Emily Faithfull, 1863.
2. Richardson, R: *Death, Dissection & the Destitute*. Chicago, Chicago University Press, 2000: 241.
3. Marshall, T: *Murdering to Dissect*. Manchester, Manchester University Press, 1995.
4. Richardson, R: *Death, Dissection & the Destitute*. Chicago, Chicago University Press, 2000: 253–4, 264. Mrs Gillard died late in 1839, the correspondence continued into 1840.
5. *Ibid.*: 239–60. See also National Archives: HO44/32–5.
6. See Buklijas, T: Cultures of Death & Politics: corpse supply in Vienna 1848–1914. *Bull.Hist.Med.*, *82* (2008) forthcoming.
7. In the twentieth century the relatives of children whose long bones and entire viscera were removed at post-mortems were regularly told that only 'tissue samples' had been taken for scientific examination, and the deception continued for years. See Richardson, R: Narratives of Compound Loss: Parents' stories from the Organ Retention Scandal. Essay in Hurwitz B, Greenhalgh T, & Skultans, V: *Narrative*

Research in Health & Illness. Oxford, Blackwell, 2004: 239–56. See also Tim Marshall's forthcoming book, *Stolen Hearts*. Nottingham, CCC Press, 2008.

8. *The Times*, 11.12.1857: 9.
9. *The Times*, 14.12.1857: 10.
10. Richardson, R: *Death, Dissection & the Destitute*. Chicago, Chicago University Press, 2000: 230, 235–8, 243–6.
11. National Archives: MH74/14: 1.9.1842: 57.
12. *Medical Times & Gazette*, 14.5.1859: 513.
13. The chapter in my own book dealing with this aspect of the Inspector's work is titled 'The Bureaucrat's Bad Dream'. See Richardson, R: *Death Dissection & the Destitute*. Chicago, Chicago University Press, 2000: chapter 10: passim.
14. *The Lancet*, 17.10.1857: 398.
15. Hospital ledgers surviving from other London hospitals show that the normal clerical process of completing entry procedures would have revealed whether a person was likely to be claimed or not, as the ledgers required data concerning marital status and addresses of next of kin.
16. Richardson, R: *Death, Dissection & the Destitute*. Chicago, Chicago University Press, 2000: 254–5.
17. An 1850s cartoon shows a chimney sweep offering his services as a joke to a man sleeping in the street inside a barrel. *Diogenes* 1853, *1*: 79. The homeless even today find doorways and other shelters to sleep in, and the journal's title looks to have triggered the cartoonist's record of this one.
18. Note the ambiguity of the phrase in stanza 2, line 4 in the above poem 'Not long before your sands had wholly run'.
19. The Gillard case never reached the newspapers.
20. It has not (as far as I am aware) been noticed before, and deserves further research.
21. A photograph of this desolate back alley, referred to as 'La Via Dolorosa' was reproduced in a St George's publication seeking rebuilding funds in the twentieth century. *St George's Hospital Rebuilding Magazine*, 1937, *4*: 6.
22. St George's Hospital Post Mortem Ledgers, 1855: 32, 33, 212, 303, 305.
23. Richardson, J: *The Annals of London*. Cassell, 2000: 1855.
24. Children under 14 years of age have been excluded. The drop recorded in the last figure does not seem to be reflected in a corresponding drop in the Inspector's figures, so this was not a London-wide phenomenon.
25. St George's Hospital Post Mortem Ledgers; National Archives: MH74/15. There is a difficulty concerning whether statistics were kept for medical school years, or for calendar years, which complicates everything, but even allowing for this, there seems to be a considerable discrepancy. The dissecting season ran from October to March.
26. Further research focusing upon this problem may yet yield more data.
27. Nicol, KE: *Henry Gray of St George's Hospital: A Chronology*. London, published by the author, 2002.
28. Concerning the Inspectors, see Richardson, R: *Death, Dissection & the Destitute*. Chicago, Chicago University Press, 2000: 256. The three people named as destined for Kinnerton Street in the St George's Post Mortem Ledgers in 1856 were: Henry Perry, who died 7.1.1856; John Crawley, who died 29.2.1856; Alexander Cotton, died 28.9.1856. All three were signed by Gray as sent to Kinnerton Street in the text, but entered as 'not examined' in the analytical index at the front of the ledger. For 1857, the only person sent to Kinnerton Street was Edward Fisher, who died 6.1.1857. He

was signed out by Gray as sent to Kinnerton Street, 'KS' in index. Sadly, the London Family Records Centre was closed while this part of the work was underway, and the new facility at the National Archives at Kew was still as yet unopened, so I was unable to access more information about this clutch of young men. My researches continue.

29. The heart of a Chelsea Pensioner, which was being dissected at the Royal College of Surgeons when Carter was studying there does not seem to have arrived with the rest of the body. See Lizars, J: Studentship Diary. 27.12.1854. The brain of a Greenwich Pensioner, likewise. See Silvester, HR: Studentship Diary. 27.7.1855. Royal College of Surgeons of England, Archives.
30. Richardson, R: *Death, Dissection & the Destitute.* Chicago, Chicago University Press, 2000: 448–9.
31. Hurren, E: A Pauper Dead-House. *Med. Hist.* 48 (1) 2004: 69–74.
32. The Human Tissue Act 1961 (also vaguely framed) never closely defined 'tissue', which allowed the scandal of Alder Hey to develop.
33. Charges to students were supposedly to defray the cost of burial, see *The Lancet,* 17.10.1857: 398. We know the sale of bodies or body parts was occurring at St George's in Gray's era, because in 1849 Carter recorded having purchased a 'full grown'. Carter Papers, Wellcome MSS 5816: 3.9.1849. Concerning commercial sale, in the song given in the text above, the working man's dead body was purchased. For dealers, see Mr Venus in Dickens's novel *Our Mutual Friend.* For Brodie and Cooper see Richardson, R: *Death, Dissection and the Destitute.* Chicago, Chicago University Press, 2000: 106, 115, 118–19. The first Inspector of Anatomy was a Brodie nominee.
34. The figures are for January to March, 1856: Poor Law 20, Bart's 11; and in 1857: Poor Law 14, Bart's 23. For both years together this totals 34 from St Bartholomew's itself, 34 from Poor Law institutions. St Bartholomew's Hospital Medical School Archives: Dissection Room Register. Five others were brought in from domestic premises, 4 stillborns/newborn, and 1 adult.
35. One body arrived from the Newington Workhouse, delivered by Mr Hogg (the undertaker who would be prosecuted in 1858) identified in the register as that of Mr George Chapman aged 75, who had died from senile apoplexy. For the condition of the London workhouses, see Richardson, R and Hurwitz, BS: Dr Joseph Rogers & the Reform of Workhouse Medicine. *History Workshop Journal,* 1997, 43: 218–25.
36. The organism which causes erysipelas also causes puerperal fever. Accidental needlestick injury can lead to septicaemia and death.
37. Bright, R: *Reports of Medical Cases.* London, Longman, 1827–31; and Sibley, SW: *Middlesex Hospital: Report on the Cholera Patients Admitted into the Hospital during the Year 1854.* London, Truscott, 1855.
38. 'An Appeal to the British Nation'. *Diogenes,* 1854, *3*: 107.
39. Barlee, E: *Friendless and Helpless.* London, Emily Faithfull, 1863. Sala, GA: 'Houseless and Hungry'. Essay in his *Gaslight and Daylight.* London, Chapman & Hall, 1859: 145–156.
40. Alexander, S: *Women's work.* London, Journeyman, 1983. Stedman Jones, G: *Outcast London.* Oxford, Oxford University Press, 1971.
41. Bright, R: *Reports of Medical Cases.* London, Longman, 1827–31, I: case LXXIII.
42. *The Times,* 4.6.1870: 11.
43. Curtis, G: *Visual Words.* Aldershot, Ashgate, 2002: 72. Curtis notes the epidemic killed 11,000 people in London alone. At St George's, none of those dying of cholera was examined after death.

44. Richardson, R: *Death, Dissection & the Destitute.* Chicago, Chicago University Press, 2000: 276–80.
45. *Household Words* 1855: 226–7. See also Richardson, R: Why was Death So Big in Victorian Britain? Chapter in Houlbrooke, R (ed.): *Death, Ritual & Bereavement.* London, Routledge, 1989: 105–17.

CHAPTER 6: CREATION

1856–1857

1. Long before he could afford to own a microscope, Carter was very likely to have possessed a camera lucida. It was a small and inexpensive article of equipment commonly sold by artists' suppliers, and (as eyepieces) by vendors of microscopes. Carter occasionally mentions making his own eye-shades, which suggests that he was using one. They are said to have been as commonly in use as cameras were before the mobile phone—see Hammond, JH & Austin, J: *The Camera Lucida in Art and Science.* Bristol, Hilger, 1987: 108. Other anatomical artists mention using them, see for example Wilson, E: *The Anatomist's Vade Mecum.* 6th ed London, Churchill, 1854: Preface. John Quekett featured the camera lucida in his book *A Practical Treatise on the Use of the Microscope.* (London, Baillière, 1848: 128–30) and they would have been in everyday use attached to a drawing board, or to a microscope, at the Royal College of Surgeons. See also: Varley, C: *A Treatise on Optical Drawing Instruments.* London, published by the Author, 1845; and the excellent work by Ivins, W: *Prints and Visual Communication.* London, Routledge, 1953. See also Erna Fiorentini: Camera Obscura vs Camera Lucida: Distinguishing early 19th century modes of seeing. ECHO open digital library 2005 <http://echo.mpiwg_berlin.mpg.de/content/optics>. preprint 307 2006 Max-Planck-Institut for the History of Science.
2. Carter Papers, Wellcome Western MSS 5818: 25.11.1855.
3. Sadly, the laudable inclusivity of the new institution was constrained by the remaining exclusion: individuals lacking testes were disqualified from candidature. See Willson, FMG: *Our Minerva.* London, Athlone, 1995: 292.
4. See Willson, FMG: *Our Minerva.* London, Athlone, 1995: 270, 276. Many English doctors went to Scotland to qualify, where there was a history of greater openness. Carter never seems to have entertained the idea, perhaps because he would not have networks of employment to sustain him financially while studying, perhaps because until quite late in his London career he was trying to make a success at St George's. To leave to attend studies in another city would have lost him his own place on the ladder there.
5. Carter first appeared in the Society's published membership listing as Carter, HV Esq., with no hospital or other institutional affiliation, a status-less and therefore suspect position for any scholar or medic, even today. Nevertheless, in the same volume he had two illustrations published, both fully credited, one of a brain tumour for JW Ogle; and other work for Athol Johnson, see below.
6. Gray, H: *The Structure and Use of The Spleen.* London, JW Parker & Son, 1854: preface.
7. *The Lancet,* 24.6.1854: 669–70.
8. Had one been forthcoming Carter would have mentioned it, either at the time, or in subsequent reflections. The lack of apology probably made the silence seem wilful.

9. Carter Papers, Wellcome Western MSS 5818: 3.6.1854.
10. Carter was not the only artist Gray failed to recognize: most illustrations appearing in Gray's work prior to this date (February 1856) were unsigned and uncredited. His lack of attribution gave the impression that they were his own, and that is certainly how others apprehended them. For example, Gray alone was accredited when Todd & Bowman reused Carter's beautiful micrographs of the spleen in the last published part (V) of their *Physiological Anatomy* in 1856: 513–4. The transverse section was credited: 'After Mr H Gray'.
11. See Chapter 2.
12. Letter from Edwards Crisp summarizing his points, 19.3.1857. Administrative records of the Astley Cooper Prize, King's College London, Archives: Guy's Hospital Records G/AD10/M1. Crisp seems to have believed that Cooper's will, which specified that the essay must be original and unpublished material, made Gray's essay 'contrary to law'.
13. Letter dated 25.3.1857. Administrative records of the Astley Cooper Prize, King's College London Archives: Guy's Hospital Records G/AD10/M1. The argument here feels rather 'behind closed doors', Babington is apparently unconcerned about the truth or falsity of Crisp's allegations, but how he can be fobbed off/silenced.
14. Administrative records of the Astley Cooper Prize, King's College London Archives: Guy's Hospital Records G/AD10/M1: Undated.
15. *Ibid.* See also Dobson, J: Dr Edwards Crisp: A Forgotten Medical Scientist. *J.Hist. Med & Allied Sciences.* 1852, 7: 384–400. Jessie Dobson says Crisp discovered valves in splenic blood vessels, not previously described.
16. Cooper Prize Archives: 1853. Gray, H: *On the Structure and Use of the Spleen*: fol 85. Manuscript in the Archives of King's College London.
17. Carter Papers, Wellcome Western MSS 5818: week ending 9.12.1855.
18. Carter Papers, Wellcome Western MSS 5810/74/1. Carter Letter to Lily, [undated, winter 1856].
19. Carter Papers, Wellcome Western MSS 5818: 22.12.1855.
20. *Ibid.*: 22.12.1855.
21. *Ibid.*: 8.1.1856.
22. It turned out later that Hewett had persuaded the medical school staff to give Carter a contribution of £50 for the remainder of that school year; Carter discovered this at the outset of the next medical school year, when Hewett had left his post at the medical school, and the arrangement ceased, leaving Carter unexpectedly £50 poorer.
23. Carter noted Gray employing others on 31.1.1856, and did his first drawing on 1.2.1856. Carter Papers, Wellcome Western MSS 5818.
24. *Transactions of the Royal Medical & Chirurgical Society*, 1856, *39*: 120–149. The paper is interesting for Gray's technique of drawing together the findings of a number of pathologists, and adding his own data and gloss. He had used specimens collected and/or written up by Brodie, Stanley, Lawrence, and Paget.
25. Carter Papers, Wellcome Western MSS 5818: 4.2.1856.
26. Nearly half Johnson's text was written by Carter. *Transactions of the Pathological Society of London*, 1856: 242–4, plate XII. *The Lancet* reported a sequel to this paper in July 1857, naming Carter as Dr Vandyke Carter, the first occasion I have found in print. *The Lancet*, 11.7.1857: 35–6.
27. The elision covers a phrase of Carter's 'and not conscientiously', meaning not according to what his own conscience would normally have allowed him to say. Carter Papers, Wellcome Western MSS 5818: 8.2.1856.

28. *Ibid.*: 9.2.1856.
29. Whether Henry Gray had attributed the micrographs correctly in his presentation at the Royal Medical and Chirurgical Society's meeting, we do not know. When his bone cancer paper was printed, Gray *did* credit Carter with the illustrations, the first occasion so far found in which he did so. Possibly Carter's words effected this, too. *Transactions of the Royal Medical & Chirurgical Society*, 1856, 39: 120–49.
30. Carter Papers, Wellcome Western MSS 5818: 10.2.1856.
31. Luther Holden, when a Demonstrator of Anatomy at St Bartholomew's Hospital, had published a *Manual of the Dissection of the Human Body* (1851, London, Highley) without a single illustration.
32. Dr Johnson's work process in the upper floor of his famous house off Fleet Street is described in Hitchings, H: *Dr Johnson's Dictionary: The extraordinary story of the book that defined the world.* Murray, 2006. My thanks to Stephanie Pickford of Dr Johnson's House.
33. The *Oxford Dictionary of National Biography* entry for Holmes says he was at Pembroke College. He graduated 42nd Wrangler and 12th Classic, in 1847.
34. Carter also ensured that no one would accuse him of intellectual theft. See my discussion of the *Medical Times* review, in Chapter 9.
35. Gray, H and Carter, HV: *Anatomy Descriptive and Surgical.* London, JW Parker & Son, 1858: List of Illustrations. Anatomists whom Carter credited most frequently as visual sources were Arnold (24) and Quain (15), with Hind (12), Goodsir (6), and Mascagni (5) following, and with Kiernan, Bowman, and Breschet following on behind (each 3). The remaining eleven named anatomists were cited once or twice each.
36. St George's Post Mortem & Case book 1856: Henry Perry, aged 28. d. 7.1.1856.
37. I thank Dr Philip Adds of St George's Hospital Medical School for this insight.
38. Gray, H and Carter HV: *Anatomy Descriptive and Surgical.* London, JW Parker and Son, 1858: 651, figure 336.
39. Carter Papers, Wellcome Western MSS 5818: 26.10.1856. One would have thought that the creation of a major textbook could not have gone along unnoticed by students, senior students, and other staff, but so far no memoir of anyone using the dissecting room at Kinnerton Street at that time has come to light which discusses it. I have traced the names of contemporary students and staff, and checked them out, but sadly so far have drawn a blank. If any reader knows of—or finds—such a memoir, or a letter or other document by a contemporary witness, I would be *most* grateful to hear of it.
40. For the *Alphabet* see Frank Buckland MSS, I: 120. Royal College of Surgeons, London. A similar text appears in Albert Smith: *The Medical Student.* London, Routledge, 1861: 83–4, so the song was probably widely known. My version marries the two texts.
41. For the specimen label, see Frank Buckland MSS I: 109, 121. Royal College of Surgeons, London.
42. Carter worked daily at Kinnerton Street for 8–9 hours at a time through midsummer warmth in June 1856. See Carter Papers, Wellcome Western MSS 5818: June 1856 *passim.* For Joe's letter home, see Carter Papers, Wellcome Western MSS. Joseph Newington Carter letters: 5813/4: 5.9.1855; 5813/6: 10.12.1856; 5813/9: 16.6.1857.
43. Engen, RK: *A Dictionary of Victorian Wood Engravers.* Cambridge, Chadwyck-Healey, 1985: 38–9. The partnership was involved in the engraving of some rather fine artworks. See note 46 below.

44. Dyson, A: *Pictures to Print.* London, Farrand Press, 1984. The Heath who did the engravings of illustrations by William Clift for Matthew Baillie's *Morbid Anatomy* (London, J Johnson, 1803) may have been a forbear.
45. Charles Butterworth continued in business under his own name from 1884 to 1910, which suggests he was a much younger man. Engen, RK: *A Dictionary of Victorian Wood Engravers.* Cambridge, Chadwyck-Healey, 1985: 38–9.
46. See *Art Journal* 1859: 57, 59, 85, 86, and on 313, the engraving after an original by Birket Foster. I say 'mistress', because Parker senior was not averse to employing women wood engravers, and is known to have commissioned work from the well-known engraving firm of Byfield, whose workshop had several women engravers, including Mary Byfield (see note 50 below). We do not yet know for certain who did the engraving for *Gray's*, and Butterworth and Heath may well have employed women.
47. Flank wood can be cut and carved into quite complex images, but is more prone to warping, and could not hold and print the detail offered by wood engraving, or serve for long print runs. Snobbery extends to prints in general, probably because its products are multiples, and not therefore 'unique'. The greatest artists have accepted them as art, regardless; see the wonderful catalogue to an Arts Council exhibition at the Victoria and Albert Museum exhibition by Ronald Pickvance: *English Influences on Van Gogh.* London, Arts Council of Great Britain, 1974/5.
48. The great exemplar of its early development was Thomas Bewick, whose exquisite engravings are justly celebrated today. Bewick's Newcastle pupils spread use of the new technique, especially to London. The workshops they established thrived, because nothing was better for clarity of line, economy, and for use in combination with text for hand or steam presses. See Bewick, T: *My Life.* (Bain, I ed.) London, Folio Society, 1981. Gill, MAV: *The Beilby and Bewick Workshop.* Newcastle, 1976. Uglow, J: *Nature's Engraver.* London, Faber, 2006.
49. My discussion of wood engraving and reverse image transfer here is based on my essay on HV Carter in Bell, GH., Credland, A., & Richardson, R: *HB Carter & Sons: Victorian Watercolour Drawing and the Art of Illustration.* Pickering, Blackthorn Press, 2006: 32–67.
50. See, for example, *The Saturday Magazine*, for 22.6.1833, which features a large wood engraving of St George's chapel Windsor, engraved at the Byfield workshop, which is a tour de force of penny engraving for its date. One join between blocks is visible, horizontally, across the centre of the image, cleverly arranged to coincide with the top edge of the chapel organ. The Byfields were a family firm, with several female engravers. See Engen, RK: *Dictionary of Victorian Wood Engravers.* Cambridge, Chadwyck Healey, 1985: 39–40.
51. The joins are sometimes visible in the printed results, though not often in *Gray's.* See Richardson, R & Thorne, RS: *The Builder Illustrations Index 1843–1883.* London, Institute of Historical Research & English Heritage, 1995: Introduction. As the Victorian era progressed, photographic sensitizing of the block would do away with the draughtsman; while 'half-tone' later dispensed with the engraver.
52. Crane, W: *An Artist's Reminiscences.* London, Methuen 1907. 45–59. The passage dates to 1858, the year *Gray's Anatomy* was published.
53. Crane, W: *An Artist's Reminiscences.* London, Methuen 1907: 45–50.
54. Carter Papers, Wellcome Western MSS 5818: 11.6.1856.
55. Bell, G (ed.): *Theakston's Guide to Scarborough.* Illustrated with thirty-eight engravings

on wood, from original drawings by Mr. HB Carter (1841). Second Edition, Pickering, Blackthorn Press, 2002. Henry Barlow Carter's close work on the detail of the Theakston engravings may have served to foster his eldest son's interest in microscopy: their minuteness may have held a peculiar charm for the shy child, to whom those tiny images contained the streets and houses of the town he knew so well, with its teeming traffic and multifarious lives. The engravings were done in London, by Stephen Sly. The bookbinding of the 6th edition, an embossed casing by the famous Victorian book designer John Leighton, also done in London, is illustrated in Edmund King's beautiful book *Victorian Decorated Trade Bindings 1830–1880*. London, British Library & Oak Knoll, 2003: 73, no.213.

56. HV Carter had tried wood engraving himself, as an adult, so he was versed in its demands. Henry Barlow Carter's artistic influence may have had differing effects upon his two sons: his aptitude in close and accurate black-and-white line-work and perhaps reverse drafting for wood engravers carried forward by Henry, while the softer-edged more impressionistic wider focus tradition of painting in colours passed to Joe. HVC is always respectful of his father's teaching, and urges Lily to persevere in her drawing, and learn as much from him as she can: it 'will make you stand out especially if you take advantage of our father's instruction'. Carter Papers, Wellcome Western Manuscripts 5810/7: 22.10.1848: HVC to Lily.
57. *The Times*, 12.4.1856.
58. In her delightful book on Dickens's illustrator Phiz, Valerie Browne Lester explains how the great illustrator felt about wood engraving. She quotes his son Edgar as saying that his father was never so successful on wood as in freehand etching: 'In drawing on wood he was obliged to use a very hard pencil, and to depend on the point alone, so that his work resembled a coarse kind of etching, and very often had to suffer from translation at the hands of the engraver, who substituted for a lively line a mechanical one, and treated spaces of shade by cutting in tint. To the end his work suffered from these drawbacks, and he suffered greatly in translation, as Dickens does himself when translated into French.' See Valerie Browne Lester: *Phiz: The Man who drew Dickens*. London, Pimlico, 2004: 170. Butterworth and Heath did occasionally use 'tint' on Carter's drawings, but not usually on the finest engravings. 'Tint' is a mechanical-looking shading, where the lines are so regularly spaced that you can see they have not been hand engraved. See also note 73 below.
59. Carter Papers, Wellcome Western MSS 5818: 28.2.1856. He chose one from Ladd of Chancery Lane. The amount he put down was £5.15 shillings.
60. *Ibid.*: undated [mid June] 1856.
61. *Ibid.*: 13.7.1856.
62. *Ibid.*: 27.7.1856. He had no evening work with pupils, school having broken up.
63. *Ibid.*: 12.10.1856.
64. *Ibid.*: 2.4.1857. His handwriting is unclear: Fugged/Fagged had similar meanings.
65. *Ibid.*: 5.4.1857.
66. *Ibid.*: 6.5.1857.
67. *Ibid.*: 6.5.1857; 31.5.1857.
68. Transcribed with permission from a copy of the original text in the Nicol Archive.
69. The Duke's family's wealth was proverbial, and their responsibility for the great cruelty of the Highland clearances was infamous. See *Oxford Dictionary of National Biography* entry for Sutherland. I thank John Hayward for his knowledge shared hereabouts.

70. Nicol, KE: *Henry Gray of St George's Hospital: A Chronology.* London, published by the author, 2002.
71. Carter Papers, Wellcome Western MSS 5818: 21.7.1857.
72. Longman Archive: JW Parker Ledgers. The payment is undated.
73. There is a possibility that Carter was also paid by the engravers for drawing directly on the wood, as his work certainly saved them labour. No evidence exists for any such payment, but it remains a clear possibility. Sums mentioned in his notebook at the time he was leaving for India add up to £396, so he may have inherited money from his mother, or earned it in some other way. Teaching cannot have given such a yield. He had to pay his own passage to India, which cost him £95. See Carter Papers, Wellcome Western MSS 5818: 23.2.1858.
74. *Ibid.*: 11.10.1857. Bethlem was also known as 'Bedlam'.

CHAPTER 7: PRODUCTION

1857–1858

1. Robin Alston, Founder of The Scolar Press, and the ESTC, and great bibliographer–bibliophile. Personal communication, British Library, August 2007.
2. Carter Papers, Wellcome Western MSS 5818: 6.11.1857.
3. *Ibid.*: 29.6.1856; 13.7.1856. These contacts may have concerned his transition from paper to woodblock drawing, which happened in early June, possibly concerning financial arrangements.
4. Much can be appreciated concerning the relationship from the respectful manner in which Henry Barlow Carter's work was credited in the finished volume. See Bell, G (ed.): *Theakston's Scarborough Guide.* [Originally published Scarborough, 1841] 2nd edition, Pickering, Blackthorn, 2002.
5. This may have been due to ill-health. He wrote to Whewell in 1846 from Harrogate, mentioning a 'sad painful and lingering' illness. Trinity College Cambridge, Archives: Whewell Papers. Add.MSS a 210 91(1). It is possible that the Parkers avoided visiting Butterworth & Heath, because their engraving workroom was situated upstairs from a medical publishing rival, Renshaw, and to visit would have been to draw attention to their flagship book. Industrial espionage could conceivably have been one reason the Parkers chose Wertheimer: Churchill, who had a large medical list, used Savill & Edwards.
6. Details of the Parker accounts for the printing of *Gray's* are reproduced in the text below.
7. Quain, J: *Elements of Anatomy.* London, Walton & Maberly, 1856. The previous price had been 40 shillings, so this lower price was an unexpected extra pressure on the Parkers. This new edition of Quain's was in *three* volumes, and edited by William Sharpey and GV Ellis.
8. The Parkers' medical list is reproduced in the text of Chapter 3.
9. Longman archive, Parker ledgers. See also Plant, M: *The English Book Trade.* London, Allen & Unwin, 1974.
10. Robin Alston, Founder of The Scolar Press. Personal communication, British Library, August 2007.
11. The document is housed in the London Metropolitan Archives. Signed Sealed and Delivered by the within named John Wertheimer and George Littlewood in the

presence of Fredk West, Solr. 3 Charlotte Row, Mansion House, London. The petition for bankruptcy had been filed on 16th December 1857. The entire document was dated 5th April 1858 and was signed and sealed 4th June, 1858, by John Wertheimer and George Littlewood. Total monies owed £6,500.00; payments agreed £2258.00. Two appointed 'Inspectors': William Caslon, of Chiswell Street, Typefounder, and Frederick Howarth Edwards, Coppice Row, Clerkenwell, Ink Manufacturer. Agreed amounts were to yield one shilling and ninepence in the pound to each creditor within three months, six months, nine months, and twelve months of the date of the document, totalling seven shillings in the pound. Caslon and Edwards to oversee these payments, and Wertheimer's entire business, until the payments schedule was complete. The whole document was endorsed by Joshua Evans, Commissioner, on the 2nd March 1858; and the last payment was due 6th March 1859. Neither Wertheimer nor Littlewood was to leave the country until the last agreed payment was made in full (it was common for people in financial straits to flee abroad to evade creditors/ avoid imprisonment. See e.g.: Strauss, R: *Sala*. London, Constable, 1942: 127). For a good guide to the complexities of the history of bankruptcy, see Lester, VM: *Victorian Insolvency*. Oxford University Press, 1995.

12. Edwards, JP: *A Few Footprints*. London, Clement's House, 1905: 25. See also *Oxford Dictionary of National Biography* entry for Passmore Edwards. RS Best, author of *The Life and Good Works of John Passmore Edwards* (Redruth, Truran Publications, 1981) calls Edwards 'Mr Greatheart' for his extensive philanthropy later in life.
13. Longman Archive: Parker Ledgers. The ledgers do not identify the paper supplier for the original edition of *Gray's*. The Parkers' most frequent suppliers were Spalding and Hodge, and John Dickinson. The ledgers do record that the latter supplied the paper for the second edition of *Gray's*. But although the papers cost the same, they differ in texture between the first and second editions, so Spalding & Hodge remain a possibility as the supplier for the first edition.
14. For the traditionally high costs of extra margin work, see Hansard, TC: *Typographia*. London, Baldwin, 1825: 782.
15. The special tool is a spoke-plane. Robin Alston, Founder of The Scolar Press. Personal communication, British Library, August 2007.
16. Carter Papers, Wellcome Western MSS 5818: 3.2.1858. Carter's examiners were Paget, Hooker, Busk, and Walshe. Carter's accounts just before he left show a figure of £12 for a microscope. His previous one had cost more than this, so my reading of the situation is that he traded it in and added the £12 to obtain a better one. *Ibid.*: 23.2.1858.
17. *Ibid.*: undated (early January) 1858.
18. For the death of Gray's brother, see Nicol, KE: *Henry Gray of St George's Hospital: A Chronology*. London, published by the author, 2002. For the proofs reaching him, see next section.
19. Neither the portfolio nor the cartridge paper was acid free, and both have deteriorated, the inner binding has collapsed in places, and several pages have become damaged and/or detached. One or two may have disappeared, as the record of the illustrations for the first edition is not complete. It is possible that pages have been used for exhibition in the past, and certainly, in one case, a piece has been cut from one of the cartridge sheets, and subsequently glued back. The proofs were exhibited at the Royal College in 2008, in an exhibition celebrating the 150th anniversary of *Gray's Anatomy*, and funding is actively being sought for their conservation.

20. Fox, C: Victorian Wood Engravers and the City. In Nadel, IB & Schwarzbach, FS (eds.) *Victorian Artists and the City.* Oxford, Pergamon, 1980: 1–13. A fine article.
21. This was probably done because the fine India paper accidentally picked up ink from lowered areas, and printed black, or bubbled up under the pressure of the press.
22. Compare, for example, proofs and 1st edition Figures 22, 23, 24, 33, 34, 38, 48, 50, 56, 89, and 103. One curiously poor group of figure numbers on page 74 looks undecided.
23. The lateral view of the spine in the book is among several other images missing from the Butterworth and Heath proofs. The present poor condition of the proofs explains why so many images are missing—see note 21 above. This is the only complete image which appears in the Butterworth and Heath proofs, and not in the book.
24. See, for example, figures 23–4; 35–6.
25. See, for example, figures 69–70; 74–5.
26. Parker Ledgers, Longman Archive.
27. Many thanks are due to Dawn Kemp and Andrew Morgan at the Royal College of Surgeons, Edinburgh and to the Librarian/Archivist for allowing me to study this important archive.
28. The Royal College of Surgeons in Edinburgh has documentation for part of the story, but it was not fully understood even at the time the donation was made, in *c.*1946.
29. The gathering labelled with the printers' mark 'RR' (pages 609–24) is not represented.
30. It is possible that the handwriting is not that of John Wertheimer, but of someone else coordinating the printing and proofing at Finsbury Circus, but the wording rules out Parker or Mr Bourn.
31. Longman Archive: MS.1393, Impression book 1865 edition. Gray's family continued to receive income for the book for many years, with a deduction for Holmes's and later editors' editorial fees.
32. The problem of word-image and author-illustrator status is important, and was changing in this era. See Gerard Curtis's fine chapter *Shared Lines*, in his *Visual Words* (Aldershot, Ashgate, 2002). See also Valerie Browne Lester's treatment of the relations between Dickens and Phiz, in her lovely book *Phiz: the Man who drew Dickens.* London, Pimlico, 2004.
33. I do not think Holmes was deeply involved in the keeping of these proofs, or if he was, it was probably for some other reason. I say this because when he became Editor of *Gray's*, Holmes did not insist on reinstating equal size type to Carter's name, or adding his new post. Holmes's own name was added below Carter's, but in letters as large as Gray's.

CHAPTER 8: PUBLICATION

1858 and on

1. Dodd, G: *Days at the Factories; or, the Manufacturing Industry of Great Britain Described.* London, Charles Knight, 1843: 363–84. See also Alexander, S: *Women's Work.* London, Journeyman Press, 1983.
2. A woodcut showing a woman at work at a sewing press is reproduced in Dodd, G: *Days at the Factories; or, the Manufacturing Industry of Great Britain Described.* London, Charles Knight, 1843: 370. The tapes were occasionally substituted by sizal string. A folder and a stitcher and a book casing maker, a gold blocker, and a standing

press man are shown at work in Secord, J: *Victorian Sensation.* Chicago, Chicago University Press, 2000: 121, 123. Secord has a rich sense of the book as an artefact. The stitcher illustration has been reproduced in an onlay, on the front board cover of Ruari McLean's beautiful book *Victorian Publishers' Book-bindings in Cloth and Leather.* London, Gordon Fraser, 1974.

3. Dodd, G: *Days at the Factories; or, the Manufacturing Industry of Great Britain Described.* London, Charles Knight, 1843: 369. Dodd's book provides a good description of the work process/production line at Westleys.
4. King, E: *Victorian Decorated Trade Bindings 1830–1880.* London, British Library and Oak Knoll Press, 2003: 247, no. 696. King also mentions an earlier Wertheimer–Burn collaboration with yellow endpapers, a Parker favourite colour, on pp.74–5, no. 216. Trying to assign unlabelled book casings to specific binders is a form of book archaeology which has yet to yield its full fruits.
5. See *Appendix* re Wertheimer's creditors.
6. Kölliker, A: *Manual of Human Microscopic Anatomy.* London, JW Parker & Son, 1860. A copy in its original binding is in the Library of the Royal College of Surgeons and Physicians of Glasgow.
7. Victorian medical journals often used their 'wrappers' for generating revenue.
8. Later advertisements, if they come to light, may help resolve this puzzle.
9. Three examples of the use of this particular rather chunky 'bead grain' bookcloth on other books are illustrated in Ruari McLean's beautiful book *Victorian Publishers' Book-bindings in Cloth and Leather.* London, Gordon Fraser, 1974: 8, 9, 74. Two of these three, like *Gray's*, are dated 1858, so it may be that it was a fashionable choice that year. That fashion's writ ran in bookbinding is clear from an examination of works on the history of bookbinding listed in the notes for Chapter 3.
10. An editor of *Quain's*, William Sharpey, told a Scots colleague, Allen Thomson, that he disliked the book's title, and he used the word 'weary' to describe it. Jacyna, LS: *A Tale of Three Cities.* London, Wellcome, 1989: 117.
11. Only another printer, probably, would have been aware of it. The fact that no subsequent commentator on *Gray's Anatomy* has so far been found to have pointed it out shows this to be the case. If I have missed a discussion of this somewhere, I would be glad to hear of it.
12. *The Lancet*, 11.9.1858: 282–3.
13. *British Medical Journal*, 13.11.1858: 949.
14. Secord, A: Botany on a Plate. *Isis*, 2002, *93*: 28–57.
15. Gray, H and Carter, HV: *Anatomy Descriptive and Surgical.* London, JW Parker and Son, 1858: 692. Needless to say, Carter's illustration shows all this in one fell swoop.
16. The image is reproduced in my *Death, Dissection and the Destitute.* Chicago, Chicago University Press, 2000: 33.
17. Jaffe, A: *Vanishing Points: Dickens, Narrative, and the Subject of Omniscience.* University of California Press, 1991. Throughout, but especially 1–25, 167–71. Lorraine Daston discusses the studied neutrality and title to authority of 'scientific objectivity'. See Daston, L: Scientific Objectivity with and without words. In Becker, P and Clark, W: *Little Tools of Knowledge.* Ann Arbor, University of Michigan Press, 2001: 259–84.
18. Carter Papers, Wellcome Western MSS 5817: 8.11.1852.
19. Ferguson, ES: The Mind's Eye: Nonverbal Thought in Technology. *Science*, 1977, *197*: 827–36.
20. Carter mentioned needing a dissecting guide in September 1849, and began work on

the illustrations for *Anatomy Descriptive and Surgical* in early February 1856. Carter Papers, Wellcome Western MSS 5816: 3.9.1849. My discussion hereabouts owes much to reading Svetlana Alpers' wonderful book *The Art of Describing* (London, John Murray and Chicago, University of Chicago Press, 1983) the finest eye-opener, mind-opener, and simple prose in art history I have ever had the good fortune to read.

21. Young, JZ: *Programs of the Brain.* Oxford, Oxford University Press, 1978: 94–5. Thompson, RF and Madigan, SA: *Memory.* Washington, DC, Joseph Henry Press, 2005: 36 discuss 'dual code theory', separate visual and verbal channels in the brain. See also Vivian Nutton's essay Representations of Memory. In Meroi, F & Pogliano, C (eds.): *Immagini per Conoscere.* Firenze, Olschki, 2001: 61–80. Kusukawa discusses the mnemonic function of early modern anatomical images in Kusukawa, S and Maclean, I: *Transmitting Knowledge: Words, Images & Instruments in Early Modern Europe.* Oxford, Oxford University Press, 2006: 77. Nick Hopwood talks about images going into the memory with repeated exposure, and the way anatomical details become more easily discernible with repeated looking. See his Visual Standards and disciplinary change: Normal plates, table and stages in vertebrate embryology. *History of Science*, 2005, 43: 239–303.
22. The *Fabrica* (1543) is probably the most significant work in the history of anatomical illustration. The artist of *Fabrica* is not named in the book. See William Ivins: What about the *Fabrica* of Vesalius? In Lambert, SW, Wiegand, W, & Ivins, W: *Three Vesalian Essays.* NY 1952, Macmillan. Ivins argues—rightly in my view—that the artist of *Fabrica* (John Stephen of Calcar) is more important to the book than is the 'author', Vesalius. 'Great claims made for Vesalius', he says, 'are based on pious professional tradition and not on critical knowledge'.
23. Histories of anatomical illustration demonstrate the truth of this statement. See, for example the illustrations in: Roberts, KB and Tomlinson, JDW: *The Fabric of the Body. European Traditions of Anatomical Illustration.* Oxford, Clarendon Press, 1992. Rifkin, BA, Ackerman, M, and Folkenberg, J: *Human Anatomy: Depicting the Body from the Renaissance to Today.* London, Thames & Hudson, 2006. Such books tend to pick aesthetically pleasing/interesting anatomical illustrations, so it may be that they are editing out others with intrinsic labels, but I do not believe so. What they show is that integral labelling on anatomical structures does not survive much beyond Vesalius, and proxy or arrowed proxy labels win out entirely until Carter's work, after which there is increased interest in intrinsic labelling.
24. Kemp, M: Gray's Greyness. In *Visualisations: The Nature Book of Art and Science.* Cambridge, Cambridge University Press, 2000: 70–1. See also Booker, P: *A History of Engineering Drawing.* London, Chatto, 1963.
25. Brewer, D: The work of the image: the plates of the Encyclopédie. In Katz, B (ed.): *The History of Book Illustration.* Metuchen, NY and London, Scarecrow, 1994: 391–411.
26. Svetlana Alpers' splendid book *The Art of Describing* (London, John Murray, and Chicago, University of Chicago Press, 1983) is enormously valuable for pondering image and word, what she calls the 'attentive eye' and the association of mapping, seeing, painting, and the culture from which they derive. She speaks beautifully about the craft of representation, the camera obscura, and the manner in which northern European art can be *at ease with inscribed words* (page 169). An entire chapter is entitled: 'With a sincere hand and a faithful eye', which I think characterizes Carter and his work on *Anatomy Descriptive and Surgical*, as well as his later work in India.
27. The only precedent so far found is that of Luther Holden (*Human Osteology.* London,

Churchill, 1855) who used the technique only on bone. I thank Andrew Baster for confirming this finding. Holden's originality was asserted by a reviewer in the *Medical Times*, when the newly illustrated version of Holden's previously pictureless *A Manual of Dissection of the Human Body*. (London, Churchill) was published in 1861. 'The student formerly stood at no time in more need of help than at the moment of his commencing the dissection of his first "part". Arrayed in sleeves and apron, new, rigid, strange, and rather too warm, with instruments in his hands, as yet unaccustomed to their use, he found himself with a *Dissector's Guide* which to him was as useful as the signpost to the man who could not read ... Now, however, thanks to Mr Luther Holden, who may almost be said to have invented a new system of teaching anatomy by diagrams, much of the old difficulty has been annihilated. [Holden's *Manual* is] enriched with diagrams and drawings which explain themselves after his favorite plan adopted in his *Human Osteology*—placing the names of the different parts either on or immediately at the side of the diagram or drawing ... not only a great saving of time is effected, but the student's mind receives a much more vivid impression.' Holden's illustrations are described as 'bold diagrams, almost as it were, transcribed from the black board'. Review in *Medical Times*, 26.10.1861. The reviewer did not mention *Gray's Anatomy*, and finished by saying that Holden was 'a scholar and a gentleman' as if in contrast to someone else. Gray had died that June. The implication in the review is that Carter's illustrations were derived from Holden's, and, like Gray's text, had failed to credit Holden with the technique. But as I have demonstrated, Carter's diagrams probably predated Holden. Blackboard technique was apparently widespread, but not yet in books, and the same cultural influences were at work on both men, see Curtis, G: *Visual Words*. Aldershot, Ashgate, 2002. It is even possible that Holden had seen some of Carter's diagrams at the Royal College of Surgeons.

28. Reproductions of mediaeval intrinsic labelling appear in Roberts, KB and Tomlinson, JDW: *The Fabric of the Body. European Traditions of Anatomical Illustrations*. Oxford, Clarendon Press, 1992: 31, 36, 51, 74, plates 4, 9, and pages 2.10; 2.13; plate 9; plate 3.5. Leonardo uses both techniques. It appears rarely in Vesalius—see Roberts & Tomlinson: plate 12—but thereafter, seems to fall out of favour. Martin Rudwick's book *The Great Devonian Controversy* (Chicago, Chicago University Press, 1985) explains the history of the development of graphic techniques for showing/explaining/sorting out the layers of strata in geological researches in the early to mid-nineteenth century, including the personality clashes involved. Gerard Curtis's *Visual Words* (Aldershot, Ashgate, 2002) is also a delight, full of fun and insight, looking at the life of typography in the largest sense, especially mapping and advertising in the early to mid-Victorian years, and especially the linkage of written/drawn lines and lettering.
29. *Diogenes*, 1854: 139, 215. The cohort that was generating much of this material was probably, like Carter, growing up in the 1830s and 1840s, when the *Penny Magazine* and Parker's *Saturday Magazine* and other illustrated periodicals, with their large wood-engraved illustrations, were circulating widely, and having great visual impact. See Curtis, G: *Visual Words*. Aldershot, Ashgate, 2002; Wicke, J: *Advertising Fictions*. New York, Columbia University Press, 1988.
30. The Carter family, and his father's cohort of fellow English artists, were great admirers of Turner, who had taken his own path.
31. Bell, J: *Anatomy of the Bones, Muscles and Joints*. London, Longman, 1810: ii.
32. *Ibid.*: ii.

33. *Ibid.*: v.
34. *Ibid.*: Plate XI. I discuss this type of dissection-room brutality elsewhere, see Richardson, R: The Dead Body. In *The Body* edited by Carol Reeve for Berg. *In Press.* forthcoming. 2009.
35. Carter Papers. Wellcome Western MSS 5819: 17.9.1853; 1.10.1854; 5819: 30.10.1853.
36. Cruveilhier, J: *Traite d'Anatomie Descriptive.* [translated into English] London, Tweedie Library of Medicine, 1842, *8*: xii.
37. Carter Papers. Wellcome Western MSS 5818: undated entry, April 1857. The church was St Mary's, Scarborough, where HV Carter was christened. Carter mentioned that his mother's grave was close to those of friends, as though the presence of benign company nearby gave him comfort. The gravestone still stands.
38. Gammon, V: Singing & Popular Funeral Practices. *Folk Music Journal,* 5 (4) 1988: 412–47.
39. The poem continues: 'Form'd for a dignity prophets but darkly name, / Lest shameless men cry "Shame!" / So rich with wealth conceal'd / That Heaven and Hell fight chiefly for this field.' Patmore, C: To the Body. In *The Unkown Eros.* London, Bell, 1878.
40. Carter, HV: On the arrangement of the cancellated osseus tissue in the foot, as illustrative of the mechanism of that organ. *Transactions of the Medical and Physical Society of Bombay,* October 1863.
41. See Alpers, S: *The Art of Describing.* London, John Murray, and Chicago, University of Chicago Press, 1983: 108.
42. This is a big subject, which deserves more space than I can give it here. See Owen, RO: *Archetype and Homologies of the Vertebrate Skeleton.* London, Voorst, 1848. See also: Kusukawa, S and Maclean, I: *Transmitting Knowledge: Words, Images & Instruments in Early Modern Europe.* Oxford, Oxford University Press, 2006: 81.
43. Gray, H and Carter, HV: *Anatomy Descriptive and Surgical.* London, JW Parker & Son, 1858: 689, fig 348. The sensibility here is not Carter's alone, as he credits Wilson with the original design (see his List of Illustrations).
44. Quain, J; *Elements of Anatomy.* London, Walton & Maberly, 1856, *3*: 397.
45. All eyes are closed, faces averted from the viewer, except where facial features have to be shown. Mouths are shown open only if the anatomical structure requires it. Scollon, R & Scollon, SW: *Discourses in Place.* Routledge, London 2003: 84 talk helpfully about 'intimate space'.

CHAPTER 9: CALAMITY

1860–1861

1. Carter Papers, Wellcome Western MSS 5818: 19.10.1858. The word in square brackets is in shorthand, Parker, my reading.
2. *Ibid.*: 16.5.1858.
3. *Ibid.*: 15.4.1858.
4. *Ibid.*: 19.10.1858.
5. *Medical Times and Gazette*, 5.3.1859: 241–244.
6. Goss, CM: *A Brief Account of Henry Gray FRS and his Anatomy, Descriptive and Surgical.* Philadelphia, Lea & Febiger, 1959: 24–5. Irving Loudon, writing in the *Oxford Dictionary of National Biography*, entry for Gray.

7. Condensed from *Medical Times & Gazette,* 5.3.1859: 241–4. Elisions have not been individually noticed.
8. Anon: *St George's Hospital Gazette,* 1908. [attrib.to CT Dent]. See also Goss, CM: *A Brief Account of Henry Gray FRS and his Anatomy, Descriptive and Surgical.* Philadelphia, Lea & Febiger, 1959: 24–5.
9. It would be instructive to look at the precise terminology utilized in other Victorian anatomical works, in order to judge more fairly how original any of these books really were. This is not the place for such an exercise, but the subject certainly calls for further research. My own view is that most anatomists probably use the work of their predecessors as a jumping-off point, and that were computer studies ever to be done, it might show much that was derived from earlier works. The discipline is accretional in content by its nature, so why it should not be so in its diction is interesting to ponder. The real problem is not the accretion, but the lack of courtesy. Entire phrases are identical between Quain and Gray, for example, Quain *1*: 223 'cotyloid ligament; but opposite the notch, where the margin of the cavity is deficient'; Gray: 170 'cotyloid ligament; but opposite the notch where the margin of this cavity is deficient'. The excision of punctuation is very much Gray's style, too.
10. Quain, J: *Elements of Anatomy.* London, Walton & Maberly, 1856, *2*: 429. Gray, H and Carter, HV: *Anatomy Descriptive and Surgical.* London, JW Parker & Son, 1858: 450–1.
11. Quain, J: *Elements of Anatomy.* London, Walton & Maberly, 1856, *1*: 79.
12. Gray, H and Carter, HV: *Anatomy Descriptive and Surgical.* London, JW Parker & Son, 1858: 68.
13. There was a dynasty of Quains, see *Oxford Dictionary of National Biography* entry for Quain, Jones.
14. There were a lot of Pollocks about, too. Two sons of Baron Pollock knew JW Parker junior, through their fathers, and their mutual interest in photography, and Henry Pollock—a contemporary of Gray's at St George's, the man Gray had thanked in *The Spleen* for assisting Mr Noad in doing chemical analyses for him, and who took a well known portrait photograph of Gray as a young man. Reproduced in Flick, CS: Henry Gray of 'Gray's Anatomy'. *The Optician,* 1949: 34–6.
15. Pollock, DG: *Presidential Address. Royal Medical and Chirurgical Society, Annual Meeting.* London, Adlard, 1888. George Pollock, incidentally, was related to Henry Pollock, too.
16. Compare, for example Wilson, E: *The Anatomist's Vade Mecum.* London, Churchill, 1854: 373, 375, and Gray, H and Carter, HV: *Anatomy Descriptive and Surgical.* London, JW Parker and Son, 1858: 392, 395 respectively.
17. Adam Sedgwick and Roderick Murchison quoted in Rudwick, M: *The Great Devonian Controversy.* Chicago, Chicago University Press, 1985: 280.
18. Quain, J: *Elements of Descriptive and Practical Anatomy for the Use of Students.* London, Simpkin Marshall, 1828: 19–20. Quain also gave footnotes to specific sources in the rest of the book.
19. Wilson, E: *A System of Human Anatomy, General and Special.* Philadelphia, Lea and Blanchard, 1844.
20. Carter's record of it in his diary was admiring: 'Call on Gray in evg.- selected first illustr. for the work—Gray at work & employing others—he's a good example for this.' Carter Papers, Wellcome Western MSS 5818: 31.1.1856.
21. See Brodie's reaction to Gray's death later in this chapter.

22. Rudwick, M: *The Great Devonian Controversy.* Chicago, Chicago University Press, 1985. Concerning Victorian anonymity see also Secord, J: *Victorian Sensation.* Chicago, Chicago University Press, 2000: 454.
23. I hope such material exists, and may yet emerge. For those who believed the review partisan, see Goss, CM: *A Brief Account of Henry Gray FRS and his Anatomy, Descriptive and Surgical.* Philadelphia, Lea & Febiger, 1959: 24–5.
24. An anonymous article in the *St George's Hospital Gazette* in 1908: 51 attributed the 'slashing criticism' of the *Medical Times* review to Quain and Sharpey, as if from contemporary knowledge, or contemporary rumour. The article is said (by a hand emendation to the old card index of the Royal College of Surgeons card catalogue) to have been written by Clinton Thomas Dent, a George's man who was a student much later than these events, but who was taught by Pickering Pick, Editor of *Gray's* directly after Holmes. Pick was a younger student at George's under Gray, and may have picked up talk around the medical school in the years he worked there, or from Holmes. Pick himself left nothing substantial in writing about this era, except for the discussion of the Langhorn photograph, but Dent clearly seems to have known something, and wrote some of it down in 1908, after Holmes's death in 1907. Dent also said that Holmes had 'polished Gray's prose' and repeated the oft-quoted statement that 'All who remember Henry Gray as a student agree in describing him as a most painstaking and methodical worker, and one who learned his anatomy by the slow by invaluable method of making dissections for himself.'
25. Sharpey had worked on *Quain's* since 1848. Sykes, AH: *Sharpey's Fibres.* York, William Sessions, 2001: 89–90.
26. A copy of Crisp's letter to the Council of the Royal Society, dated 18.3.1857, is in the Archives of the Astley Cooper Prize. The Royal Society had been reproached previously by *The Lancet* about the questionable manner in which its medals and honours were conferred. See *The Lancet,* 9.5.1846: 535. Sharpey was Secretary of the Royal Society 1852–1866, and had an ongoing feud with *The Lancet.* He looks to have been angered by Gray's taking advantage of the Royal Society's funding to line his own pocket, and for abusing the Society's generosity. Sharpey would have known of the many cases of really poor scientists who could genuinely have done with the Society's £100 grant. This in addition to the lack of courtesy in *Anatomy, Descriptive & Surgical.* The *Medical Times* review was payback time.
27. Huxley had been prevented from taking the second MB by poverty. Desmond, A: *Huxley.* London, Michael Joseph, 1994: 34.
28. Quain, J: *Elements of Anatomy.* London, Walton and Maberly, 1856, *3*: 213, 216.
29. Sykes, AH: *Sharpey's Fibres.* York, William Sessions, 2001: 66.
30. *Ibid.*: 46–47. One of course wonders about the long delay in the publication of the review, and whether this represented friction, or editorial indecision.
31. This looks like what Latour and Woolgar call 'deindexicalization'. Quoted in Harré, R: Some Narrative Conventions in Scientific Discourse. In Nash, C (ed.): *Narrative in Culture.* London, Routledge, 1990: 98.
32. Liston, R: *Practical Surgery.* London, Churchill, 1846: v–vi.
33. Halford, G: Gray's Descriptive and Surgical Anatomy. *The Lancet,* 26.3.1859.
34. Gray, H and Carter, HV: *Anatomy Descriptive & Surgical.* London, JW Parker & Son, 1860: 148. Gray's list includes Cloquet, Cruveillier, Bourgery, Boyer, Henle, Kolliker, Owen, Todd and Bowman, FO Ward, and Holden. Gray added further such paragraphs, for example, on pages 202, 325, 422, and 502.

35. Crisp, E: The Carmichael Prize. *The Lancet*, 1.10.1859: 346.
36. The title page motto on the US edition was 'QUAE PROSUNT OMNIBUS' meaning (Those things) which are useful to all/which benefit everyone. I thank Christian Forsdyke for his translation.
37. Several of the really large illustrations still extended into the margins, but the overall sense of inclusion was improved by the alteration. The printing of the illustrations in the US edition was not as black as the London edition, and the detail not as sharp. This is probably not because the stereotypes were poor, but something to do with paper or ink quality. It may be that the edition I have been able to study in the UK came from the end of a long run, but the strong blacks and whites of Wertheimer's work seem to be lacking in the Philadelphia version.
38. Carter Papers, Wellcome Western MSS 5820/16 News-cutting from *Home News* 3.8.1865.
39. Carter Papers, Wellcome Western MSS 5818: undated, November 1861.
40. *Ibid.*
41. *Ibid.*
42. *Ibid.*: undated January 1862. Carter entirely rejected the idea of taking a native mistress as other men did in Bombay, essentially a vow of self-denial. 'Bad as [I] am, cannot take the measures others adopt ... will do no wrong to any woman, or anything to confirm her in vice.' *Ibid.*: 9.1.1862.
43. Parker's death certificate, dated 9.11.1860, attributed his death to an illness of a month's duration caused by congestion and abscess of the left lung, and effusion into both lungs.
44. Dunn, WH: *James Anthony Froude.* Oxford, Oxford University Press, 1961, 1: 286. See also Kingsley, C: *His Letters and memories of his Life.* [Edited by his Wife.]. London, HS King, 1877, *2*: 105. For the funeral see Pollock, Sir Frederick: *Personal Remembrances of Sir F Pollock.* London, Macmillan, 1887, *2*: 86–7. The funeral was on Friday 16th November 1860.
45. [Parker, JW jun. (ed.)]: *Essays and Reviews.* London, JW Parker & Son, 1860.
46. The book was reviewed in the *Westminster Review* by the young Frederic Harrison. See Ashton, R: *142 Strand.* Chatto & Windus, 2006: 278, 291.
47. Ellis, I: *Seven Against Christ.* Leiden, Brill, 1980: 116–17. See also Francis, M: The Origin of Essays and Reviews. *Historical Journal*, 1974, 17(4): 797–811, and next note.
48. The Parkers' book has recently been issued in a comprehensive new edition, with essays discussing its importance and impact. See Shea, V and Whitla, W: *Essays and Reviews.* Charlottesville, University of Virginia Press, 2000.
49. *The Bookseller* (26.1.1861: 2) says the partnership was created after young Parker's death. But the fact of his name being kept in the house's title suggests the legal agreement may have been earlier. See also *Oxford Dictionary of National Biography* entry for JW Parker.
50. The lines of the lettering and shadow-engraving are more slender, which makes labels stand out less. The engraver uses the long s—showing a different hand is at work.
51. For the printing of the third edition, a number of copies of the second edition were disposed of at a discount in the June previous to publication. See Parker Ledgers, Longman Archive.
52. *The Lancet*, 9.2.1861: 140.

53. Gray's Preface is dated December, 1860. Kölliker's genuine thanks to the Parkers for their efforts in making his book the fine thing it was appeared in 1860, and offers a contrast to Gray's silence. One wonders if the size problem, the title page emendations and/or the *Medical Times* review had strained the relationship.
54. Condensed from [Froude, JA]: Mr John William Parker Jun. Publisher, of West Strand. Obituary. *Gentleman's Magazine*, 1861, 210: 221–4.
55. Dean, DR: *Gideon Algernon Mantell.* Delmar NY, Scholar's Facsimiles, 1998: 261.
56. The Hospital was named for a local benefactor, Jamsetjee Jeejeebhoy, who endowed several public institutions and other public works in Bombay.
57. Carter, HV: Studentship Diary MS0134: Feb 1854, Royal College of Surgeons, Archive, London, re: cedar fungus from the Physic Garden. The tree had been planted by Sir Hans Sloane, in 1683, and is shown in old prints of the Garden.
58. A further influence upon Carter may have been Richard Owen's discovery of a mould in the lung of a flamingo, in 1832, showing the existence of entophyta as well as entozoa. See Ainsworth, GC: *Medical and Veterinary Mycology.* Cambridge, CUP, 1986: 5.
59. Carter could not identify the fungus, and had little success in growing it. The identification and artificial culture of the fungus turned out to be highly complex, not fully sorted out until recently, and even now, perhaps, not fully understood. Carter was working at the edge of known science at the time, and even now much remains to be known about these deep mycoses, of which mycetoma was the first to be discovered. See Keddie, FM: Medical Mycology 1841–1870. In Poynter, FNL: *Medicine and Science in the 1860s.* London, Wellcome, 1968: 137–49. Keddie says the first use of fungus culture techniques dates from 1868, which is certainly wrong, as Kuchenmeister (Kuchenmeister, F: *Animal and Vegetable Parasites of the Human Body.* [Translated by Edwin Lankester] New Sydenham Society, 1857) mentions the use of several media, and Carter and a colleague in the UK to whom he sent material were using rice paste as a culture medium in 1862. See also Ainsworth, GC: *Medical and Veterinary Mycology.* Cambridge, CUP, 1986. The frontispiece in this work shows Carter's illustration of black mycetoma of the foot, but has missed Carter's 1860 Bombay publication. An essay by W Symmers does justice to Carter's work on Mycetoma. See Symmers, W St C: *Curiosa.* London, Bailliere Tindall, 1974: 128–43. It turns out that more than one fungus or mycobacterium can cause mycetoma, but that the symptoms appear similar. Those affected are usually the poorest people in the world, who work in the fields without shoes.
60. Carter Papers, Wellcome Western MSS 5818: 21.4.1859.
61. Carter, HV: On a new and striking form of fungus disease affecting the foot. *Transactions of the Bombay Medical and Physical Society*, 1860, 6 (NS). Bombay, 1861: 104–42. For a modern paper on mycetoma, see: Ahmed, AOA et al: Mycetoma: A neglected infectious burden. *The Lancet Infectious Diseases,* 2004: 568–74.
62. *The Lancet*, 15.6.1861: 600, 601.
63. *The Lancet* was in error in stating Gray's age as 36, and others have followed the mistake. Gray was 34 years old at the time of his death. There has always been confusion about his age, and the suspicion arises that the wrong date of birth was given when he signed up for medical school, so he might attend sooner. His memorial card and grave both give his mother's record of his year of birth, which was 1827. Other students at St George's did not begin medical school until after their seventeenth birthday.
64. John Simon wrote soon afterwards to *The Times* alerting the public to the problem of a fatal form of smallpox emerging especially amongst children in districts with low

take-up of vaccination. He was discussing the new strain, and the failure of what we know as 'herd immunity'. *The Times*, 24.6.1861: 10.

65. The Medical Officer of Health in Westminster later observed that the 1861 epidemic (in which Gray died) was 'one of the greatest attacks of smallpox since the days of Jenner'. The case fatality rate in 1863 was 17%. See Hardy, A: *The Epidemic Streets.* Oxford, Oxford University Press, 1993: 113–21. Gray's death was a tragedy that demonstrated vaccination immunity was not necessarily life-long. Gould. T and Uttley, D: *A Short History of St George's Hospital.* Athlone Press, 1997: 75 seem to be writing in ignorance when they say: 'it is a telling commentary on the state of medicine at the time ... that even medical men with strong connections to Jenner's old teaching hospital disregarded the benefits of routine vaccination at their peril.' Few people believed revaccination necessary, although there had been voices raised in favour. See *Medical Times & Gazette*, 21.5.1859: 530.
66. The election was due on the 21st June. *British Medical Journal*, 15.6.1861: 639, 658.
67. Benjamin Brodie, MSS Letter to Charles Hawkins, 15.6.1861. From a copy in The Royal College of Surgeons Archives, London. My thanks are due to Matthew Derrick, for helping me decipher Brodie's script. The letter is said to be one of the last Brodie wrote, as he was nearly blind, and died the following year.
68. Lee had been an undergraduate at King's before he came to train at George's. Holmes wrote warmly of Lee after his death in 1898. Holmes, T: In Memoriam Henry Lee. *St George's Hospital Gazette*, 1898: 112–16.
69. *Medical Times and Gazette*, 22.6.1861: 656.
70. Perhaps too, he might there have mentioned news of the early death of John Quekett, his teacher at the Royal College, who died soon afterwards at the young age of 46. Gray died on 12th June 1861, Quekett on 20th August 1861. See *Oxford Dictionary of National Biography* entry for John Quekett. A letter from Quekett's brother in law, announcing his death, is held in the Frank Buckland MSS, Royal College of Surgeons Archive, London, (*r.* 35) with a printed appeal issued in 1862 for financial support for his widow and four young sons, signed by Owen, Busk, Buckland, and many others, including Paget, Ogle, Lister, and Angela Burdett Coutts. The appeal says that Quekett was 'beloved for his kind and obliging disposition, and the liberality with which he placed the stores of his knowledge at the disposal of all who asked his assistance' (*r.* 36)
71. A doctor's letter conveys how horrific the disease was: see *The Lancet*, 15.9.1858: 315.
72. Carter's usual reaction was not an uncommon pattern among Victorians. 'Great steadying shock' are Richard Monkton Milnes's words, describing his reaction to the death of Arthur Hallam, the man for whom Tennyson wrote *In Memoriam.* Friends at the time are reported to have said to one another: 'We must be more earnest workers since the labourers are fewer.' See Brookfield, FM: *The Cambridge Apostles.* London, Pitman, 1906: 155, 226. I am reminded of a fine song by the American singer-songwriter Holly Near, which says: 'I'll keep doing the things you were doing as if I were two.'
73. Carter Papers, Wellcome Western MSS 5810/89. HVC Letter to Lily. February 1862.
74. *Ibid.*
75. Carter Papers, Wellcome Western Mss. 5818; undated, March 1862.
76. *Ibid.*: undated, [December 1858–February 1859].
77. Carter Papers, Wellcome Western MSS 5816: 14.6.1850.
78. Carter Papers, Wellcome Western MSS 5810/102: undated [1860s].

CHAPTER 10: FUTURITY

After 1861

1. HV Carter would have been pleased, as he favoured female medical education.
2. Richardson, R and Hurwitz, BS: Donors' Attitudes towards Body Donation for Dissection. *The Lancet*, 346, 1995: 277–79.
3. I owe the notion of ritual pillage to the historian Carlo Ginzburg. For Victorian dreams of cooperation in the dissecting room, see Richardson, R and Hurwitz, BS: Celebrating New Year in Bart's Dissecting Room. *Clinical Anatomy*, 9, 1996: 408–13.
4. Charles Butterworth continued in business on his own 1884–1910. The Tenniel work was for *The Mirage of Life*, 1859. See Engen, RK: *A Dictionary of Victorian Wood Engravers*. Cambridge, Chadwyck-Healey, 1985: 38–9.
5. See Richardson, R & Thorne, RS: *The Builder Illustrations Index 1843–1883*. Institute of Historical Research & English Heritage, 1995: Introduction.
6. The company (known as Williams Lea) is now owned by Deutsche Post.
7. Parker Ledgers, Longman Archive.
8. Holmes, T: *A System of Surgery*. London, JW Parker, JW Parker, Son & Bourn, and Longman, 1860–1864, 4 vols.
9. *The Lancet*, 1.2.1862: 135.
10. *Plarr's Lives of the Fellows of the Royal College of Surgeons of England*. D'Arcy Power et al (eds). London, Royal College of Surgeons of England, 1930. Holmes's two laments for Parker read: 'nor could the scheme have ever been realised but for the energy and liberality of Messrs Parker. The Editor, and all connected with it, have to lament, in the premature death of Mr. Parker Junior, one who was warmly interested in the success of this undertaking, whose friendship the Editor was happy enough to form in consequence of their common connection with it, and from whom he constantly received valuable advice and encouragement' (vol 1, pvii). 'The heavy pecuniary risks connected with it were undertaken by the late Mr. JW Parker junior, mainly in consequence of his interest in a profession in which he had many friends; and the Editor cannot conclude his task without renewed expressions of regret for the loss of one whom he had learned to esteem, and who was esteemed by all (and they were many) with whom he was connected' (vol 4, pii).
11. Holmes, T: In Memoriam Henry Lee. *St George's Hospital Gazette* 1898: 112–16.
12. See Appendix 1 *Gray's Anatomy*: Editions 1858–2008.
13. Pick, P: The Men of My Time. *St George's Hospital Gazette*. 1893: 37–41.
14. Illustrators are listed separately, under Acknowledgements. My own contribution was the Historical Introduction.
15. There is also now a smaller separate students' edition.
16. For the publishers, the human chain would be even shorter, from Elsevier, back to Churchill Livingstone, to Longman, and the Parkers. See Appendix 3: *Gray's Anatomy*: Editions, 1858–2008.
17. *Annual Report 1859–60, Grant Medical College, Bombay*. Bombay, Education Society Press, 1860: 7–8.
18. Carter, HV: On the condition of the nerve trunks in anaesthetic leprosy. *Transactions of the Pathological Society of London*, 1861–62, *13*: 13–16. Delivered 1.4.1862, by Mr T.Holmes, for Dr HV Carter.
19. Rolleston, H: *St. George's and the Progress of Physic*. 1909, Reprint in BL: 1–16. [Originally published in *St George's Hospital Gazette*, VIII: 103–8; 119–29].

20. Carter endowed the Reay Lectureship at Grant Medical College, Bombay, named for Donald James Mackay, 11th Baron Reay, a reforming Governor of Bombay.
21. Henry Vandyke Carter Mss. Letter to John Chatto: 19.3.188[7]. Royal College of Surgeons, London, Archive. Chatto died in 1887, so it looks as though Carter missed seeing him after all.
22. Carter was eventually honoured with the courtesy title of Honorary Surgeon to Queen Victoria, 'in consideration of his eminent services to Medical Science', but only after he had been retired for more than two years. HVC retired August 1888, his honour was announced November 1890. See Crawford, DG: *Roll of the Indian Medical Service, 1615–1930*. London, Thacker, 1930.
23. The plaque, sponsored by The Scarborough Civic Society and Coulson's, a local company, was unveiled in July 2008 by Brian Hurwitz, Professor of Medicine & the Arts at King's College, London.

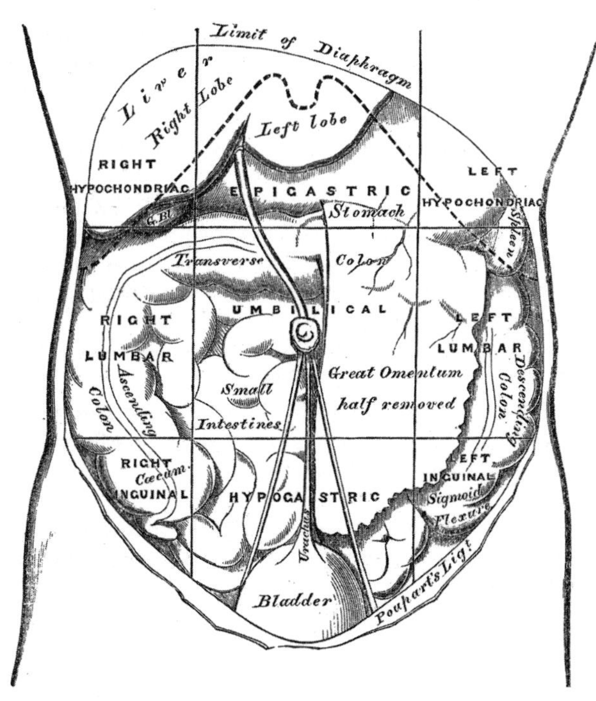

INDEX

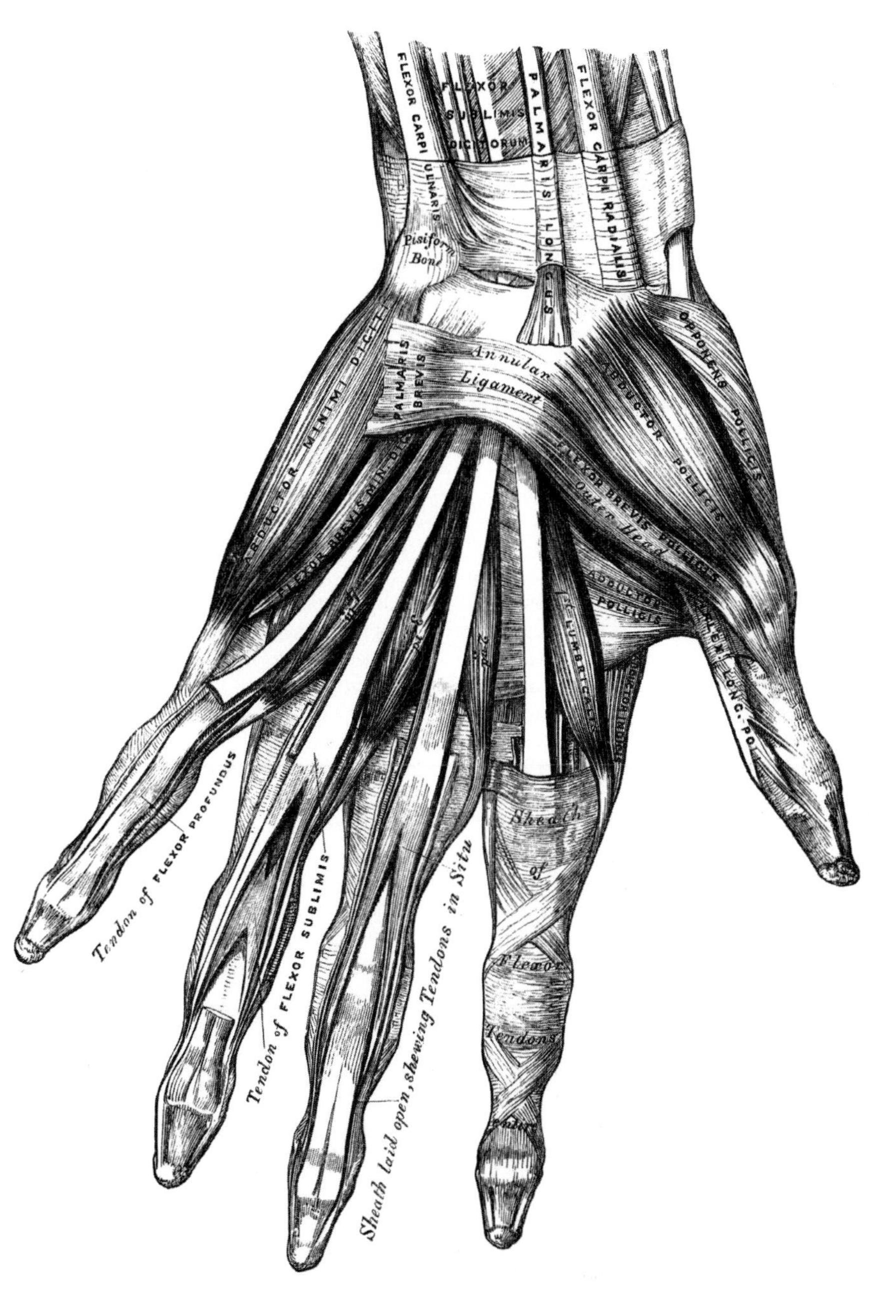
FLEXOR CARPI ULNARIS
FLEXOR SUBLIMIS DIGITORUM
PALMARIS LONGUS
FLEXOR CARPI RADIALIS
Pisiform Bone
Annular Ligament
PALMARIS BREVIS
ABDUCTOR MINIMI DIGITI
OPPONENS POLLICIS
ABDUCTOR POLLICIS
FLEXOR BREVIS POLLICIS
Outer Head
FLEXOR BREVIS MIN. DIG.
ADDUCTOR POLLICIS
LUMBRICALES
FLEX. LONG. POL.
Tendon of FLEXOR PROFUNDUS
Tendon of FLEXOR SUBLIMIS
Sheath laid open, showing Tendons in Situ
Sheath of Flexor Tendons

KEY

§ Parker author
‡ person probably dissected

NB: Many London placenames (including most hospitals and workhouses) are classified under London.

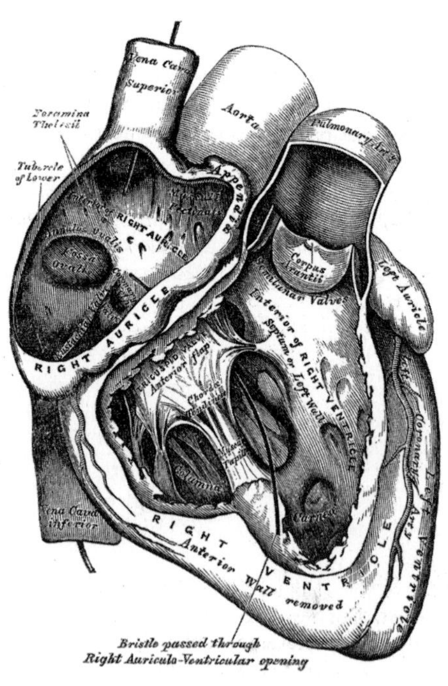
Aorta
Corpus Arantii
Left Auricle
RIGHT AURICLE
Right Anterior Wall removed
Bristle passed through Right Auriculo-Ventricular opening